Body Composition in Children

Body Composition in Children

Editors

Odysseas Androutsos
Antonis Zampelas

MDPI

Basel • Beijing • Wuhan • Barcelona • Belgrade • Novi Sad • Cluj • Manchester

Editors

Odysseas Androutsos
Lab of Clinical Nutrition and Dietetics (CND-lab),
Department of Nutrition and Dietetics
University of Thessaly
Trikala
Greece

Antonis Zampelas
Department of Food Science and Human Nutrition
Agricultural University of Athens
Athens
Greece

Editorial Office
MDPI
St. Alban-Anlage 66
4052 Basel, Switzerland

This is a reprint of articles from the Special Issue published online in the open access journal *Children* (ISSN 2227-9067) (available at: www.mdpi.com/journal/children/special_issues/Body_Composition_in_Children).

For citation purposes, cite each article independently as indicated on the article page online and as indicated below:

Lastname, A.A.; Lastname, B.B. Article Title. *Journal Name* **Year**, *Volume Number*, Page Range.

ISBN 978-3-7258-0294-4 (Hbk)
ISBN 978-3-7258-0293-7 (PDF)
doi.org/10.3390/books978-3-7258-0293-7

Contents

About the Editors

Odysseas Androutsos

Dr. Odysseas Androutsos is Associate Professor and Director of the Lab of Clinical Nutrition and Dietetics at the Department of Nutrition-Dietetics, University of Thessaly. He has published 146 scientific papers in peer-reviewed journals and >130 oral or poster presentations in national and international scientific conferences. Since 2008, he has participated as the Principal Investigator, Project Manager or Member of research groups in national and international research programmes. Dr. Androutsos is a member of the Editorial Board and reviewer of several scientific journals and reviewer of international conferences, books and research proposals. He has provided support as an external Scientific Consultant to the Ministry of Health of Malta and the National Institute of Health of Estonia for the implementation of research programmes for the prevention of childhood obesity. His scientific work has received national and international awards, such as the John M. Kinney Award for one of his scientific papers in the field of Pediatric Nutrition in 2019. He is an elected member of the ESDN Obesity of EFAD.

Antonis Zampelas

Prof. Antonis Zampelas is President of the Management Board of the Hellenic Food Authority, Professor of Human Nutrition at the Department of Food Science and Human Nutrition at the Agricultural University of Athens and Member of the Management Board of the European Food Safety Authority. He is also Honorary Professor, Division of Medicine, University College London, and Visiting Professor at the University of Nicosia, Cyprus.

Prof. Zampelas served as the President of the Hellenic Society of Lipidology, Atherosclerosis and Vascular Diseases; as the Vice President of the Committee for the approval of Dietary Supplements at the National Drug Administration; as a Member of the National Nutrition Policy Committee of the Ministry of Health; and as Member of the Executive Committee of the European Atherosclerosis Society. He is Associate Editor of the international scientific journal *Clinical Nutrition*, member of the Editorial Board of the international scientific journal *Nutrients* and Editor in Chief of the section "Nutritional Policies" of *Nutrients*. He also served as Associate Editor of the international scientific journal *Atherosclerosis*. He has been a reviewer of scientific research protocols for international (European Union, New Zealand, the Netherlands, and Cyprus) and Greek institutions. Prof. Zampelas has been invited to give more than 120 lectures at national and international conferences.

For his contribution in the field of Human Nutrition, he was awarded the honorary distinction of an Honorary Member of the Cyprus Association of Dietitians–Nutritionists (2021). In 1994, he received the Young Scientist of the Year Award by the British Nutrition Foundation, and in 2020 and 2023, he was ranked in the Top 2% worldwide in an international database in health sciences (University of Stanford).

Preface

Body composition analysis focuses on the quantification of different compartments of the human body (e.g., fat, muscle and/ or bone mass and water). Based on the assessment of certain tissues, health professionals can diagnose diseases, such as obesity and osteoporosis, and develop intervention plans to support the diagnosed individuals. Moreover, body composition tools are used to explore the prevalence of non-communicable diseases, such as obesity, in wide population groups and to support policy making, accordingly.

Various methods of body composition analysis have been previously proposed, ranging from simple, non-invasive and low-cost methods (e.g., anthropometric measurements and bioelectrical impedance analysis) to more expensive and sophisticated methods (e.g., dual-energy X-ray absorptiometry and air displacement plethysmography). Still, novel, valid and easily applicable body composition methods are required to optimize health care and to screen wider population groups on a systematic basis.

Body composition is a determinant of human health and disease progress in all age groups. It is influenced by a wide range of genetic and non-genetic factors such as age, gender, ethnicity, nutrition, physical activity and the presence of diseases. Unravelling the underlying mechanisms linking body composition and health as well as understanding how each of the aforementioned factors influence body composition are essential for developing effective interventions to prevent non-communicable diseases and to support patients' care and therapies' effectiveness. Furthermore, it is important to understand how certain types of interventions can improve individuals' body composition in order to use the most effective of these strategies to improve public health, especially in vulnerable population groups such as children and adolescents.

This Special Issue, "Body Composition in Children", collates 16 articles that aim to shed light in these fields, including 1 systematic literature review and 15 original articles from Asia (Korea, Malaysia and Taiwan), America (Mexico) and Europe (Austria, Greece, Poland, Serbia and Spain). The authors have provided novel insights about the prevalence of obesity, the influence of certain determinants (e.g., nutrition, physical activity and puberty) on body composition, the role of body composition on children's health indices and quality of life in health or disease state, and the use of certain body composition techniques in children and adolescents.

Odysseas Androutsos and Antonis Zampelas
Editors

children

MDPI

Editorial

Body Composition in Children: What Does It Tell Us So Far?

Odysseas Androutsos [1,*] and Antonis Zampelas [2]

1 Laboratory of Clinical Nutrition and Dietetics, Department of Nutrition and Dietetics, University of Thessaly, Thessaly, 42132 Trikala, Greece
2 Laboratory of Dietetics and Quality of Life, Department of Food Science and Human Nutrition, Agricultural University of Athens, 11588 Athens, Greece
* Correspondence: oandroutsos@uth.gr

The Special Issue "Body Composition in Children" of the journal "CHILDREN" aimed to publish both original and review articles focusing on the prevalence and determinants of obesity across childhood, the role of body composition in children's health, new approaches to assessing body composition, and interventions aiming to improve body composition in children and adolescents.

Body composition plays an important role in children's health and influences their energy requirements. It is determined by several factors, including genetic predisposition, age, gender, ethnicity, perinatal factors, energy balance (nutrition, physical activity, and sedentary behavior), and health status, while it can be assessed using various complex techniques (e.g., the four-compartment model) or more surrogate methods (anthropometry, bioelectrical impedance analysis (BIA), etc.). Developing new tools, standardizing the assessment methods, and evaluating the validity and applicability of existing or novel methods of assessing body composition in pediatric populations would help to optimize nutritional assessment and enable the scientific community to overcome important barriers related to its applicability in public health actions and in clinical practice.

Furthermore, understanding the underlying mechanisms linking body composition and health is essential. Studies exploring the pathways through which adiposity induces changes in health indices are required in order to tackle the effects of obesity on children's metabolic profile and quality of life, and on the development of chronic diseases across the lifespan. On the other hand, exploring the effectiveness of lifestyle interventions in improving children's body composition is essential for the prevention of obesity.

To shed further light on this field, fifteen articles following different designs and methodologies were included in the present Special Issue, and their major findings are given in Table 1.

The cross-sectional study by Markovic et al. reported prospective changes in the prevalence of obesity within a national representative sample of primary schoolchildren (n = 6105) in Serbia [1]. Data were collected in 2015 and 2019, and the children were categorized according to the International Obesity Task Force (IOTF) and the WHO criteria. The results of this study showed that the prevalence of overweight/obesity increased by 4.1% based on the WHO criteria, or by 7.2% based on the IOTF criteria, indicating that the monitoring and surveillance of pediatric malnutrition should be included in Serbia's public health agenda.

Citation: Androutsos, O.; Zampelas, A. Body Composition in Children: What Does It Tell Us So Far?. *Children* **2022**, *9*, 1199. https://doi.org/10.3390/children9081199

Received: 27 July 2022
Accepted: 8 August 2022
Published: 10 August 2022

Table 1. Short description and main findings of the studies published in the Special Issue "Body Composition in Children" of the journal "CHILDREN".

n/n	Authors	Title	Study Population	Study Design	Type of Intervention/Exposure/Data Collection Method	Main Findings/Outcomes
1	Markovic et al. [1]	Childhood Obesity in Serbia on the Rise	Primary school children aged 6.00 to 8.99 years in 2015 and 6.00 to 9.99 in 2019. In both rounds, nationally representative samples were selected through cluster sampling, with the primary school as the primary sampling unit. Schools were selected randomly from the list of public primary schools provided by the Serbian Ministry of Education. Since less than 1% of the target children were enrolled in private or special schools, these schools were excluded from the sampling frame. Stratification by the region, district, and level of urbanization was applied. The minimal planned sample size consisted of 2400 children per round, as recommended by the WHO European Office.	Cross-sectional study (2015 & 2019)	WHO COSI-design anthropometric measurements (body weight and height). Calculation of BMI and conversion to BMI categories (IOTF criteria and WHO criteria).	Prevalence of overweight/obesity in 7–9-year-old children increased by 4.1% (2015: 30.7%, 2019: 34.8%, $p < 0.05$) based on the WHO criteria, or 7.2% (2015: 22.8%, 2019: 30.0%, $p < 0.05$) based on the IOTF criteria.
2	Georgiou et al. [2]	Do Children and Adolescents with Overweight or Obesity Adhere to the National Food-Based Dietary Guidelines in Greece?	Children and adolescents aged 2–18 years (n = 1467) attending the out-patient clinic for the prevention and management of overweight and obesity in childhood and adolescence were recruited between October 2014 and March 2017.	Cross-sectional study	Anthropometric measurements (body weight and height). Calculation of BMI and conversion to BMI categories (IOTF criteria). FFQ & Questionnaire (SES, medical issues)	• The consumption of dairy products, fruits, vegetables, legumes, and fish by children/adolescents with overweight or obesity was lower than the national recommendations (ranging from a minimum of 39.5% for fish to a maximum of 75.5% for cereals/potatoes/rice). • The consumption of meat/poultry was found to exceed the national recommendation (estimated coverage of 131.3%). • A large proportion of participants regularly consumed various unhealthy foods/beverages.

Table 1. *Cont.*

n/n	Authors	Title	Study Population	Study Design	Type of Intervention/Exposure/Data Collection Method	Main Findings/Outcomes
3	Magriplis et al. [3]	Dietary Sugar Intake and Its Association with Obesity in Children and Adolescents	Hellenic National Nutrition and Health Survey (2013–2015). Multi-stage stratified design aimed at achieving representativeness in six groups (0–19 years, 20–65 years, and 65+, in males and females), across four geographical regions of Greece. For this study, data from 1165 children & adolescents aged ≥2–18 years (66.8% males) were used (n = 781; aged 2–11 years) and adolescents (n = 384; aged 12–18 years). Those who reported energy intakes of >6000 kcal/day were excluded to avoid extreme over-reporting.	Cross-sectional study	Anthropometric measurements (body weight and height). Calculation of BMI and conversion to BMI categories (IOTF criteria). Total of 2 × 24 h recalls (1st: interview & 2nd: phone) on non-consecutive days (Automated Multipass Method). Nutrient analysis with the Nutrition Data System for Research and Greek food composition tables (+newly marketed foods). FDA definition for total sugars. Lifestyle: Pre-Physical Activity Questionnaire (PAQ) and IPAQ-A Questionnaire (screen time, SES).	• 18.7% of children and 24.5% of adolescents had >10% of their total energy intake from added sugars. • The main sources of added sugars in both age groups were sweets (29.8%) and processed/refined grains and cereals (19.1%). • In adolescents, the third main contributor was sugar-sweetened beverages (20.6%). • Children/adolescents with >10% of their total energy intake from added sugars were more likely to be overweight or obese compared to their peers with <10% of their total energy intake from added sugars. • The predicted probability of becoming obese was also significant with higher total and added sugar consumption.

Table 1. *Cont.*

n/n	Authors	Title	Study Population	Study Design	Type of Intervention/Exposure/Data Collection Method	Main Findings/Outcomes
4	Karatzi et al. [4]	The Impact of Nutritional and Lifestyle Changes on Body Weight, Body Composition and Cardiometabolic Risk Factors in Children and Adolescents during the Pandemic of COVID-19: A Systematic Review	Electronic search in: PUBMED, COCHRANE, Google Scholar, and SCOPUS (up to 31 October 2021). Total of 15 studies included.	Systematic literature review	PUBMED, COCHRANE, Google Scholar, and SCOPUS RoB 2.	• COVID-19 measures negatively influenced children's and adolescents' diets and lifestyles. • Children's and adolescents' body weight and central fat increased. • Parental presence and control resulted in better glycemic control in children with type-1 diabetes mellitus. • The impact of the pandemic on the glycemic control of children with type-2 diabetes mellitus is controversial. • Dietary and lifestyle changes had a differential impact on children's hypertension prevalence.
5	Ferrer-Santos et al. [5]	Moderate-to-Vigorous Physical Activity and Body Composition in Children from the Spanish Region of Aragon	CALINA Cohort Study 308 children 7 years old (161 boys) recruited in September 2016 to September 2017.	Cross-sectional study	Assessment of PA levels (accelerometry). Body composition analysis (DXA). Questionnaires (lifestyle, perinatal factors, parental BMI, SES.	• Higher percentage of boys (69.6%) met the WHO PA recommendations compared to girls (40.9%). • A negative association was observed between MVPA and subtotal fat and abdominal fat. • Subtotal body fat, abdominal fat, and fat mass index (FMI) were significantly lower in those classified as active. • MVPA was associated with body fat.

Table 1. *Cont.*

n/n	Authors	Title	Study Population	Study Design	Type of Intervention/Exposure/Data Collection Method	Main Findings/Outcomes
6	Rusek et al. [6]	Changes in Children's Body Composition and Posture during Puberty Growth	Total of 464 children/adolescents 6–16 years old (234 boys/230 girls) living in Trzebownisko Municipality, a rural area in southeastern Poland, 8 randomly selected schools (5 primary & 3 secondary).	Cross-sectional study	Tanner stage assessment of body posture (Zebris system). Body composition analysis (TANITA MC 780 MA): measurement of body weight, adiposity, lean mass, muscle mass, total body water, BMR, compartmental body composition analysis. Measurement of body height Calculation of BMI.	• Pelvic obliquity was lower in older children. • Age played a significant role in differences in the height of the right pelvis, and the difference in the height of the right shoulder. • The content of adipose tissue (FAT%) increased with BMI and decreased with increasing weight, age, and height. • FAT% was lower in boys compared to girls. • Older children (puberty) had greater asymmetry in the right shoulder blade and right shoulder. • Younger children (who were still prepubescent) had greater anomalies in the left trunk inclination as well as in the pelvic obliquity. • Girls in puberty were characterized by greater asymmetry on the right side, including the shoulders, the scapula, and the pelvis. • In boys, the problem related only to the asymmetry of the shoulder blades. • Girls were characterized by a greater increase in adipose tissue and boys by muscle tissue. • Significant differences appeared in children's body posture. • Greater asymmetry within the scapulas and shoulders were seen in children during puberty.

Table 1. *Cont.*

n/n	Authors	Title	Study Population	Study Design	Type of Intervention/Exposure/Data Collection Method	Main Findings/Outcomes
7	Hsu et al. [7]	Can Anthropometry and Body Composition Explain Physical Fitness Levels in School-Aged Children?	Total of 360 schoolchildren (180 boys/180 girls; mean age: 10 ± 0.7 years; mean BMI score 0.336). Data collected between 2013–2016 from the hospital of Chang Gung, Linkou Main Branch, Taoyuan, Taiwan.	Cross-sectional study	Questionnaires (demographics) Anthropometric measurements (body weight & height), calculation of BMI Body composition analysis Assessment of fitness level (800 m run, sit-and-reach, 1 min sit-ups, and standing long jump)	• BF% was associated with the 800 m run. • Sex, age, and BMI z-score were independently related to the sit-and-reach. • Age, BF%, and muscle weight were associated with the 1 min sit-ups.
8	Tengku H et al. [8]	Oral Diseases and Quality of Life between Obese and Normal Weight Adolescents: A Two-Year Observational Study	Total of 397 adolescents (195 adolescents with overweight/obesity and 202 adolescents with normal weight, matched by age = 14 years at baseline and gender, were followed up for two years (2015–2017) in Malaysia).	Prospective observational cohort study	Anthropometric measurements (body weight & height), calculation of BMI and BMI weight status according to WHO criteria. Clinical examination (caries according to WHO criteria; periodontal assessment). Questionnaire (demographics, oral health related behaviors) Oral health related quality of life (OHRQoL) using the short form of the Malaysian version of Oral Health Impact Profile OHIP(M)-14.	• The prevalence of dental caries was higher in the normal-weight group. • Higher significant caries index was reported in the overweight/obese (OW/OB) group. • The prevalence of gingivitis was high in all groups. • At the 2-year follow up, the authors recorded: (1) reduction in gingival bleeding index prevalence, (2) increase in oral health-related quality of life (OHRQoL) prevalence in the OW/OB group. • The findings of this study suggest that obesity status did not have an influence on the burden of oral diseases and OHRQoL.

Table 1. *Cont.*

n/n	Authors	Title	Study Population	Study Design	Type of Intervention/Exposure/Data Collection Method	Main Findings/Outcomes
9	Jeong et al. [9]	Positive Associations between Body Mass Index and Hematological Parameters, Including RBCs, WBCs, and Platelet Counts, in Korean Children and Adolescents	Korea National Health and Nutrition Examination Survey (KNHANES) (nationally representative survey). Children and adolescents 10–18 years old ($n = 7997$; 4259 girls and 3738 boys) recruited between 2007–2018.	Cross-sectional study	Hematological parameters (including white blood cells (WBCs), red blood cells (RBCs), hemoglobin (Hb), hematocrit (Hct), and platelets). Anthropometric measurements (body weight, height, WC), calculation of BMI. Assessment of blood pressure (SBP, DBP). Questionnaire (lifestyle, alcohol consumption, smoking, SES, previous diagnoses).	• Significantly higher mean levels of WBCs, RBCs, Hb, Hct, and platelets were recorded in the obese compared to the normal weight group. • BMI SDS had significant positive associations with the levels of WBCs, RBCs, Hb, Hct, and platelets. • Higher BMI was associated with elevated WBC, RBC, Hb, Hct, and platelet counts in children and adolescents.
10	Hsu et al. CN (2021) [10]	Fat Mass Index Associated with Blood Pressure Abnormalities in Children with Chronic Kidney Disease	Total of 63 children and adolescents (8–18 years old) with chronic kidney disease (CKD) stage G1–G4 were recruited from the Kaohsiung Chang Gung Memorial Hospital, Taiwan, in December 2016–October 2018.	Prospective cohort study	Medical history and examination. Biochemical indices. Anthropometric measurements (body weight, height) and calculation of BMI and BMI percentiles. Body composition measurement (DEXA/calculation of fat mass index—FMI; android to gynoid fat ratio—A/G ratio). Cardiovascular assessments: 24 h ambulatory blood pressure monitoring (ABPM), arterial stiffness index, echocardiography.	• Up to 63.5% of CKD children had abnormal changes in BP detected by ABPM. • CKD children with abnormal ABPM were older, had higher numbers of CKD stage G2 to G4, hyperuricemia, obesity, and higher FMI z-score and A/G ratio compared to individuals with normal ABPM. • Among these factors, only the FMI z-score showed an independent association with abnormal ABPM.

Table 1. *Cont.*

n/n	Authors	Title	Study Population	Study Design	Type of Intervention/Exposure/Data Collection Method	Main Findings/Outcomes
11	Ofenheimer et al. [11]	Using Body Composition Groups to Identify Children and Adolescents at Risk of Dyslipidemia	LEAD Cohort Study *n* = 1394 children aged 6–<18 years from the LEAD study (2011–2019) in Austria.	Cross-sectional study	Body composition analysis (DEXA). Blood samples (8 h fasting; HDL-c, LDL-c, TG + their z-scores, total cholesterol). Anthropometric measurements (body weight and height), calculation of BMI fitness level (handgrip strength). Appendicular lean mass (ALMI) = sum of the lean mass of 4 limbs (Kg) divided by height$^{3.5}$ ($m^{3.5}$). Fat mass index (FMI) = fat mass (Kg) divided by height$^{2.5}$ ($m^{2.5}$).	• Different body composition groups, which are not distinguishable by BMI, exist. • Children with high appendicular lean mass (ALMI) and high fat mass index (FMI) showed higher triglycerides and LDL-c, but lower HDL-c levels. • Levels of these indices did not differ between those with high FMI but low (or normal) ALMI, and other body composition groups. • These findings suggest that children/adolescents with high FMI who have concomitantly high ALMI should be followed closely in future studies to investigate whether they are at increased risk of cardiovascular problems.
12	Valencia-Sosa et al. [12]	Percentile Reference Values for the Neck Circumference of Mexican Children	Total of 1059 (52.9% girls) normal-weight schoolchildren aged 6–11 years from six different schools located in Acatlán de Juárez and Villa Corona, Jalisco, Mexico.	Cross-sectional study	Anthropometric measurements (body weight and height, WC, neck circumference), calculation of BMI. Body composition (skinfolds: triceps & subscapular/Slaughter equation).	• Weight, height, and BMI values were higher for males (not statistically significant). • The 50th percentile for females was 24.6 cm at six years old and 28.25 cm at 11 years old. • The 50th percentile for males was 25.75 cm at six years old and 28.76 cm at 11 years old. • Both males and females displayed a pronounced increase in neck circumference between 10 and 11 years of age. • The greatest variability was found in the 11-year-old group, with an increase of 5.5 cm for males and 5.4 cm for females.

Table 1. *Cont.*

n/n	Authors	Title	Study Population	Study Design	Type of Intervention/Exposure/Data Collection Method	Main Findings/Outcomes
13	Thajer et al. [13]	Comparison of Bioelectrical Impedance-Based Methods on Body Composition in Young Patients with Obesity	Total of 123 children and adolescents (6–18 years) with obesity BMI: 21–59 kg/m^2).	Cross-sectional study	BIA (TANITA and BIACORPUS) in all subjects. BOD POD and DEXA for a small patient group.	• TANITA overestimated body fat percentage and fat mass relative to BIACORPUS and underestimated fat-free mass. • A Bland–Altman plot indicated little agreement between methods, which produced clinically relevant differences for all three parameters. • Gender-specific differences were observed with both methods, with body fat percentage being lower and fat-free mass higher in males than in females.
14	Bielec et al. [14]	Changes in Body Composition and Anthropomorphic Measurements in Children Participating in Swimming and Non-Swimming Activities	Two groups of children (11–12 years old) from Olsztyn, Poland, who attended swimming classes (n = 46) or participated in training activities in other sports, including, but not limited to, football, basketball, and athletics (n = 42).	Quasi-experimental, two-group study (non-random sampling)	Anthropometric measurements (body height and height). Calculation of BMI BIA (Tanita BC 418 MA analyzer). Students individually reported their rate of perceived exertion during training using the Pictorial Children's Effort Rating Table (PCERT) scale.	• The weekly volume of training was higher in the group of swimmers than in that playing other sports. • After 12 weeks of training, body height and weight increased in both groups. However, the BMI value and adipose tissue content only increased in the group of non-swimmers. • Swimmers perceived greater exertion during training than non-swimmers.

Table 1. *Cont.*

n/n	Authors	Title	Study Population	Study Design	Type of Intervention/Exposure/Data Collection Method	Main Findings/Outcomes
15	Trajkovic et al. [15]	Effects of After-School Volleyball Program on Body Composition in Overweight Adolescent Girls	Total of 42 overweight adolescent girls were randomly divided into a volleyball group (VG) ($n = 22$; age: 15.6 ± 0.5 years) and control group (CG) ($n = 20$; age: 15.5 ± 0.7 years).	Randomized study comparing an after-school volleyball program with traditional physical education classes in overweight female adolescents.	Description of volleyball program: Volleyball activities, played as small-sided games, two times a week after school. The experimental program lasted for 12 weeks. Volleyball was played on a court and the number of players was modified from 3 vs. 3 to 5 vs. 5 players. Duration & Context: • Warm-up (10–12 min). • Main part: 40 min of volleyball and 5 min cool-down. • Intensity of training monitored (rate of perceived exertion) • CG undertook their regular PE classes. • In both groups, volleyball was excluded from regular PE classes. Assessments: - BIA (InBody Co.). - Anthropometrics (body weight, height), calculation of BMI.	• A significant interaction between groups (intervention vs. control) and time (pre- vs. post) was observed for weight and BMI. • Significant main effect of time was found for body fat (kg) and body fat (%). • The results of the current study show that a 12-week after-school volleyball program, including 2 session/week, can improve body composition in overweight adolescent girls.

Georgiou et al. conducted a cross-sectional study to record the nutritional habits and to explore the level of adherence to the Greek food-based guidelines among children and adolescents classed as overweight or obese (n = 1467, 2–18 years old) [2]. According to the findings of this study, the consumption of various core food groups (dairy products, fruit, vegetables, legumes, and fish) was lower than the national recommendations, while the consumption of meat/poultry exceeded them. Moreover, a high number of the participants were found to consume unhealthy foods and beverages (e.g., regular soft drinks, sweets, salty snacks, etc.) regularly. The findings of this study indicate that a large proportion of overweight/obese children and adolescents in Greece do not adhere to the national food-based recommendations; therefore, public health initiatives must be carried out to improve their dietary habits and prevent further increases in their body weights.

The studies by Magriplis et al., Karatzi et al., and Ferrer-Santos et al. aimed to provide more evidence regarding the determinants of obesity in children and adolescents [3–5]. More specifically, Magriplis et al. conducted a cross-sectional study to record sugar intake in Greek children and adolescents and to explore its association with overweight/obesity [3]. The study sample included 1165 children/adolescents aged $\geq$ 2–18 years, who participated in the Hellenic National Nutrition and Health Survey (HNNHS). According to the results of this study, a large proportion of the participants (18.7% of children and 24.5% of adolescents) exceeded the recommendation of 10% of total energy intake from added sugars, which may be attributed to their high consumption of foods rich in sugar, such as sweets and processed/refined grains and cereals in the case of children, or sugar-sweetened beverages in the case of adolescents. Furthermore, individuals with an intake of $\geq$10% of their total energy from added sugars were more likely to be overweight/obese compared to their peers with an intake of <10%, even after adjusting for the intake of other foods or macronutrients. In their systematic review, Karatzi and her colleagues explored the impact of the COVID-19 pandemic on children's and adolescents' lifestyles and cardiovascular risk factors [4]. The electronic search resulted in 15 studies, which met the eligibility criteria set by the authors. It was revealed that the prolonged measures taken to fight against the COVID-19 pandemic had negative effects on children's and adolescents' dietary and lifestyle behaviors (e.g., an increase in the consumption of fats, fast-foods, processed foods, sweet and salty snacks, and sugar-sweetened beverages, as well as an increase in screen-time and a decrease in physical activity). However, on the other hand, a few studies also reported some improvements (e.g., an increase in the frequency of breakfast consumption and in fruit and vegetable intake, and a decrease in soft drinks consumption). Such differences could be attributed to the different lockdown periods addressed by each study and the diversity of the measures applied across different countries, as well as the participants' adherence to these measures. This study also reported that the COVID-19 period negatively affected children's and adolescents' weight status and body composition. On the other hand, the more significant presence of parents during the home confinement lead to greater support for type-1 diabetic children, who ultimately achieved better glycemic control. The study by Ferrer-Santos et al. aimed to assess the association between moderate-to-vigorous physical activity (MVPA) and body composition in 308 schoolchildren (aged 7 years old) in Aragon, Spain [5]. Gold standard techniques were used to assess both the participants' body composition (i.e., dual-energy X-ray absorptiometry) and their physical activity (i.e., accelerometers) and included several potential covariates (e.g., lifestyle, body weight gain during the first year of life, parental BMI, smoking status, etc.). Based on the results of this study, MVPA was inversely associated with subtotal and abdominal fat, indicating that the level and intensity of physical activity constitute important determinants of childhood obesity.

Rusek et al. conducted a study based on 464 schoolchildren aged 6–16 years in Poland to examine the alterations in their body composition and posture during puberty growth [6]. The results of this study showed that children with higher BMIs tended to have higher percentages of body fat, while younger children had lower percentages of body fat. Moreover, girls experienced a greater increase in adipose tissue (both in the total and segmental body composition analyses), and boys in lean mass, muscle mass, and total body water. Pubertal status also appeared to influence the children's posture. The findings of this study indicate the necessity of screening children periodically to assess their body composition and posture, and of early intervening, wherever needed, to prevent any adverse consequences of these factors on their health status.

The studies by Hsu CU et al., Tengku et al., Jeong et al., and Hsu CN et al. examined the effects of body composition on health indices and quality of life among healthy or ill children [7–10]. More specifically, Hsu CU et al. aimed to shed more light on the impact of body composition on children's level of physical fitness [7]. In total, 360 children were recruited from elementary schools in Northern Taiwan. Their body composition was assessed using BIA and their fitness level using different tests and outcomes (800 m run to assess cardiorespiratory fitness; sit-and-reach exercise to assess flexibility; 1 min sit-ups to assess speed/agility; and long-jump to assess lower body power). The findings of this study showed that the % of body fat was independently associated with cardiorespiratory fitness, and the BMI z-score was associated with flexibility. Based on these observations, the authors concluded that weight management and physical fitness should be addressed simultaneously in pediatric populations. Similarly, Tengku et al. conducted a prospective study focusing on the association between the oral disease burden and oral health related quality of life in adolescents, based on their weight status [8]. In total, 195 adolescents classed as overweight/obese and 202 adolescents classed as a normal weight, who were matched by age (14 years at the baseline) and gender, were followed up for two years (2015–2017) in Malaysia. No significant differences were observed between the two groups regarding the burden of oral diseases and indices of oral health related quality of life, thus indicating that obesity in adolescence does not appear to negatively affect these health indices. Jeong et al. compared hematological indices among children and adolescents of different weight categories [9]. The subjects participated in the Korea National Health and Nutrition Examination Survey (KNHANES) in 2007–2018, and, for the purposes of this study, 7997 children/adolescents (4259 boys) were included. The obese participants were found to have higher levels of white blood cells, red blood cells, and platelets compared to their normal-weight peers. Moreover, positive associations were observed between BMI standard deviation scores and white blood cells, red blood cells, hemoglobin, hematocrit, and platelets. As shown in this study, body composition influences hematological parameters and, therefore, children and adolescents with obesity should be periodically screened to assess their levels of these indices. Hsu CN et al. conducted a prospective cohort study based on 63 children and adolescents (8–18 years old) from Taiwan with chronic kidney disease to examine the association of body composition with cardiovascular risk factors [10]. The fat mass index (FMI) z-score was found to be associated with ambulatory blood pressure monitoring, indicating that body composition may influence the levels of certain cardiovascular risk factors and could be considered as a new method for identifying children and adolescents with early-stage chronic kidney disease, who are at a high risk of developing cardiovascular disease.

The studies of Ofenheimer et al., Valencia-Sosa et al., and Thajer et al. focused on the use and utility of certain body composition indicators in healthy or ill children [11–13]. In particular, Ofenheimer et al. explored serum lipid profiles in relation to body composition in a large cohort of 1394 children and adolescents (6–<18 years old), who participated in the LEAD study in 2011–2019 [11]. The body composition analysis was performed with dual X-ray absorptiometry (DXA), and the appendicular lean mass (ALMI) and fat mass (FMI) indices were calculated. It was observed that children with high ALMI and high FMI had abnormal levels of certain blood indices, including high levels of triglycerides and LDL cholesterol and low levels of HDL cholesterol. The results of this study also suggest that BMI does not adequately reflect body composition/adiposity. Consequently, more precise methods should be in use to assess body composition and to identify the children who are at a higher risk of developing cardiovascular disease.

Valencia-Sosa et al. provided reference values for neck circumference among Mexican children, a method which was introduced about a decade ago as a new technique for screening for obesity, especially central obesity, and cardiometabolic risk indices in pediatric populations [12]. A number of 1059 schoolchildren aged 6–11 years old with normal weight participated in this study. Overall, the authors recorded an age-dependent increase in the values of neck circumference, as well as a gender-specific difference, since boys had higher values compared to girls. Considering that neck circumference is a valid, simple, low-cost, and easy-to-apply method of screening for obesity and cardiometabolic risk factors, the publication of the percentile reference values for Mexican children is expected to contribute to the early identification of high-risk children and improve public health in Mexico.

Thajer et al. compared different methods of BIA in 123 children and adolescents (6–18 years old) with obesity in Austria [13]. More specifically, two BIA devices, namely TANITA and BIACORPUS, were used to assess body fat in all the study participants, while gold standard techniques (i.e., dual X-ray absorptiometry and air displacement plethysmography) were also applied in a subsample of volunteers. The agreement between the measurements taken by the two BIA methods and the agreement between each BIA method and each gold-standard method were tested using Bland–Altman analyses. The study showed that the BIA methods produced different results in the same individuals, with TANITA overestimating the % of body fat and the fat mass compared to BIACORPUS and underestimating the fat-free mass. Gender-specific differences were also identified. These findings suggest that both BIA devices can be used in clinical practice and research. However, it is important to use the same type of device in order to be able to compare a series of measurements taken at the individual level for clinical assessment and to compare measurements taken by different individuals in population-based studies.

The last two studies by Bielec et al. and Trajkovic et al. examined the impact of physical activity on children's and adolescents' body composition [14,15]. In brief, Bielec et al. compared body composition and anthropometric indices between children 11–12 years old who participated regularly in swimming activities (e.g., football, basketball, athletics, etc.) ($n = 46$) or other sports, including non-swimming activities ($n = 42$) for 5–12 h/week over a period of 12 weeks [14]. It was observed that participants' body weights and heights increased in both groups after the 12-week period; however, BMI and body fat percentage increased only among the non-swimmer groups. The authors noted, however, that the group of swimmers engaged in higher volumes of training compared to their non-swimmer peers, which may explain the study findings. It was concluded that the engagement for >10 h/week in vigorous exercise, such as swimming, may prevent excess body fat increase in children.

Trajkovic et al. evaluated the effects of an after-school volleyball program on body composition indices among overweight adolescent girls in Serbia [15]. The participants were randomly allocated to the intervention group ($n = 22$) or the control group ($n = 20$) and attended the program for 12 weeks. Both groups followed their standard PE activities, and the intervention group also engaged in small-sided games two times/week. Body composition was analyzed using BIA. The study showed that the after-school volleyball program improved body composition among the intervention group compared to the control group, indicating that it may be a useful strategy for tackling overweight/obesity in adolescent girls.

In conclusion, the studies included in the Special Issue "Body Composition in Children" of the journal "CHILDREN" showed that the prevalence of childhood obesity is increasing in certain areas of Europe. This increase can be attributed, to a large extent, to the adoption of unhealthy dietary and lifestyle behaviors by children and adolescents, which may lead to poor health status and quality of life and pose an economic burden for the public health systems. New methods, such as the neck circumference technique, may be used to identify children/adolescents at a high risk of developing cardiovascular disease and guide them within the health care system so that they can be treated accordingly. In parallel, effective strategies and interventions to promote healthy lifestyles were proposed, and these may be incorporated into future obesity prevention/treatment actions targeting children and adolescents.

Author Contributions: O.A. wrote and edited the manuscript. A.Z. edited and supervised the work. All authors have read and agreed to the published version of the manuscript.

Funding: This research received no external funding.

Conflicts of Interest: The authors declare no conflict of interest.

References

1. Marković, L.; Đorđić, V.; Trajković, N.; Božić, P.; Halaši, S.; Cvejić, D.; Ostojić, S.M. Childhood Obesity in Serbia on the Rise. *Children* **2021**, *8*, 409. [CrossRef] [PubMed]
2. Georgiou, A.; Androutsos, O.; Chouliaras, G.; Charmandari, E. Do Children and Adolescents with Overweight or Obesity Adhere to the National Food-Based Dietary Guidelines in Greece? *Children* **2022**, *9*, 256. [CrossRef] [PubMed]
3. Magriplis, E.; Michas, G.; Petridi, E.; Chrousos, G.P.; Roma, E.; Benetou, V.; Cholopoulos, N.; Micha, R.; Panagiotakos, D.; Zampelas, A. Dietary Sugar Intake and Its Association with Obesity in Children and Adolescents. *Children* **2021**, *8*, 676. [CrossRef] [PubMed]
4. Karatzi, K.; Poulia, K.-A.; Papakonstantinou, E.; Zampelas, A. The Impact of Nutritional and Lifestyle Changes on Body Weight, Body Composition and Cardiometabolic Risk Factors in Children and Adolescents during the Pandemic of COVID-19: A Systematic Review. *Children* **2021**, *8*, 1130. [CrossRef] [PubMed]
5. Ferrer-Santos, P.; Iglesia, I.; Muñiz-Pardos, B.; Miguel-Berges, M.; Flores-Barrantes, P.; Moreno, L.; Rodríguez-Martínez, G. Moderate-to-Vigorous Physical Activity and Body Composition in Children from the Spanish Region of Aragon. *Children* **2021**, *8*, 341. [CrossRef] [PubMed]
6. Rusek, W.; Baran, J.; Leszczak, J.; Adamczyk, M.; Baran, R.; Weres, A.; Inglot, G.; Czenczek-Lewandowska, E.; Pop, T. Changes in Children's Body Composition and Posture during Puberty Growth. *Children* **2021**, *8*, 288. [CrossRef] [PubMed]
7. Hsu, C.-Y.; Chen, L.-S.; Chang, I.-J.; Fang, W.-C.; Huang, S.-W.; Lin, R.H.; Ueng, S.W.-N.; Chuang, H.-H. Can Anthropometry and Body Composition Explain Physical Fitness Levels in School-Aged Children? *Children* **2021**, *8*, 460. [CrossRef] [PubMed]
8. Tengku, H.T.N.N.; Peh, W.Y.; Shoaib, L.A.; Baharuddin, N.A.; Vaithilingam, R.D.; Saub, R. Oral Diseases and Quality of Life between Obese and Normal Weight Adolescents: A Two-Year Observational Study. *Children* **2021**, *8*, 435. [CrossRef] [PubMed]
9. Jeong, H.R.; Lee, H.S.; Shim, Y.S.; Hwang, J.S. Positive Associations between Body Mass Index and Hematological Parameters, Including RBCs, WBCs, and Platelet Counts, in Korean Children and Adolescents. *Children* **2022**, *9*, 109. [CrossRef] [PubMed]
10. Hsu, C.-N.; Lu, P.-C.; Tain, Y.-L. Fat Mass Index Associated with Blood Pressure Abnormalities in Children with Chronic Kidney Disease. *Children* **2021**, *8*, 621. [CrossRef] [PubMed]
11. Ofenheimer, A.; Breyer-Kohansal, R.; Hartl, S.; Burghuber, O.C.; Krach, F.; Franssen, F.M.E.; Wouters, E.F.M.; Breyer, M.-K. Using Body Composition Groups to Identify Children and Adolescents at Risk of Dyslipidemia. *Children* **2021**, *8*, 1047. [CrossRef] [PubMed]
12. Valencia-Sosa, E.; Chávez-Palencia, C.; Vallarta-Robledo, J.R.; Romero-Velarde, E.; Larrosa-Haro, A.; Vásquez-Garibay, E.M.; Ramos-García, C.O. Percentile Reference Values for the Neck Circumference of Mexican Children. *Children* **2021**, *8*, 407. [CrossRef] [PubMed]
13. Thajer, A.; Skacel, G.; Truschner, K.; Jorda, A.; Vasek, M.; Horsak, B.; Strempfl, J.; Kautzky-Willer, A.; Kainberger, F.; Greber-Platzer, S. Comparison of Bioelectrical Impedance-Based Methods on Body Composition in Young Patients with Obesity. *Children* **2021**, *8*, 295. [CrossRef] [PubMed]
14. Bielec, G.; Gozdziejewska, A.; Makar, P. Changes in Body Composition and Anthropomorphic Measurements in Children Participating in Swimming and Non-Swimming Activities. *Children* **2021**, *8*, 529. [CrossRef] [PubMed]
15. Trajković, N.; Lazić, A.; Trkulja-Petković, D.; Barišić, V.; Milić, V.; Nikolić, S.; Sporiš, G. Effects of After-School Volleyball Program on Body Composition in Overweight Adolescent Girls. *Children* **2021**, *9*, 21. [CrossRef] [PubMed]

children

MDPI

Communication

Childhood Obesity in Serbia on the Rise

Lidija Marković [1], Višnja Đorđić [1], Nebojša Trajković [2,*], Predrag Božić [3,4], Szabolcs Halaši [5], Dragan Cvejić [6] and Sergej M. Ostojić [1]

1 Faculty of Sport and Physical Education, University of Novi Sad, 21101 Novi Sad, Serbia; markoviclidija169@gmail.com (L.M.); djordjicvisnja@gmail.com (V.Đ.); sergej.ostojic@chess.edu.rs (S.M.O.)
2 Faculty of Sport and Physical Education, University of Niš, 18000 Niš, Serbia
3 Serbian Institute of Sport and Sports Medicine, 11030 Beograd, Serbia; predrag.bozic@rzsport.gov.rs
4 Faculty of Sport and Physical Education, University of Montenegro, 81400 Nikšić, Montenegro
5 Teacher Training Faculty in the Hungarian Language in Subotica, University of Novi Sad, 24000 Subotica, Serbia; halasi.sabolc@gmail.com
6 Faculty of Education in Sombor, University of Novi Sad, 25101 Sombor, Serbia; dcvejic@yahoo.com
* Correspondence: nele_trajce@yahoo.com

Abstract: The aim of the study was to examine changes in obesity prevalence among primary school children in Serbia between 2015 and 2019 rounds of the national WHO European Childhood Obesity Surveillance Initiative (COSI-Serbia). Cross-sectional studies were conducted in 2015 and 2019. The nationally representative samples of primary school children were measured for body height and weight, following the COSI protocol. Body Mass Index was calculated, and the IOTF and WHO definitions were used to classify children as overweight or obese. Participants were children of both sexes aged 7.00–8.99 years (n = 6105). Significant differences in overweight (obesity included) prevalence between two COSI rounds were identified regardless of definitions applied. According to the WHO definitions, prevalence of overweight and obesity combined increased in 7–9-year-old children in Serbia from 30.7% in 2015 to 34.8% in 2019 ($z = -3.309$, $p < 0.05$), and according to the IOTF standards, the increase from 22.8% to 30% was registered ($z = -6.08$, $p = 0.00$). The childhood overweight/obesity rate is increasing in Serbia, which places monitoring and surveillance of children's nutritional status high on the public health agenda.

Keywords: childhood obesity; overweight; Serbia; WHO definitions; IOTF definitions

check for updates

Citation: Marković, L.; Đorđić, V.; Trajković, N.; Božić, P.; Halaši, S.; Cvejić, D.; Ostojić, S.M. Childhood Obesity in Serbia on the Rise. *Children* **2021**, *8*, 409. https://doi.org/10.3390/children8050409

Academic Editor: Odysseas Androutsos

Received: 19 April 2021
Accepted: 14 May 2021
Published: 18 May 2021

1. Introduction

Obesity is an alarming global health problem, with more than 107 million children and 603 million adults affected [1]. It is defined as abnormal or excessive fat accumulation that presents a risk to health [2]. Many studies suggest that the prevalence of obesity depends on different factors such as gender, ethnicity, socioeconomic status, and parent's level of education [3–5]. In previous decades, the prevalence of obesity has increased both in children and adults, yet in many countries, the increase is higher in children. As for the European countries, the number of children with obesity seems to have stabilized in the last decades, with substantial regional differences [6]. The highest prevalence rates of overweight and obesity were reported in the Mediterranean region, while in the Atlantic region, the rates were the lowest.

According to available World Health Organization data, overweight (including obesity) prevalence in primary-school children in 35 European countries varies from 9% to 42% among boys and from 5% to 43% among girls, and the obesity prevalence reaches 2% to 21% among boys and 1% to 19% among girls [7]. In Serbia, the overweight and obesity combined rate was reported to be as high as 23% in 6–9-year-old children, with an obesity rate of 6.9%. Therewith, overweight (including obesity) prevalence in Serbian 6–9-year-old children varies from 22.1% to 24.6% among boys, and from 23.1% to 24.3% among girls [8].

Associations between individual health-risk behaviors related to the frequency of consuming food and physical activity and childhood obesity in European countries appeared to be inconsistent, with some health-risk behaviors showing positive and other negative associations with obesity or overweight [9]. However, on the other hand, the authors stated that a combination of health-risk behaviors was consistent in association with obesity or overweight, especially in multiple physical activity-related risk behaviors. Oppositely, physical activity in children, spending time outdoors and engaging in activities such as play improve children's health and wellbeing [10].

Considering the scale of the problem and numerous adverse health effects of overweight and obesity [11–16], it is of vital importance to develop policies concerning prevention, monitoring, and treatment of childhood obesity. In order to provide a European-wide harmonized surveillance system for childhood obesity, the World Health Organization (WHO) Regional Office for Europe launched the WHO European Childhood Obesity Surveillance Initiative (COSI) in 2006. The COSI targets primary school children (6–9 years) since quality data on the nutrition status of this population group are lacking. The initial main outcomes of interest of COSI implementation were anthropometric outcome measures, such as BMI. Family record form implies data on the children's characteristics and lifestyle behavior, household characteristics, and parental socioeconomic status (educational and occupational level). It was optional and provided by children's parents or caregivers. School administrators filled in a school record form considering data on a school, classes, environment, and organization in school, also physical activity, nutrition, and promotion of active lifestyle behavior.

Serbia joined the COSI for Implementation Round 4 (2015/2016), intending to develop a surveillance system that would provide reliable, objective, and internationally comparable data on overweight and obesity prevalence among primary-school children. In Round 5 (2018/19) the Serbian national team collected data for the second time in order to identify obesity prevalence changes over four years. Further, obesity trends could be used to inform the public, professionals, and policymakers and helped them to develop new policies for obesity prevention in children.

2. Materials and Methods

The research was instigated following the ethical principles, and all procedures were in accordance with ethical standards of the responsible committee on human experimentation (institutional and national) and with the Helsinki Declaration of 1975, as revised in 2008. The present study was performed under the COSI protocol. The study was conducted in accordance with the Declaration of Helsinki, and all procedures were approved by World Health Organization (No. 2018/873491-0). Parents, teachers, and school administrators were fully informed about all study procedures, and active informed written consent was requested from the parents. Verbal consent from the child to participate in the study was obtained on the measurement day.

The first and the second COSI data collection rounds in Serbia were conducted in 2015 and 2019, respectively. Following the COSI protocol [17], cross-sectional design was employed in both rounds, targeting primary school children aged 6.00 to 8.99 years in 2015, and 6.00 to 9.99 in 2019. In both rounds, nationally representative samples were selected through cluster sampling, with the primary school as the primary sampling unit. Schools were selected randomly from the list of public primary schools provided by the Serbian Ministry of Education. Since less than 1% of the target children were enrolled in private or special schools, these schools were excluded from the sampling frame. Stratification by the region, district, and level of urbanization was applied. The minimal planned sample size consisted of 2400 children per each round, as recommended by the WHO European Office.

Trained field examiners performed standardized anthropometric measurements in participating schools. Children were asked to take off their shoes and to remove any heavy objects like wallets and mobile phones before measurement. They were mostly measured in everyday light clothes, with bodyweight additionally corrected for an average weight of the clothes worn (gym clothes −0.15 kg, light clothing −0.35 kg, heavy clothing −0.5 kg). Bodyweight was measured to the nearest 0.1 kg using portable digital scales (Omron BF214, Kyoto, Japan), while height was measured to the nearest 0.1 cm by a portable stadiometer (Seca 213, Hamburg, Germany).

Body Mass Index (BMI) was calculated as weight (kg) divided by height squared (m^2). In order to identify overweight and obese status in children, the revised international IOTF cut-offs [18] and the WHO cut-offs [19] for school-age children and adolescents were applied. The WHO AnthroPlus software (Nutritional Survey module) for 2007 WHO growth reference was used to calculate the Z-scores for BMI-for-age [20]. The z score test for two population proportions was applied to test differences in overweight/obesity prevalence in COSI national round 1 and 2. The data were analyzed using the statistical package (SPSS Statistics 21.0, IBM Corporation, Chicago, IL, USA).

3. Results

In the first data collection round in 2015, we recruited 5102 children aged 6.00–8.99 years from 42 public primary schools. After a quality check evaluation, 214 children were excluded due to missing or inaccurate data. The final sample consisted of 4861 participants (2386 girls) who entered the final analysis.

In the second national round in 2019, 3920 children aged 6.00–9.99 years from 57 public elementary schools were approached. Only two schools refused to participate, and 55 schools from all statistical regions in Serbia, and 26 out of 29 districts participated in the study. However, some children were absent on the measurement day, and in some cases, parental consents were not provided, therefore, 563 children were excluded. The next 178 children were ruled out because of missing or inaccurate data, so the final sample consisted of 3179 participants (1506 girls). The basic characteristics of the 2015 and 2019 samples are presented in Table 1. In order to compare childhood overweight and obesity prevalence in 2015 and 2019 rounds, only data of overlapping age groups: 7.00–7.99 and 8.00–8.99 years were used. Prevalence rates for overweight and obesity in 2015 and 2019, using the IOTF and WHO definitions, are presented in Table 2.

Table 1. Basic characteristics of the final samples in Serbian COSI 2015 and 2019 rounds.

	2015	2019
Final sample size	4861	3179
Girls	49.1% (2386)	47.4% (1506)
Boys	50.9% (2475)	52.6% (1673)
6-year-olds	16.9% (824)	0.7% (22)
7-year-olds	48.2% (2341)	31.1% (989)
8-year-olds	34.9% (1696)	33.9% (1079)
9-year-olds	-	34.2% (1089)
Age (years)	7.7 ± 0.6	8.5 ± 0.8
Weight (kg)	28.4 ± 6.3	31.7 ± 7.9
Height (cm)	129.7 ± 6.8	134.1 ± 7.8
Body mass index (kg/m^2)	16.7 ± 2.6	17.4 ± 3.0

Table 2. Prevalence of overweight and obesity among Serbian children aged 7–9 years in national COSI round 1 (2015) and 2 (2019) using the WHO and IOTF definitions.

	WHO (2015)				WHO (2019)				IOTF (2015)				IOTF (2019)			
	Overweight		Obese		Overweight		Obese		Overweight		Obese		Overweight		Obese	
	%	95% CI	%	95% CI	%	95% CI	%	95% CI	%	95% CI	%	95% CI	%	95% CI	%	95% CI
Girls																
7–8 years	19.1	16.9–21.5	9.7	8.0–11.5	18.8	15.3–22.7	10.7	8.0–14.0	17.0	14.9–19.2	7.3	5.9–8.9	19.2	15.7–23.2	9.4	6.9–12.5
8–9 years	19.5	16.9–22.5	7.9	6.1–10.0	23.8	20.2–27.6	13.2	10.4–16.4	16.5	14.0–19.3	5.3	3.8–7.0	23.2	19.6–27.0	10.9	8.4–13.9
Boys																
7–8 years	18.7	16.5–21.0	13.9	12.0–16.0	20.5	17.2–24.1	15.1	12.2–18.4	15.9	13.9–18.1	6.2	4.9–7.8	18.3	15.1–21.8	9.4	7.1–12.2
8–9 years	17.8	15.3–20.4	15.6	13.3–18.2	19.2	16.0–22.7	17.1	14.0–20.4	15.9	13.5–18.4	6.9	5.4–8.8	19.7	16.5–23.3	9.9	7.5–12.7
All	18.8	17.6–20.0	11.9	10.9–12.9	20.6	18.9–22.4	14.2	12.7–15.8	16.3	15.2–17.5	6.5	5.8–7.4	20.1	18.4–21.9	9.9	8.7–11.3

WHO, WHO obesity/overweight definition; IOTF, IOTF obesity/overweight definition; 95% CI, 95% confidence interval; y, years.

Overall obesity and overweight ratio have increased between two rounds in 8–9-year-old girls regardless of the definition applied, and boys of both ages when the IOTF criteria were used. At the same time, similar trends were observed in 7–8-year-old girls, and boys of both ages according to the WHO criteria (Table 2). The rates of overweight and obesity are generally higher when the WHO definitions are applied in comparison to the IOTF. This particularly refers to the estimates of obesity, with the WHO numbers being up to nine percentage points higher.

According to the WHO and IOTF cut-offs, the lowest obesity rate was observed in 8–9-year-old girls in 2015 (7.9% and 6.2%, respectively). However, the highest obesity estimates differ depending on the criteria applied. According to the WHO criteria, the highest obesity rate was observed in 8–9-year-old boys in 2019 (17.1%), while the IOTF standards pointed out 8–9-year-old girls measured in the same round (10.9%).

As for the sex differences, the tendency of higher obesity prevalence in boys was identified in both rounds, when the WHO definitions are applied. For example, in the 2015 round, the prevalence of obesity was almost two times higher in 8–9-year-old boys in comparison to the girls of the same age (15.7% and 7.9%, respectively). In the 2019 round, the sex differences were less pronounced. On the other hand, when IOTF definitions are used to estimate childhood obesity, the pattern no longer exists. As for the overweight prevalence, the numbers are higher in girls in almost all cases, regardless of the criteria employed.

Finally, z-test for proportions revealed significant differences in overweight (obesity included) prevalence between two COSI rounds. According to the WHO cut-off points, the prevalence of overweight and obesity combined increased in 7–9-year-old children in Serbia from 30.7% in 2015 to 34.8% in 2019 ($z = -3.309$, $p < 0.05$). Provided that the IOTF standards are used, the increase in overweight and obesity rate is even more pronounced, from 22.8% in the first round to 30% in the second round ($z = -6.08$, $p = 0.00$).

4. Discussion

Serbia joined COSI in 2015, at the fourth data collection round, and also participated in the fifth round in 2019. The increase in both overweight and obesity in Serbian children aged 7–9 years has been identified, while in most of the COSI participating countries, a decrease or stable prevalence were reported within a ten-year time frame [7]. A recent meta-analysis suggests that although the prevalence of childhood overweight and obesity is very high in European countries, in most cases, trends have stabilized from 1999 to 2016 [6]. However, the increase in the prevalence of overweight and obesity was recorded in some Mediterranean countries [6]. Contrary to our findings, the same study pointed out that the prevalence of overweight and obesity is higher in girls in most European countries. In accordance with our results are the COSI data collected in the 2015/2017 round, where the prevalence of obesity was higher in boys in nearly all countries, and the prevalence of overweight was higher in girls in two-thirds of countries [7]. This was confirmed in the neighboring country, using COSI methodology, showing that Croatia has one of the highest prevalence of childhood overweight and obesity among European countries, which was more frequent among boys [21]. One more neighboring country (Hungary) showing the COSI results from 2016 is facing overweight and obesity as emerging problems [22].

The unfavorable trend in childhood obesity might be related to the lack of health-promoting lifestyle habits in Serbian children and youth, where only 8% of early adolescents in Serbia choose a diet low in fats, 17% limit use of sugars and sweets, and 20% eat recommended servings of vegetables each day [23]. Only 45% of Serbian adolescents follow a planned exercise program, 43% exercise vigorously for 20 or more minutes at least three times a week, 40% take part in light to moderate physical activity, and 29% take part in recreational physical activity [23].

The high prevalence of childhood overweight and obesity in Serbia confirmed in the current study, along with the previous findings that overweight (obesity included) rates are lowest among children from well-developed communities [8], suggest that Serbia is in the third stage of obesity transition, like the USA and many European countries [24]. However, it should be noted that there were small differences in the reference system, especially for estimating obesity, with the WHO numbers being up to nine percentage points higher. The difference between IOTF and WHO reference system was documented recently in 5–17 years old children [25]. The authors explained these differences with population variations in the pattern of BMI with gender and age between nations and also with the time of data collection.

According to Jaacks et al. [24], reducing the prevalence of adulthood obesity might be related to reducing obesity in children in the first place, therefore, regular monitoring and assessment of children's nutritional status and nutritional environments should be considered as a public health priority. This was confirmed in Spain that reached a stabilization situation regarding obesity by the implementation of public measures to prevent childhood obesity [26]. Moreover, some countries with the highest prevalence that were involved in COSI project, showed a decrease of the BMI z-score with public health measures taken after the first round [27]. However, stabilization in nutritional status could be transitory and could appear easy again in the future. Therefore, schools should integrate educational procedures relying on critical thinking, nutrition, and physical activity via diet and school curriculum and provide suitable areas for recreation, sport, and outdoor play, parental and community participation, and restrain commercialization of processed food [28]. Moreover, besides schools, coordinated efforts from multiple governmental sectors and institutions are essential in order to decrease, stabilize, or prevent obesity.

5. Conclusions

The childhood overweight/obesity rate is increasing in Serbia, which should place monitoring and surveillance of children's nutritional status high on the public health agenda. Comparing the incidence of obesity in the world and in our country, we see that our country follows the trend of increasing obesity. Based on the presented results of the research, we see that it every third child in our country is overweight or obese, which is very worrying because, according to research, this trend of obesity in children will grow even more. Serbia, as a developing country, follows the trend of obesity in all structures of society. Therefore, the entire state and numerous organizations should be involved, and a lot more should be invested in order to stop this trend of obesity worldwide.

Author Contributions: Conceptualization, S.M.O. and V.Đ.; methodology, D.C.; software, S.H.; validation, S.M.O.; formal analysis, P.B.; investigation, L.M.; data curation, L.M., S.H.; writing—original draft preparation, N.T.; writing—review and editing, N.T., V.Đ.; visualization, D.C., S.H.; supervision, S.M.O.; project administration, P.B. All authors have read and agreed to the published version of the manuscript.

Funding: The study was supported by the World Health Organization, the Serbian Ministry of Education, Science and Technological Development, the Faculty of Sport and Physical Education, University of Novi Sad and the Center for Health, Exercise and Sport Sciences (CHESS).

Institutional Review Board Statement: This study was conducted according to the guidelines laid down in the Declaration of Helsinki and all procedures involving research study participants were approved by the Ethics Committee of the Faculty of Sport and Physical Education in Novi Sad (46-06-01/2020-1).

Informed Consent Statement: Written informed parental consent was obtained from all subjects.

Data Availability Statement: The data presented in this study is available on request from the corresponding author.

Acknowledgments: The authors thank the staff and participants of COSI-Serbia 2015 and 2019 for their invaluable contributions.

Conflicts of Interest: The authors declare no conflict of interest.

References

1. Fernandes, J.C. The GBD 2015 Obesity Collaborators. Health Effects of Overweight and Obesity in 195 Countries over 25 Years. *N. Engl. J. Med.* **2017**, *377*, 13–27. [CrossRef]
2. World Health Organization. Obesity and Overweight. Facts about Overweight and Obesity [Monograph on the Internet]. Fact Sheet No. September 2006. Available online: http://www.who.int/mediacentre/factsheets/fs311/en/index.html (accessed on 20 January 2009).
3. Brown, C.L.; Halvorson, E.E.; Cohen, G.M.; Lazorick, S.; Skelton, J.A. Addressing Childhood Obesity: Opportunities for Prevention. *Pediatr. Clin. N. Am.* **2015**, *62*, 1241–1261. [CrossRef] [PubMed]
4. Datar, A.; Chung, P.J. Changes in Socioeconomic, Racial/Ethnic, and Sex Disparities in Childhood Obesity at School Entry in the United States. *JAMA Pediatr.* **2015**, *169*, 696–697. [CrossRef] [PubMed]
5. Parikka, S.; Mäki, P.; Levälahti, E.; Lehtinen-Jacks, S.; Martelin, T.; Laatikainen, T. Associations between parental BMI, socioeconomic factors, family structure and overweight in Finnish children: A path model approach Disease epidemiology. *Chronic. BMC Public Health* **2015**, *15*, 271.
6. Garrido-Miguel, M.; Cavero-Redondo, I.; Álvarez-Bueno, C.; Rodríguez-Artalejo, F.; Moreno, L.A.; Ruiz, J.R.; Ahrens, W.; Martínez-Vizcaíno, V. Prevalence and trends of overweight and obesity in European children from 1999 to 2016: A systematic review and meta-analysis. *JAMA Pediatr.* **2009**, *173*, 903–915. [CrossRef] [PubMed]
7. World Health Organization, Regional Office for Europe. COSI Factsheet: Childhood Obesity Surveillance Initiative: Highlights 2015–2017. 2018. Available online: http://www.euro.who.int/__data/assets/pdf_file/0006/372426/WH14_COSI_factsheets_v2.pdf?ua=1 (accessed on 23 January 2020).
8. Djordjic, V.; Radisavljević, S.; Milanović, I.; Bozic, P.; Grbic, M.; Jorga, J.; Ostojic, S.M. WHO European Childhood Obesity Surveillance Initiative in Serbia: A prevalence of overweight and obesity among 6–9-year-old school children. *J. Pediatr. Endocrinol. Metab.* **2016**, *29*, 1025–1030. [CrossRef] [PubMed]
9. Wijnhoven, T.M.A.; van Raaij, J.M.A.; Yngvee, A.; Sjoberg, A.; Kunesova, M.; Duleva, V.; Petrauskiene, A.; Breda, J. WHO European Childhood Obesity Surveillance Initiative: Health-risk behaviours on nutrition and physical activity in 6–9-year-old schoolchildren. *Public Health Nutr.* **2015**, *18*, 3108–3124. [CrossRef]
10. Dankiw, K.A.; Tsiros, M.D.; Baldock, K.; Kumar, S. The impacts of unstructured nature play on health in early childhood development: A systematic review. *PLoS ONE* **2020**, *15*, e0229006. [CrossRef]
11. Reilly, J.J.; Bonataki, M.; Leary, S.D.; Wells, J.; Davey-Smith, G.; Emmett, P.; Steer, C.; Ness, A.R.; Sherriff, A. Progression from childhood overweight to adolescent obesity in a large contemporary cohort. *Pediatr. Obes.* **2011**, *6*, e138–e143. [CrossRef]
12. Martinez, S.M.; Blanco, E.; Burrows, R.; Lozoff, B.; Gahagan, S. Mechanisms linking childhood weight status to metabolic risk in adolescence. *Pediatr. Diabetes* **2020**, *21*, 203–209. [CrossRef]
13. Killedar, A.; Lung, T.; Petrou, S.; Teixeira-Pinto, A.; Tan, E.J.; Hayes, A. Weight status and health-related quality of life dur-ing childhood and adolescence: Effects of age and socioeconomic position. *Int. J. Obes.* **2020**, *44*, 637–645. [CrossRef] [PubMed]
14. Halasi, S.; Lepeš, J.; Đorđić, V.; Stevanovic, D.; Ihasz, F.; Jaksic, D.; Zivkovic-Vukovic, A.; Cvetkovic, A.; Milic, Z.; Stajer, A.; et al. Relationship between obesity and health-related quality of life in children aged 7–8 years. *Health Qual. Life Outcomes* **2018**, *6*, 149–156. [CrossRef]
15. Rankin, J.; Matthews, L.; Cobley, S.; Han, A.; Sanders, R.; Wiltshire, H.D.; Baker, J.S. Psychological consequences of child-hood obesity: Psychiatric comorbidity and prevention. *Adoles. Health Med. Ther.* **2016**, *7*, 125–146.
16. Puhl, R.M.; Latner, J.D.; O'Brien, K.; Luedicke, J.; Forhan, M.; Danielsdottir, S. Cross-national perspectives about weight-based bullying in youth: Nature, extent and remedies. *Pediatr. Obes.* **2015**, *11*, 241–250. [CrossRef] [PubMed]
17. Regional Office for Europe of the World Health Organization. Childhood Obesity Surveillance Initiative (COSI) Proto-col: October 2016. 2017. Available online: http://www.euro.who.int/__data/assets/pdf_file/0018/333900/COSI-protocol-en.pdf?ua=1 (accessed on 13 December 2019).
18. Cole, T.J.; Bellizzi, M.C.; Flegal, K.M.; Dietz, W.H. Establishing a standard definition for child overweight and obesity worldwide: International survey. *BMJ* **2000**, *320*, 1240. [CrossRef] [PubMed]
19. de Onis, M.; Onyango, A.W.; Borghi, E.; Siyam, A.; Nishida, C.; Siekmann, J. Development of a WHO growth reference for school-aged children and adolescents. *Bull. World Health Org.* **2007**, *85*, 660–667. [CrossRef] [PubMed]
20. World Health Organization. WHO AnthroPlus for Personal Computers Manual: Software for Assessing Growth of the World's Children and Adolescents. 2009. Available online: http://www.who.int/growthref/tools/en/ (accessed on 25 May 2019).
21. Musić Milanović, S.; Lang Morović, M.; Bukal, D.; Križan, H.; Buoncristiano, M.; Breda, J. Regional and sociodemographic determi-nants of the prevalence of overweight and obesity in children aged 7–9 years in Croatia. *Acta Clin. Croat.* **2020**, *59*, 303–311.
22. Erdei, G.; Bakacs, M.; Illés, É.; Nagy, B.; Kaposvári, C.; Mák, E.; Nagy, E.S.; Cserháti, Z.; Kovács, V.A. Substantial variation across geo-graphic regions in the obesity prevalence among 6–8 years old Hungarian children (COSI Hungary 2016). *BMC Pub. Health.* **2018**, *18*, 1–9. [CrossRef]
23. Đorđić, V.; Božić, P.; Milanović, I.; Radisavljević, S.; Batez, M.; Jorga, J.; Ostojić, S.M. Guidelines-driven educational interven-tion promotes healthy lifestyle among adolescents and adults: A Serbian national longitudinal study. *Medicina* **2019**, *55*, 39. [CrossRef]
24. Jaacks, L.M.; Vandevijvere, S.; Pan, A.; McGowan, C.J.; Wallace, C.; Imamura, F.; Mozaffarian, D.; Swinburn, B.; Ezzati, M. The obesity transition: Stages of the global epidemic. *Lancet Diabetes Endocrinol.* **2019**, *7*, 231–240. [CrossRef]

25. Valerio, G.; Balsamo, A.; Baroni, M.G.; Brufani, C.; Forziato, C.; Grugni, G.; Licenziati, M.R.; Maffeis, C.; Del Giudice, E.M.; Morandi, A.; et al. Childhood obesity classification systems and cardiometabolic risk factors: A com-parison of the Italian, World Health Organization and International Obesity Task Force references. *Ital. J. Pediatr.* **2017**, *43*, 19. [CrossRef] [PubMed]
26. Pastor-Fajardo, M.T.; Bosch-Giménez, V.M.; Larqué, E.; Navarro, C.S.; Fuentes-Castelló, M.Á.; Pastor-Rosado, J. Prevalence and secular trend of childhood overweight and obesity in a Mediterranean area of Southeast Spain. *Child. Adolesc. Obes.* **2020**, *3*, 136–149. [CrossRef]
27. Wijnhoven, T.M.; van Raaij, J.M.; Spinelli, A.; Starc, G.; Hassapidou, M.; Spiroski, I.; Rutter, H.; Martos, É.; Rito, A.I.; Hovengen, R.; et al. WHO European Childhood Obesity Surveillance Initiative: Body mass index and level of overweight among 6–9-year-old children from school year 2007/2008 to school year 2009/2010. *BMC Pub. Health* **2014**, *14*, 806. [CrossRef] [PubMed]
28. Andrade De Medeiros Moreira, R.; Ricardo Moreira, T.; Dias Da Costa, G.; Vidigal Castro, L.C.; Minardi Mitre Cotta, R. Mul-tilevel analysis of factors that influence overweight in children: Research in schools enrolled in northern Brazil School Health Program. *BMC Pediatr.* **2020**, *20*, 188. [CrossRef] [PubMed]

children

MDPI

Article

Do Children and Adolescents with Overweight or Obesity Adhere to the National Food-Based Dietary Guidelines in Greece?

Alexandra Georgiou [1,†], Odysseas Androutsos [2,†], Giorgos Chouliaras [1] and Evangelia Charmandari [1,3,*]

1 First Department of Paediatrics, School of Medicine, National and Kapodistrian University of Athens, "Aghia Sophia" Children's Hospital, 11527 Athens, Greece; alsgeorgiou@gmail.com (A.G.); georgehouliaras@msn.com (G.C.)
2 Department of Nutrition and Dietetics, School of Physical Education, Sport Science and Dietetics, University of Thessaly, 42132 Trikala, Greece; oandroutsos@uth.gr
3 Division of Endocrinology and Metabolism, Center of Clinical, Experimental Surgery and Translational Research, Biomedical Research Foundation of the Academy of Athens, 11527 Athens, Greece
* Correspondence: evangelia.charmandari@googlemail.com; Tel.: +30-21-3201-3384
† These authors contributed equally to this work.

Abstract: Childhood obesity increases the risk for metabolic disorders, but is also related to nutritional deficiencies, such as anemia and hypovitaminosis D. Although children/adolescents with overweight/obesity may have higher energy intake, their diet quality and diversity may be low. The present study aimed to evaluate the consumption of foods against the national food-based dietary guidelines in children and adolescents with overweight or obesity in Greece. Sociodemographic, anthropometric and lifestyle data were collected from a sample of 1467 children 2–18 years old (962 obese, 505 overweight, 51.2% females) in 2014–2017. The results of this study show that the consumption of dairy products, fruit, vegetables, legumes and fish by children/adolescents with overweight or obesity was lower than the national recommendations (ranging from a minimum of 39.5% for fish, to a maximum of 75.5% for cereal/potato/rice). Only the consumption of meat/poultry was found to exceed the national recommendation (estimated coverage of 131.3%). Moreover, a large proportion of participants regularly consumed various unhealthy foods/beverages. The present findings indicate that the majority of children/adolescents with overweight/obesity do not comply with the national food-based dietary guidelines in Greece. The implementation of new strategies to promote healthy diets among children/adolescents with overweight/obesity are urgently required.

Keywords: obesity; nutrition; national guidelines; adherence

Citation: Georgiou, A.; Androutsos, O.; Chouliaras, G.; Charmandari, E. Do Children and Adolescents with Overweight or Obesity Adhere to the National Food-Based Dietary Guidelines in Greece? *Children* **2022**, 9, 256. https://doi.org/10.3390/children9020256

Academic Editor: George Antonogeorgos

Received: 16 December 2021
Accepted: 10 February 2022
Published: 15 February 2022

1. Introduction

Childhood obesity has significantly increased over the past decades [1]. Previous studies have shown that the prevalence of overweight and obesity among children and adolescents in Greece is high compared with other European countries. More specifically, the ToyBox study revealed that 20.6% of preschool children in Greece were overweight/obese, while the Healthy Growth Study and the ENERGY project showed that this figure almost doubles at the age of 10–12 [2–4].

The etiology of obesity is complex and attributed to a wide range of genetic, perinatal, socioeconomic and lifestyle factors that have been associated with this disease [5,6]. It may have detrimental effects on health, with significant short-term implications for children, including metabolic abnormalities, musculoskeletal and breathing problems, and psychosocial effects [1]. Furthermore, individuals with overweight or obesity in childhood are at higher risk of being overweight or obese in adulthood and, consequently, at higher risk for the development of non-communicable diseases, such as type 2 diabetes mellitus, hypertension and certain types of cancer [7–9]. Beyond its health consequences, obesity also causes financial burden to the national health systems due to the high expenses required for the management of its complications [10].

It is suggested that the main cause of obesity is attributed to an energy imbalance maintained for long periods, resulting from a higher energy intake compared with energy expenditure [1]. Furthermore, previous studies have highlighted the so-called "double burden of disease", according to which childhood obesity is not only associated with metabolic abnormalities, but also with the development of certain nutritional deficiencies, such as vitamin D insufficiency or deficiency, and anemia [11]. It might be speculated that high energy intake and poor diet quality determine these health conditions in childhood obesity. However, the findings from the National Health and Nutrition Examination Survey (NHANES) (2009–2014) showed that diet quality does not significantly differ between children with overweight/obesity and their normal-weight peers [12]. Similarly, the GRECO study in Greece showed that adherence to a Mediterranean diet, as assessed with valid indices, such as the KIDMED, did not differ significantly between normal-weight and overweight or obese schoolchildren [13]. The findings regarding potential differences in caloric intake between children with overweight or obesity and their normal-weight peers are contradictory, possibly due to methodological barriers (e.g., self-reported data; targeting the general population, thus having a small number of children with overweight/obesity; etc.) [14,15].

To better understand the determinants of childhood obesity and assess children's nutrition, previous studies have also explored children's adherence to food-based dietary guidelines. According to the findings of two European studies, the EsKiMo II study, which included a nationwide, representative sample of 1353 volunteers aged 12–17 years in Germany, and the HELENA study, which included 1593 volunteers aged 12.5–17.5 years from ten European cities, adolescents' consumption of several core foods (e.g., fruit, vegetables, milk/dairy products) was below the recommendations, whereas the consumption of meat/meat products exceeded the recommendations [16,17]. The ToyBox study, which included 3301 preschool children from six European countries, showed that about 35% and 55% of children with normal weight did not comply with the recommendations on fruit and vegetables consumption, respectively [18]. In Greece, the national food-based dietary guidelines were developed and became available in 2014. According to these guidelines, children and adolescents are advised to consume foods from all food groups (i.e., fruit, vegetables, milk and dairy products, bread, cereals and potatoes, legumes, meat, egg, fish and seafood, nuts and oils) on a daily or weekly basis, with the recommended servings varying according to children's age and gender. Still, no previous study has examined if children and adolescents with overweight/obesity in Greece comply to these guidelines.

The aim of the present study was to explore if children and adolescents with overweight or obesity in Greece adhere to the national food-based dietary guidelines, using a large-scale cohort consisting only of children/adolescents with overweight/obesity.

2. Materials and Methods

2.1. Study Design and Participants

The present study was a cross-sectional study. Children and adolescents aged 2–18 years attending the out-patient clinic for the prevention and management of overweight and obesity in childhood and adolescence were recruited between October 2014 and March 2017. The following inclusion criteria were applied: living in Greece, being able to complete the study questionnaire in Greek language, child/adolescent aged 2–18 years with overweight/obesity and providing a consent form. For all age groups of children and adolescents, the parents were initially approached and informed about the study purpose and protocol, and they were asked to provide a written consent to enroll their child to the study. This common recruitment approach was followed since all children/adolescents were brought to the clinic by their parents. Additionally, all children and adolescents gave their verbal assent to proceed with the measurements.

The study adhered to the Declaration of Helsinki and the conventions of the Council of Europe on human rights and biomedicine. The study was approved by the Committee on the Ethics of Human Research of 'Aghia Sophia' Children's Hospital (Approval Number: EB-PASCH-MoM: 28 November 2013, Re: 10290-14 May 2013). Written informed consent was obtained in all cases by a parent, and assent was given by children older than 7 years. Further data regarding the whole study design can be found in detail elsewhere [19]. Children and adolescents were classified as "overweight" or "obese" according to the International Obesity Task Force (IOTF) cut-off points.

2.2. Procedure

All participants were admitted to the Endocrine Unit early in the morning on the day of the assessment, and a detailed medical history and clinical examination, including standard anthropometric measurements (weight, height, waist circumference, hip circumference) were obtained by one trained researcher. A complete interview with the accompanying parents/caregivers was performed regarding demographic and socioeconomic data, and data regarding the family composition (number of children, married/divorced parents, the participant's line of birth between the family's offspring) and children's/adolescents' consumption of foods.

2.3. Instruments and Variables

Data regarding children's/adolescents' usual food and beverage habits over the last year before the assessment were collected from the parents/caregivers by completing a food frequency questionnaire (FFQ), which has been validated in Greek subjects [20–22]. In brief, this FFQ comprised questions regarding the frequency of consumption of thirty-seven food and beverage items. Response categories were: "never or less than once per month", "1–3 days per month", "1 day per week", "2–4 days per week", "5–6 days per week" and "every day". Next, the average consumption of each food or beverage category per day was asked. Parents/caregivers were asked to indicate the portion size category that best fit the daily portion of their child, between specific response categories depending on the food item. A list of common standard measures was given as examples for each food or beverage category. The average daily intake for each food and beverage category was estimated by multiplying the number of days per week with the average consumption per day and then dividing the result by 7. The portions per day that were consumed by each child/adolescent were calculated by dividing the result obtained in the previous step by the portion size, which is described for children and adolescents in the Greek food-based national recommendations. Children who had no data on both frequency and portion size for all thirty-seven items were excluded from the study. The frequency of consumption of the three main meals (breakfast, lunch and dinner) was assessed based on three relevant questions asking how many days per week children eat each main meal, with the responses varying in a six-point scale ranging from "(almost) never" to "every day". The FFQ also included four questions regarding the frequency at which the child consumes a food or beverage between the main meals (with a response range between "never or less than once per month" to "every day"), and what kind of food and/or beverage they usually choose.

The Greek national food-based guidelines were used to compare participants' food consumption. From the list of 37 food items included in the FFQ, all kinds of fruit (fresh fruit, dried fruit, canned fruit and freshly squeezed/home-made fruit juice) were grouped into one category (i.e., "fruit"). Similarly, all kinds of vegetables (i.e., fresh vegetables and cooked vegetables) were grouped into one category (i.e., "vegetables"). Plain milk, plain yoghurt, fruit/sugared/aromatized yoghurt and cheese were grouped into one category ("dairy"). Unsweetened breakfast cereals, sweetened breakfast cereals, white bread and other bakery products, brown or wholemeal bread and other bakery products, pasta, rice, deep-fried potato products and potatoes were grouped into one category ("cereal/potato/rice"). Legume consumption corresponds to the category "legumes". The portion that was consumed by each child/adolescent per day was compared with the age-specific recommended portion, which is described in the Greek food-based national guidelines for children and adolescents, leading to three categories of consumers: "according to recommendation", "less than recommended" and "more than recommended".

The food items which could not be matched to the food groups of the national food guidelines are presented as single food items. No data were recorded regarding added lipids/oils/nuts or eggs, so, for these two food groups, no comparisons against the national food recommendations were possible.

2.4. Statistical Analysis

Continuous data are presented as mean ± standard deviation, and median (interquartile range), whereas categorical variables are presented as absolute and relative frequencies. Pie charts are used to present graphically categorical outcomes. Student's t-test was used for comparison of continuous data between two groups and one-way analysis of variance (ANOVA) with Bonferroni corrections for post hoc t-tests implemented for comparisons between more than two groups. Comparisons of percentage of recommended consumption and absolute deviation from recommended consumption were performed between obese and overweight children, males and females, and age class (2–6, 6–12 and 12–18 years). All statistical analyses were performed in Stata 11.2 statistical software (StataCorp, College Station, TX, USA).

3. Results

A total of 1540 children were recruited and 1467 (962 obese, 505 overweight, 51.2% females) with a mean age of 10.4 ± 3.0 years (range: 2–18) were included in the analysis. Descriptive demographic and socioeconomic data are summarized in Table 1. Approximately 25% (394 children) had been, previously, subjected to some type of intervention, aiming at reducing body weight. The most frequent was dietary counseling (88.3%), whereas other reported interventions included psychological support (6.4%), physical activity programs (1.7%) and a combination of interventions (3.6%). In more than half of the participating children (54.4%), both parents were classified as overweight/obese (BMI > 25 Kg/m^2), whereas in 37.1%, only one parent was overweight/obese. Only 8.5% of the included children had both parents with BMI < 25 Kg/m^2.

Table 1. Socioeconomic and demographic data.

Parameter	N	%
Origin		
Greek	1358	92.6%
Other	109	7.4%
Residency		
Major urban center	1294	88.2%
Rural areas	173	11.8%
Single-parent family, yes	224	15.3%
Stay with relatives, yes	362	24.7%
Systemic exercise, yes	839	57.2%
Sports at school, yes	1277	87.1%
Frequency of physical activity (days per week)		
1–2	459	31.3%
3	565	38.5%
>3	443	30.2%
Smoking, yes	11	0.77%
Alcohol, yes	8	0.56%
Drugs, yes	0	0%

Table 2 illustrates the average consumption of food groups presented as absolute values, percentage of recommended consumption and absolute deviation from recommended consumption. It can be seen that for all food groups, with the striking exception of meat/poultry, the consumption was lower than the national recommendations. The average coverage ranged from a minimum of 39.5% for fish, to a maximum of 75.5% for cereal/potato/rice. Meat/poultry was the only food group that, on average, was consumed in excess of the national recommendations, with an estimated coverage of 131.3%.

Table 2. Average consumption of food groups expressed as absolute values of consumption, percentage of recommended consumption and absolute deviation from recommended consumption.

	Average Consumption	Average Consumption as Percentage (%) of Recommendation	Average Absolute Deviation from Recommended Consumption
Dairy (daily servings)	2.1 ± 1.1, 2.0 (1.4, 2.7)	73.2 ± 32.5, 74.3 (49.7, 100.0)	−0.7 ± 1.0, −0.7 (−1.4, 0)
Vegetables (daily servings)	0.8 ± 0.5, 0.7 (0.3, 1.1)	45.3 ± 31.2, 38.9 (19.3, 68.6)	−1.1 ± 0.8, −1.0 (−1.6, −0.5)
Meat/poultry (weekly servings)	3.8 ± 1.8, 3.8 (2.9, 5.1)	131.3 ± 54.2, 127.7 (100.0, 170.8)	1.0 ± 1.5, 0.8 (0, 2.1)
Fish (weekly servings)	0.9 ± 0.8, 0.7 (0.4, 0.9)	39.5 ± 30.1, 36.5 (20.1, 47.9)	−1.2 ± 0.6, −1.3 (−1.6, −1.0)
Fruit (daily servings)	1.5 ± 1.1, 1.3 (0.8, 2.1)	75.0 ± 45.9, 75.9 (44.1, 100.0)	−0.5 ± 0.9, −0.4 (−1.2, 0)
Legumes (weekly servings)	1.7 ± 1.5, 1.1 (0.8, 2.0)	52.7 ± 40.3, 40.0 (25.0, 75.0)	−1.3 ± 1.2, −1.6 (−2.2, −0.7)
Cereal/potato/rice (daily servings)	3.9 ± 2.1, 3.5 (2.4, 5.0)	75.5 ± 35.4, 71.8 (49.5, 100.0)	−1.2 ± 1.9, −1.3, (−2.3, 0)

Data are presented as mean ± standard deviation, median (interquartile range) of daily or weekly servings.

The classification of the participants according to whether the consumed quantities followed the national recommendations (i.e., less, as suggested or more) is presented in Figure 1.

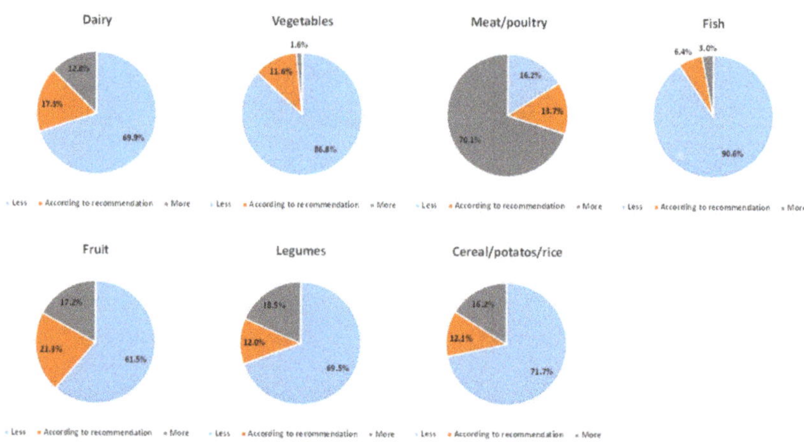

Figure 1. Compliance (less, as suggested or more) of the average consumed quantity of food groups to national recommendations.

The frequency of consumption of non-recommended food groups is shown in Table 3.

Table 3. Frequency of consumption of non-recommended food groups.

	Never or Less than 1 per Month	1–3 Times per Month	1 Time per Week	2–4 Times per Week	5–6 Times per Week	Everyday
Sugared milk	75.5%	11.6%	4.2%	3.6%	1.3%	3.8%
Soft drinks	38.1%	24.7%	18.5%	13.3%	2.4%	3.0%
Light soft drinks	67.1%	13.5%	8.8%	7.5%	1.0%	2.1%
Bottled fruit juice	23.3%	23.3%	18.4%	24.7%	5.0%	5.3%
Tea	68.7%	15.3%	6.4%	7.1%	1.1%	1.4%
Smoothies	82.5%	8.3%	4.3%	3.7%	0.4%	0.8%
Chocolate	9.3%	19.6%	25.0%	31.5%	7.5%	7.1%
Chocolate spread	23.8%	22.2%	19.9%	24.5%	4.5%	5.1%
Milk-based desserts	77.3%	10.5%	5.4%	4.6%	0.6%	1.6%
Cakes	18.0%	36.1%	21.7%	19.2%	2.5%	2.5%
Biscuits	14.1%	25.9%	22.7%	26.4%	5.7%	5.2%
Pastries	30.4%	28.1%	20.1%	15.9%	2.1%	3.4%
Sugar-based desserts	41.9%	22.6%	12.4%	15.1%	3.8%	4.2%
Salty snacks	36.2%	30.3%	18.9%	11.4%	1.9%	1.3%
Processed meat products	14.3%	8.1%	8.9%	33.4%	14.1%	21.2%

4. Discussion

The present study is the first study in Greece that describes the compliance of children with overweight/obesity with the national food-based guidelines (Table S1). The findings of this study reveal that the majority of participating children/adolescents with overweight/obesity consume less quantities of food groups than those proposed in the Greek food-based recommendations for these age groups, except for meat/poultry, which is over-consumed. Additionally, participants were found to frequently consume unhealthy foods and beverages. It is noted that two of the food categories which are included in the Greek national food-based guidelines (i.e., added lipids/oils/nuts and eggs) were not recorded in the present study. Therefore, no conclusions can be reached regarding the compliance of the study sample against these specific food categories.

Low diet quality may lead to significant health problems in youth. Childhood and adolescence comprise two main periods of growth and the dietary requirements of energy, macro- and micro-nutrients are high during this stage of life. Inadequate consumption of core foods and/or overconsumption of unhealthy foods may lead to malnutrition, metabolic abnormalities and insufficiency or deficiency of micronutrients, such as minerals and vitamins. Moreover, the adoption of unhealthy dietary patterns in these age groups increases the likelihood of retaining them in the next stages of life and, consequently, the risk for developing specific non-communicable diseases [1,7–9].

The findings of the present study are in line with those of the study by Brettschneider et al., which showed that the majority of adolescents (all weight categories) did not comply with the German national dietary guidelines [16]. Specifically, their consumption of fruit, vegetables, milk/dairy products and starchy foods was below the recommended quantity, but their consumption of meat/meat products exceeded the recommendations. Similarly, the HELENA study showed that European adolescents did not meet the food-based guidelines "Optimized Mixed Diet" and "Food Guide Pyramid", since they consumed a lesser quantity of fruit, vegetables and milk/milk products, and a greater quantity of meat/meat products [17]. In younger age groups, the findings of previous studies are also similar to those of the current study. In a sample of 521 children aged 5–10 years (all weight categories), Kunin-Batson et al. showed that only 14% of children met the guideline of five servings of fruit/vegetables per day, and 42% met the guidelines regarding the consumption of sugar-sweetened beverages, while the percentage of those meeting both guidelines was even lower (8%) [23]. Still, no differences between overweight/obese children and their normal weight peers were identified. The ToyBox study, which focused on a large number of European preschool children from six European countries, and used the same FFQ as the one used in the present study, showed that no clear differences were observed in diet quality assessed by the DQI index between overweight/obese and normal-weight children [24]. However, based on the same study sample, Cardon et al. showed that the mean consumption of soft drinks and fruit was significantly higher in overweight/obese preschool children compared with their normal-weight peers [18]. Similarly, the present study showed that >37% of the participants consumed soft drinks (with added sugar) at least once per week. Further studies are needed to confirm if the differences between children with normal weight and overweight/obesity are focused only on specific foods, and to elucidate if the overconsumption of these foods may impact overweight/obese children's nutritional status. Future initiatives should aim to improve the adherence of children/adolescents with overweight/obesity to the food-based guidelines and lower their consumption of unhealthy foods. In this direction, a series of multilevel actions is required. A recent study in Greek schoolchildren aged 10–12 years and two systematic reviews suggested that low health literacy is associated with unhealthy dietary behavior (e.g., less frequent consumption of breakfast, healthy snacks, main meals and family meals) and childhood obesity [25–27]. Therefore, it is important to increase health literacy in the general population, to raise their awareness on the detrimental effects of unhealthy nutrition in childhood and adulthood. Considering that low socioeconomic groups may have lower health literacy, these vulnerable subgroups of the population should be prioritized and targeted [28]. Furthermore, actions to promote healthy eating and avoid unhealthy food consumption should be taken. These should include practices related to the improvement of the school food environment and the implementation of new communication strategies to reach children/adolescents. The study of Lien et al. showed that only 46% of the primary schools in Greece that participated in the ENERGY project prohibited children from bringing unhealthy foods/drinks to school [29]. This highlights the necessity of adopting policies and programs to promote healthy eating at school, which, according to FAO, is an essential element to foster healthy diets at school and has been previously shown to combat unhealthy food consumption [30,31]. In addition, previous studies suggested that healthy diets could be promoted through communication channels that are appealing to children and adolescents (e.g., social media and gamification), in order to reach broader population groups, and especially those who live with overweight and obesity [32]. For example, a recent systematic literature review and meta-analysis by Suleiman-Martos showed that gamification could be an effective method to improve knowledge of healthier nutritional habits in children and adolescents [32]. Moreover, the systematic literature review by Chau et al. suggested that social media is a promising feature for nutrition interventions in adolescents [33].

The findings of the current study should be interpreted under the light of its strengths and limitations. Regarding its strengths, the study included a large sample size, while all study procedures were harmonized, following specific protocols, validated questionnaires and equipment. Additionally, all measurements were taken by trained personnel and FFQs were completed during interviews with the parents. On the other hand, the study limitations include the study sample, which consists only of children with overweight/obesity, as well as the cross-sectional design of the study, which does not allow the establishment of causal relationships between adherence to the dietary guidelines and risk for overweight/obesity. Moreover, the study sample is not representative. It should also be mentioned that dietary data were self-reported, which may be subject to recall bias and socially desirable answers. Another limitation of the present study is that data regarding children's/adolescents' consumption of eggs and of added lipids/oils/nuts was not recorded; therefore, it was not feasible to report the degree of their compliance against the national food recommendations. Future studies are needed to elucidate if Greek children/adolescents with overweight or obesity comply with the national food recommendations for added lipids/oils/nuts and eggs consumption.

5. Conclusions

The present study showed that a large number of the participating children and adolescents with overweight or obesity did not comply with the national food-based dietary guidelines. Considering the detrimental effect of poor diet quality on health, urgent actions are needed to promote healthy nutrition and treat overweight and obesity in childhood and adolescence.

Supplementary Materials: The following are available online at https://www.mdpi.com/article/10.3390/children9020256/s1, Table S1: Recommended dietary intake according to the Greek national guidelines for children and adolescents.

Author Contributions: Conceptualization, E.C. and O.A.; methodology, E.C. and O.A., statistical analysis, G.C.; data curation, A.G.; writing—original draft preparation, O.A. and G.C.; writing—review and editing, E.C. and A.G.; supervision, E.C. and O.A. All authors have read and agreed to the published version of the manuscript.

Funding: This work was supported by national funds through Research Funding of the National and Kapodistrian University of Athens (UOA) Medical School, Athens, Greece.

Institutional Review Board Statement: The study was conducted according to the guidelines of the Declaration of Helsinki and approved by the Committee on the Ethics of Human Research of "Aghia Sophia" Children's Hospital. (Approval Number: EB-PASCH-MoM: 28 November 2013, Re: 10290-14 May 2013).

Informed Consent Statement: Informed consent was obtained from all subjects involved in the study.

Data Availability Statement: The data presented in this study are available on request from the corresponding author. The data are not publicly available due to data protection issues.

Acknowledgments: The authors would like to thank the study participants.

Conflicts of Interest: The authors declare no conflict of interest.

References

1. WHO. Available online: https://www.who.int/news-room/fact-sheets/detail/obesity-and-overweight#:~{}:text=lived%20in%20Asia.-,Over%20340%20million%20children%20and%20adolescents%20aged%205%2D19%20were,just%20over%2018%25%20in%202016 (accessed on 2 July 2021).
2. Manios, Y.; Androutsos, O.; Katsarou, C.; Vampouli, E.A.; Kulaga, Z.; Gurzkowska, B.; Iotova, V.; Usheva, N.; Cardon, G.; Koletzko, B.; et al. Prevalence and sociodemographic correlates of overweight and obesity in a large Pan-European cohort of preschool children and their families: The ToyBox study. *Nutrition* **2018**, *55–56*, 192–198. [CrossRef] [PubMed]
3. Moschonis, G.; Tanagra, S.; Vandorou, A.; Kyriakou, A.E.; Dede, V.; Siatitsa, P.E.; Koumpitski, A.; Androutsos, O.; Grammatikaki, E.; Kantilafti, M.; et al. Social, economic and demographic correlates of overweight and obesity in primary-school children: Preliminary data from the Healthy Growth Study. *Public Health Nutr.* **2010**, *13*, 1693–1700. [CrossRef] [PubMed]
4. Brug, J.; van Stralen, M.M.; Te Velde, S.J.; Chinapaw, M.J.; De Bourdeaudhuij, I.; Lien, N.; Bere, E.; Maskini, V.; Singh, A.S.; Maes, L.; et al. Differences in weight status and energy-balance related behaviors among schoolchildren across Europe: The ENERGY-project. *PLoS ONE* **2012**, *7*, e34742. [CrossRef] [PubMed]
5. Silventoinen, K.; Jelenkovic, A.; Sund, R.; Hur, Y.M.; Yokoyama, Y.; Honda, C.; vB Hjelmborg, J.; Möller, S.; Ooki, S.; Aaltonen, S.; et al. Genetic and environmental effects on body mass index from infancy to the onset of adulthood: An individual-based pooled analysis of 45 twin cohorts participating in the Collaborative project of Development of Anthropometrical measures in Twins (CODATwins) study. *Am. J. Clin. Nutr.* **2016**, *104*, 371–379.
6. Androutsos, O.; Moschonis, G.; Ierodiakonou, D.; Karatzi, K.; De Bourdeaudhuij, I.; Iotova, V.; Zych, K.; Moreno, L.A.; Koletzko, B.; Manios, Y.; et al. Perinatal and lifestyle factors mediate the association between maternal education and preschool children's weight status: The ToyBox study. *Nutrition* **2018**, *48*, 6–12. [CrossRef]
7. Singh, A.S.; Mulder, C.; Twisk, J.W.; van Mechelen, W.; Chinapaw, M.J. Tracking of childhood overweight into adulthood: A systematic review of the literature. *Obes. Rev.* **2008**, *9*, 474–488. [CrossRef]

8. Reilly, J.J.; Kelly, J. Long-term impact of overweight and obesity in childhood and adolescence on morbidity and premature mortality in adulthood: Systematic review. *Int. J. Obes.* **2011**, *35*, 891–898. [CrossRef]
9. Lobstein, T.; Jackson-Leach, R. Estimated burden of paediatric obesity and co-morbidities in Europe. Part2. Numbers of children with indicators of obesity-related disease. *Int. J. Pediatr. Obes.* **2006**, *1*, 33–41. [CrossRef]
10. Tremmel, M.; Gerdtham, U.-G.; Nilsson, P.M.; Saha, S. Economic Burden of Obesity: A Systematic Literature Review. *Int. J. Environ. Res. Public Health* **2017**, *14*, 435. [CrossRef]
11. Manios, Y.; Moschonis, G.; Chrousos, G.P.; Lionis, C.; Mougios, V.; Kantilaftim, M.; Tzotzolam, V.; Skenderi, K.P.; Petridou, A.; Tsalis, G.; et al. The double burden of obesity and iron deficiency on children and adolescents in Greece: The Healthy Growth Study. *J. Hum. Nutr. Diet.* **2013**, *26*, 470–478. [CrossRef]
12. Thomson, J.L.; Landry, A.S.; Tussing-Humphrey, L.M.; Goodman, M.H. Diet quality of children in the United States by body mass index and sociodemographic characteristics. *Obes. Sci. Pract.* **2020**, *6*, 84–98. [CrossRef]
13. Farajian, P.; Risvas, G.; Karasouli, K.; Pounis, G.D.; Kastorini, C.M.; Panagiotakos, D.B.; Zampelas, A. Very high childhood obesity prevalence and low adherence rates to the Mediterranean diet in Greek children: The GRECO study. *Atherosclerosis* **2011**, *217*, 525–530. [CrossRef]
14. Garipağaoğlu, M.; Sahip, Y.; Budak, N.; Akdikmen, O.; Altan, T.; Baban, M. Food types in the diet and the nutrient intake of obese and non-obese children. *J. Clin. Res. Pediatr. Endocrinol.* **2008**, *1*, 21–29. [CrossRef]
15. Patricia, A.; Stefanik, M.S.; Felix, P.; Heald, J.R.; Mayer, J. Caloric Intake in Relation to Energy Output of Obese and Non-Obese Adolescent Boys. *Am. J. Clin. Nutr.* **1959**, *7*, 55–62.
16. Brettschneider, A.K.; Lage Barbosa, C.; Haftenberger, M.; Lehmann, F.; Mensink, G.B.M. Adherence to food-based dietary guidelines among adolescents in Germany according to socio-economic status and region: Results from Eating Study as a KiGGS Module (EsKiMo) II. *Public Health Nutr.* **2021**, *24*, 1216–1228. [CrossRef]
17. Diethelm, K.; Jankovic, N.; Moreno, L.A.; Huybrechts, I.; De Henauw, S.; De Vriendt, T.; Gonzalez-Gross, M.; Leclercq, C.; Gottrand, F.; Gilbert, C.C.; et al. HELENA Study Group Food intake of European adolescents in the light of different food-based dietary guidelines: Results of the HELENA (Healthy Lifestyle in Europe by Nutrition in Adolescence) Study. *Public Health Nutr.* **2012**, *15*, 386–398. [CrossRef]
18. Cardon, G.; De Bourdeaudhuij, I.; Iotova, V.; Latomme, J.; Socha, P.; Koletzko, B.; Moreno, L.; Manios, Y.; Androutsos, O.; De Craemer, M.; et al. Health related behaviours in normal weight and overweight preschoolers of a large Pan-European sample: The ToyBox-Study. *PLoS ONE* **2016**, *11*, e0150580. [CrossRef]
19. Genitsaridi, S.M.; Giannios, C.; Karampatsou, S.; Papageorgiou, I.; Papadopoulos, G.; Farakla, I.; Koui, E.; Georgiou, A.; Romas, S.; Terzioglou, E.; et al. A Comprehensive Multidisciplinary Management Plan Is Effective in Reducing the Prevalence of Overweight and Obesity in Childhood and Adolescence. *Horm. Res. Paediatr.* **2020**, *93*, 94–107. [CrossRef]
20. Mouratidou, T.; Mesana-Graffe, M.I.; Huybrechts, I.; De Decker, E.; De Craemer, M.; Androutsos, O.; Manios, Y.; Galcheva, S.; Lateva, M.; Gurzkowska, B.; et al. Reproducibility and relative validity of a semiquantitative food frequency questionnaire in European preschoolers: The ToyBox study. *Nutrition* **2019**, *65*, 60–67. [CrossRef]
21. Mouratidou, T.; Miguel, M.L.; Androutsos, O.; Manios, Y.; De Bourdeaudhuij, I.; Cardon, G.; Kulaga, Z.; Socha, P.; Galcheva, S.; Iotova, V.; et al. Tools, harmonization and standardization procedures of the impact and outcome evaluation indices obtained during a kindergarten-based, family-involved intervention to prevent obesity in early childhood: The ToyBox-study. *Obes. Rev. Off. J. Int. Assoc. Study Obes.* **2014**, *15* (Suppl. S3), 53–60. [CrossRef]
22. Gonzalez-Gil, E.M.; Mouratidou, T.; Cardon, G.; Androutsos, O.; De Bourdeaudhuij, I.; Gozdz, M.; Usheva, N.; Birnbaum, J.; Manios, Y.; Moreno, L.A.; et al. Reliability of primary caregivers reports on lifestyle behaviours of European pre-school children: The ToyBox-study. *Obes. Rev. Off. J. Int. Assoc. Study Obes.* **2014**, *15* (Suppl. S3), 61–66. [CrossRef]
23. Kunin-Batson, A.S.; Seburg, E.M.; Crain, A.L.; Senso, M.M.; Langer, S.L.; Levy, R.L.; Sherwood, N.E. Household factors, family behavior patterns, and adherence to dietary and physical activity guidelines among children at risk for obesity. *J. Nutr. Educ. Behav.* **2015**, *47*, 206–215.e1. [CrossRef]
24. Pinket, A.S.; De Craemer, M.; Huybrechts, I.; De Bourdeaudhuij, I.; Deforche, B.; Cardon, G.; Androutsos, O.; Koletzko, B.; Moreno, L.; Socha, P.; et al. Diet quality in European pre-schoolers: Evaluation based on diet quality indices and association with gender, socio-economic status and overweight, the ToyBox-study. *Public Health Nutr.* **2016**, *19*, 2441–2450. [CrossRef]
25. Kanellopoulou, A.; Notara, V.; Antonogeorgos, G.; Chrissini, M.; Paola Rojas-Gil, A.; Kornilaki, E.N.; Lagiou, A.; Panagiotakos, D.B. Inverse Association Between Health Literacy and Obesity Among Children in Greece: A School-Based, Cross-Sectional Epidemiological Study. *Health Educ. Behav.* **2021**, *15*, 1090198120982944. [CrossRef]
26. Chrissini, M.K.; Panagiotakos, D.B. Health literacy as a determinant of childhood and adult obesity: A systematic review. *Int. J. Adolesc. Med. Health* **2021**, *33*, 9–39. [CrossRef]
27. Michou, M.; Panagiotakos, D.B.; Costarelli, V. Low health literacy and excess body weight: A systematic review. *Cent. Eur. J. Public Health* **2018**, *26*, 234–241. [CrossRef]
28. Vaillancourt, R.; Cameron, J.D. Health Literacy for Children and Families. *Br. J. Clin. Pharmacol.* **2021**. epub ahead of print. [CrossRef]
29. Lien, N.; van Stralen, M.M.; Androutsos, O.; Bere, E.; Fernandez-Alvira, J.M.; Jan, N.; Kovacs, E.; van Lippevelde, W.; Manios, Y.; Te Velde, S.J.; et al. The school nutrition environment and its association with soft drink intakes in seven countries across Europe-the ENERGY project. *Health Place* **2014**, *30*, 28–35. [CrossRef]

30. Available online: http://www.fao.org/school-food/areas-work/food-environment/en/ (accessed on 2 July 2021).
31. Driessen, C.E.; Cameron, A.J.; Thornton, L.E.; Lailand, S.K.; Barnett, L.M. Effect of changes to the school food environment oneating behaviours and/or body weight in children: A systematic review. *Obes. Rev.* **2014**, *15*, 968–982. [CrossRef]
32. Suleiman-Martos, N.; García-Lara, R.A.; Martos-Cabrera, M.B.; Albendín-García, L.; Romero-Béjar, J.L.; Cañadas-De laFuente, G.A.; Gómez-Urquiza, J.L. Gamification for the Improvement of Diet, Nutritional Habits, and Body Composition in Children andAdolescents: A Systematic Review and Meta-Analysis. *Nutrients* **2021**, *13*, 2478. [CrossRef]
33. Chau, M.M.; Burgermaster, M.; Mamykina, L. The use of social media in nutrition interventions for adolescents and young adults—A systematic review. *Int. J. Med. Inform.* **2018**, *120*, 77–91. [CrossRef] [PubMed]

children

MDPI

Article

Dietary Sugar Intake and Its Association with Obesity in Children and Adolescents †

Emmanuella Magriplis [1,*], George Michas [1], Evgenia Petridi [2], George P. Chrousos [3], Eleftheria Roma [3], Vassiliki Benetou [4], Nikos Cholopoulos [5], Renata Micha [6,7], Demosthenes Panagiotakos [8] and Antonis Zampelas [1]

1 Department of Food Science and Human Nutrition, Agricultural University of Athens, Iera Odos 75, 118 55 Athens, Greece; gv.michas@gmail.com (G.M.); azampelas@aua.gr (A.Z.)
2 Department of Life and Health Sciences, University of Nicosia, Makedonitisas Avenue, Nicosia CY1700, Cyprus; evipetridi@hotmail.com
3 First Department of Pediatrics, Medical School, National and Kapodistrian University of Athens, Mikras Asias 75, 115 27 Athens, Greece; chrousos@gmail.com (G.P.C.); roma2el@otenet.gr (E.R.)
4 Department of Hygiene, Epidemiology and Medical Statistics, School of Medicine, National and Kapodistrian University of Athens, 115-27 Athens, Greece; vbenetou@med.uoa.gr
5 Department of Medicine, School of Health Sciences, Aristotle University of Thessaloniki, University Campus, 54 124 Thessaloniki, Greece; n.cholopoulos@gmail.com
6 Department of Food Science & Human Nutrition, University of Thessaly, 382 21 Volos, Greece; Renata.Micha@tufts.edu
7 Friedman School of Nutrition Science and Policy, Tufts University, Boston, MA 02155, USA
8 Department of Nutrition and Dietetics, School of Health Science and Education, Harokopio University, Athens, Eleftheriou Venizelou 70, 176 76 Athens, Greece; dbpanag@hua.gr
* Correspondence: emagriplis@aua.gr
† Contributors: Ioannis Dimakopoulos [1], Dimitra Karageorgou [1], Anastasia-Vasiliki Mitsopoulou [1], Ioanna Bakogianni [1], Triantafyllia Ntouroupi [1], Sophia-Maria Tsaniklidou [1], Kostantina Argyri [1], Evangelia Fappa [1], Eleni-Maria Theodoraki [1], Eirini Trichia [1], Theodora-Eirini Sialvera [1], Aggeliki Varytimiadi [1], Eleni Spyreli [1], George Karlis [1], Stauroula Zacharia [1], Anna Papageorgiou [1]; Advisory Committee: George P. Chrousos [3], Georgios Dedoussis [8], George Dimitriadis [1], Yiannis Manios [8], Eleftheria Roma [3].

Citation: Magriplis, E.; Michas, G.; Petridi, E.; Chrousos, G.P.; Roma, E.; Benetou, V.; Cholopoulos, N.; Micha, R.; Panagiotakos, D.; Zampelas, A. Dietary Sugar Intake and Its Association with Obesity in Children and Adolescents. *Children* **2021**, *8*, 676. https://doi.org/10.3390/children8080676

Academic Editor: Niels Wedderkopp

Received: 10 June 2021
Accepted: 30 July 2021
Published: 3 August 2021

Abstract: Sugar intake has been associated with increased prevalence of childhood overweight/obesity; however, results remain controversial. The aim of this study was to examine the probability of overweight/obesity with higher sugar intakes, accounting for other dietary intakes. Data from 1165 children and adolescents aged ≥2–18 years (66.8% males) enrolled in the Hellenic National Nutrition and Health Survey (HNNHS) were used; specifically, 781 children aged 2–11 years and 384 adolescents 12–18 years. Total and added sugar intake were assessed using two 24 h recalls (24 hR). Foods were categorized into specific food groups to evaluate the main foods contributing to intakes. A significant proportion of children (18.7%) and adolescents (24.5%) exceeded the recommended cut-off of 10% of total energy intake from added sugars. Sweets (29.8%) and processed/refined grains and cereals (19.1%) were the main sources of added sugars in both age groups, while in adolescents, the third main contributor was sugar-sweetened beverages (20.6%). Being overweight or obese was 2.57 ($p = 0.002$) and 1.77 ($p = 0.047$) times more likely for intakes ≥10% of total energy from added sugars compared to less <10%, when accounting for food groups and macronutrient intakes, respectively. The predicted probability of becoming obese was also significant with higher total and added-sugar consumption. We conclude that high consumption of added sugars increased the probability for overweight/obesity among youth, irrespectively of other dietary or macronutrient intakes.

Keywords: total sugars; added sugars; children; adolescents; overweight and obesity; dietary intake

1. Introduction

The prevalence of obesity nearly tripled from 4% in 1975 to over 18% in 2016 worldwide [1] with 124 million children and adolescents aged 5–19 years old being classified as obese and 213 million as overweight in 2017 [2]. These numbers rank childhood obesity as one of the major public health issues of the 21st century [3] as it is generally associated with obesity in adulthood. Childhood obesity is also associated with premature onset of cardiovascular disease risk factors such as hypertension, dyslipidemia, insulin resistance, and glucose intolerance [4].

Sugar consumption has been associated with reduced diet quality, increased energy intake [5], and with increased prevalence of obesity. These findings, however, remain controversial, especially with regards to energy consumption [6,7]. The controversy is further enhanced, by recent findings from a large national US survey, which showed that total sugar intake among adults had decreased in the past years, to levels consumed in 1975, while the prevalence of obesity remained increased [8]. Interestingly, in children and adolescents, however, 16% of their total energy intake was obtained from added sugars, which is 6% higher than the current recommendations [8]. Although most public health professionals agree on limiting added sugars, this is mostly recommended in an attempt to minimize empty calory intake [9], and increase nutrient density; it is not based on sufficient evidence for obesity reduction, since data remain inconclusive. The World Health Organization (WHO), however, strongly emphasizes on reducing free sugars intake to <10% of total energy intake, for preventing and controlling excess body weight [1]. Reduction to <5% has been acknowledged as a recommendation to prevent dental caries alone [1].

The association, therefore, between obesity rates and added sugars and/or total sugar intake remains under investigation, especially among children. Several studies have investigated the association of sugar intake and weight status in children [10–20]. The majority of them have investigated potential effects of added sugars and SSB's on Body Mass Index (BMI). Specifically, Wang et al. (2015), indicated that liquid but not solid added sugars were positively associated with higher BMI for age, BMI z-scores, and body fat, although both were associated with higher energy consumption [21]. Other studies showed that an increased intake of SSBs was associated with an increase in BMI in preschool and school aged children [10–13]. In these studies, however, the main characteristic of the sampled children was frequent, often daily SSB consumption. A study that investigated the effect of SSB as well as artificially sweetened beverage consumption, on obesity, in school aged children in the UK, found a significant association in both cases [11]. This raises the question whether it is the whole dietary pattern, the sugar content consumed or the interaction of these, granted that artificially sweetened beverages do not provide added sugars or calories. Other studies have failed to observe an association between SSBs and body weight status in children [16–20]. Furthermore, studies investigating specifically total, added, or free sugars, depending on definition chosen, have been inconclusive with regards to their effect on weight in children and adolescents [14–16], despite the fact that in some populations sugar intakes were tabulated above the recommended levels among school aged children [14,17]. Finally, a recent systematic review has shown that dietary patterns which include sweets and other potential obesogenic foods may be contributing to childhood obesity via an interaction of foods and nutrients [22]. Consequently, an investigation of total sugars, SSBs, and added sugar intakes, in relation to dietary patterns, and their associations on childhood obesity may help to resolve these discrepancies.

Therefore, the purpose of this study was to investigate the association between total and added sugar intakes with overweight/obesity in children and adolescents, accounting for other dietary intakes, using nationally representative data. In addition, the study aimed to examine the major foods or food groups that contributed to total sugar intake by age group and sex.

2. Methods

2.1. Study Design and Population Sampling

The Hellenic National Nutrition and Health Survey (HNNHS) is a population-based survey launched between September 2013 and May 2015. The study was designed to assess the health and nutritional status of Greek residents, including children over 6 months and adults. Individuals residing in various institutions, members of the armed forces, those with progressed mental disabilities, and pregnant and lactating women were excluded. In more detail, HNNHS is a cross-sectional observational survey. Responders' selection was performed with a random stratified design based on the 2011 census data. The sampling frame included a multi-stage stratified design and we aimed in achieving representativeness in six groups (0–19 years, 20–65 years, and 65+; in males and females), across four geographical regions of Greece. A total of 4574 individuals (42.5% males). A random selection of more than one individual per household was possible but no more than one individual from the same age group could be enrolled in the study. If households had children < 6 years of age, one (if more were present) was randomly selected to be included in the study, upon consent. For this study data from 1165 children and adolescents aged ≥2–18 years (66.8% males) were used. More details of the study have been published elsewhere [23].

2.2. *Final Study Population and Factors Evaluated*

For the purposes of this study, all children and adolescents ≥2–18 years that had provided at least one dietary 24 h recall (24 hR) were included. This resulted in a total sample of 1165 (66.8% males) of children (n = 781; aged 2–11 years) and adolescents (n = 384; aged 12–18 years). Children and adolescents that reported energy intakes >6000 kcal/ day (>25,080 kJ), were excluded to avoid extreme over-reporters. Due to the nature of the study sample, a specific value for extreme under-reporters was not set. A total of eight children, with no significant sex differences, had extremely overreported. The final sample consisted of 1165 children (≥2–11 years) and adolescents (≥12–18 years).

Parental or main guardian consent was signed for all underaged individuals, and by adolescents themselves if they were 18 years of age. The study was approved by the Ethics Committee of the Department of Food Science and Human Nutrition of the Agricultural University of Athens in 2013. The study was also approved by the Hellenic Data Protection Authority. It was conducted in accordance with the 1964 Helsinki declaration and its later amendments or comparable ethical standards. All members of the staff signed confidentiality agreements.

2.3. *Data Collection and Dietary Assessment*

Interpersonal Computer Assisted Personal Interview (CAPI) were performed by trained personnel, to obtain information on anthropometric, sociodemographic and lifestyle parameters. For dietary intake 24 hR's were obtained using the Automated Multipass Method (AMPM) [24], in order to reduce reporting bias [25]. This method has been found by a recent systematic review to provide the most accurate energy consumption data for children aged 0–18 years when compared to doubly labelled water, using parents as proxy reporters [26]. The goal was to obtain two 24 hR, in non-consecutive days, 8–20 days apart: the first via interview and the second via phone. Parents or the primary guardian was used as a proxy responder for all children <12 years. Those ≥12 self-responded with the parent or guardian having an aiding action. Age-specific food atlases and specific grids and household volume measures such as cups, glasses, plates, and spoon sizes were used in order to achieve accurate portion sizes. All foods were disaggregated into specific food groups, 42 in total as the primary value as shown in Supplementary Table S1. For nutrient analysis the Nutrition Data System for Research (NDSR) (developed by the University of Minnesota) as well as Greek food composition tables for traditional Greek recipes (e.g., baklavas) [27] were used.

2.3.1. Sugar Intake Assessment and Major Food Contributors

Total sugar, added sugar, as well as specific mono- and disaccharides were estimated for each food item reported in the 24 hR, using the Nutrition Data System for Research (NDSR) database, adapted with traditional and other newly marketed foods. The definition of the Food and Drug Administration (FDA) was used for total sugars: "the sum of all free mono and disaccharides" which would include glucose, fructose, galactose, and lactose as well as sucrose and maltose [7]. Moreover, the FDA defines added sugar as the "sugars and syrups that are added to foods during processing or preparation" excluding sugars naturally found in foods, such as fruits or dairy products. The sum of each mono- and di-saccharide was calculated and then the average intake per child per day was estimated and used for further analysis. Each of these intakes were further investigated in relation to children's weight status.

Foods were also categorized into specific food groups differentiating SSBs from solid sugars, but also natural sources from processed foods containing added sugars. Major food groups contributing to total sugar intake were then investigated using specific methodological approaches. Specifically, the percentage contribution made by each food group to total- and added sugar was estimated for the total sample by sex, but also for children and adolescents, separately. This was calculated by dividing the total nutrient provided by a specific food group by the total nutrient provided by all food groups and multiplying this by 100 [28]. Children and adolescents were grouped based on % of total and added sugar intakes for further analysis. Specifically, for total sugars, the median of the population was used for comparative reasons, since to date no specific cut-off is recommended. For added sugars the 10% cutoff of total energy from added sugar consumption as recommended by the US Department of Health and Human Services and US Department of Agriculture [29].

2.3.2. Other Nutritional Intakes

As per sugar intake, other macronutrients including total animal protein, plant protein, total fat, fiber (in grams), non-sugars carbohydrates, and energy were calculated for each child. In addition, specific food groups were derived, including non-milk dairy, plant protein, non-processed animal protein, processed animal protein, fruit, and vegetables. Grouping details can be seen in supplements (Supplementary Table S1), where the preliminary 42 food groups were regrouped for study purposes.

2.4. Anthropometric and Lifestyle Data

2.4.1. Weight Status Assessment

Body weight and height were reported by the parent and guardian for children <12 of age and by the adolescents ($\geq$12). For missing information on weight or height (n = 107 in total), multiple imputation was performed, using missing at random (MAR) process (STATA Corp LLC., College Station, TX, USA). Body Mass Index (BMI) was derived by body weight in kg divided by height in meters [2] and was used for the evaluation of weight status. All children and adolescence were categorized using the extended International Obesity Task Force (IOTF) tables. These tables correlate the children's BMI by month after the 2nd year of age, to the respective adult's BMI; these can be expressed as BMI centiles as well, for direct comparison with other child BMI references [30]. The descriptive table (Table 1) contains healthy weight children (none were found to be undernourished), overweight, and obese; overweight and obese children were grouped for further analysis, in order to have adequate power of analysis.

Table 1. Anthropometric, dietary and lifestyle characteristics of children and adolescents, with main guardian sociodemographic information.

		Age Category		
	Total	**Children**	**Adolescents**	***p* Value**
n	1165	781	384	-
Weight, kg, mean (SD)	40.7 (17.1)	31.9 (10.6)	58.6 (13.6)	<0.001
Height, m, mean, (SD)	1.44 (0.23)	1.33 (0.19)	1.66 (0.11)	<0.001
BMI, kg/m2, mean, (SD)	18.5 (3.3)	17.3 (2.4)	21.1 (3.4)	<0.001
Weight status *, *n* (%)				<0.001
Healthy weight	1000 (85.8)	699 (89.5)	301 (78.4)	
Overweight	142 (12.2)	66 (8.5)	76 (19.8)	
Obese	23 (2.0)	16 (2.1)	7 (1.8)	
Total sugar				
Grams, median (IQR)	43.7 (32.5, 70.0)	41.2 (41.2, 66.8)	55.7 (38.1, 80.6)	<0.001
Total calories from total sugars, median (IQR)	175 (165, 280)	165 (165, 267)	222 (153, 319)	<0.001
Total % energy intake, mean, (SD)	15.6 (6.7)	16.3 (6.3)	14.3 (7.4)	<0.001
≥median, *n* (%)	587 (41.8)	303 (38.8)	184 (47.9)	0.003
Added sugar				
Total grams, median (IQR)	28.4 (17.7, 34.5)	28.4 (22.7, 28.4)	24.5 (13.1, 44.6)	<0.001
Total calories from added sugar, median (IQR)	114 (71, 138)	114 (91, 114)	98 (52, 178)	<0.001
Total % energy intake median (IQR)	9.4 (4.5, 9.4)	9.4 (6.3, 9.4)	5.5 (3.5, 9.8)	<0.001
≥median, *n* (%)	259 (22.2)	156 (20.0)		0.008
≥10% of total energy, *n* (%)	240 (20.6)	146 (18.7)	94 (24.5)	0.022
Total METS, median (IQR)	0 (0, 3953)	0 (0, 2016)	3228 (1477, 6628)	<0.001
Total screen time (hours), median 1 (IQR)	3 (1.5, 3.75)	3.75 (1.6, 3.75)	2.45 (1.3, 4.0)	<0.001
Primary guardian Educational level, *n* (%)				<0.001
≤6 years	29 (3.11)	8 (1.2)	21 (7.6)	
>6–12 years	599 (64.2)	457 (69.6)	142 (51.5)	
≥12 years	305 (32.7)	192 (29.2)	113 (40.9)	
Primary Guardian Professional Status, *n* (%)				<0.001
Employed	727 (78.8)	561 (86.0)	166 (61.3)	
Unemployed/Homeworkers	157 (17.0)	86 (13.2)	71 (26.2)	
Pension	30 (4.2)	5 (0.8)	34 (12.6)	

Significance level set at 5%; Student *t*-test for normally distributed values and Kruskal–Wallis test for skewed numerical variables (two group comparison); chi square test or Fisher's exact test for categorical variables. * Weight status categorized based on IOTF criteria. [1] Total screen time includes total TV-viewing, computer, phone, and other screens that do not involve active movement. IQR: Interquartile range.

2.4.2. Lifestyle and Other Variables

Details on other lifestyle and socioeconomic variables, such as physical activity, screen time, and parental educational level, known to be associated with adiposity, were also collected [11,31]. Specifically, physical activity was evaluated by two different questionnaires based on age groups (i) Pre-Physical Activity Questionnaire (PAQ) Home Version for children ≥2–<12 years old [32] and (ii) the International Physical Questionnaire–Adolescents (IPAQ-A), for those ≥12–<18 years old [33]. PAQ questionnaire estimated only specific activities performed over 7-days; IPAQ-A is also a 7-day recall instrument, used to estimate total metabolic equivalents (METS), based on information obtained on specific activities performed at leisure time, during transport, at school (physed and recess), and at home. These have been previously validated and PAQ-A specifically, has also been tested against doubly labelled water and was found to have a strong correlation with energy expenditure for groups [34]. The measure of physical activity was estimated by multiplying the time spent with each activity with the corresponding MET-value per total minutes of activity as per various protocols [34]. Physical activity quartiles were derived to describe the children's physical activity status.

Screen time was also obtained by calculating the average time spent in front of any type of screen, on weekday and weekends, excluding time spent using gaming platform that requires movement. Total screen time was categorized using the upper recommended level of 2 h per day.

The children's or adolescent's SES cannot differ from that of their parent hence parental educational level was used to categorize them accordingly and adjust in the models. If this was not performed systematic errors would prevail. The primary guardian's educational level and type of profession were also accounted for, as socioeconomic proxies. Education was categorized into elementary (≤6 years), middle school (≥6–11 years), and higher level of education (≥12 years). Professional status was categorized as employed, unemployed and/or homeworker, and pension. The latter characterize mostly grandparents.

2.4.3. Sensitivity Analysis

Extra statistical analysis was performed by reporting status in order to assess association differences and further decrease reporting error, since this was not controlled for. Specifically, it has been shown that dietary under-reporting prevalence is higher compared to over-reporting among children and adolescents [35]. Normal misreporters, including over- and under-reporters as identified based on sex, age, BMR, and IPAQ level, were therefore not excluded, in order to avoid a shift to the right which means a decrease in the estimated energy and macronutrients intake. A sensitivity analysis, however, per reporting status, was performed to investigate potential effect differences of added and total sugar on weight status. For this specific analysis, over- and under- reporters were identified by dividing total energy intake by calculated basal metabolic rate (BMR) using the Scholffield equation, including for IPAQ, and then using validated Goldberg cut-offs [36] for adolescents and those reported for children by age group and sex [35].

2.5. Statistical Analysis

Distribution plots were examined for continuous variables using k density kernels and P-P plots. All variables are summarized in total and by sex; numerical variables following normal distribution are depicted as means (±sd) and those skewed as medians and interquartile range (IQR) and categorical variables are presented as relative frequencies. Group differences were examined using parametric or non-parametric methods for numerical data, accordingly and chi-square test for categorical. Significance was assessed at $\alpha = 5\%$, using two-sided test ($p < 0.05$). Mixed effects logistic regression models were used to account for the skewed distribution of the data. The likelihood of being overweight/obese compared to healthy weight was explored as per (i) the median of total sugar intake (adjusted for energy) and (ii) added sugar intake above 10% (>10%) of total energy intake, and (iii) median intake of added sugar intake. Two models were used following preliminary analysis of intakes (i) food groups including animal- and plant-protein, non-milk dairy, fruit and vegetables; and (ii) macronutrient (fat, protein, other carbohydrates) and dietary fiber intakes. All models were adjusted for children's' lifestyle, activity level (IPAQ), total screen time, and the primary guardian's socioeconomic status. Marginal effect sizes were also conducted to derive probability of overweight/obese status for higher total- and added- sugar intakes. The statistical software package STATA 12.0 was used for statistical calculations (STATA Corp LLC., College Station, TX, USA).

3. Results

Table 1 summarizes the anthropometric and lifestyle characteristics, the dietary intake of the children and the adolescents, as well as the main guardian sociodemographic information. In the present study, 10.6% of the children and 21.6% of the adolescents were overweight/obese. Median total sugar daily intake in relation to total energy consumption, was 15.6% (SE: 6.7), with a significantly higher consumption observed in children compared to adolescents (16.3% and 14.3% respectively; $p < 0.001$). The main food contributing to this higher intake was milk (Figure 1). The percentage of energy from added sugars was

9.5% in children and 5.5% in adolescents, with 18.7% of the children and 24.5% of the adolescents exceeding the 10% of total energy intake recommended upper limit (Table 1). As per basic lifestyle characteristics, physical activity levels differed between children and adolescents, with adolescents being more physically active. Overall, only 200 out of 781 children reported of performing a specific activity at least once a week (hence the median of "0"). In contrast although both children and adolescents spent more time in front of a screen than recommended (>2 h per day), children's total screen time was higher than adolescents. Finally, primary guardian education level was between 6 and 12 years, and the majority of them were employed.

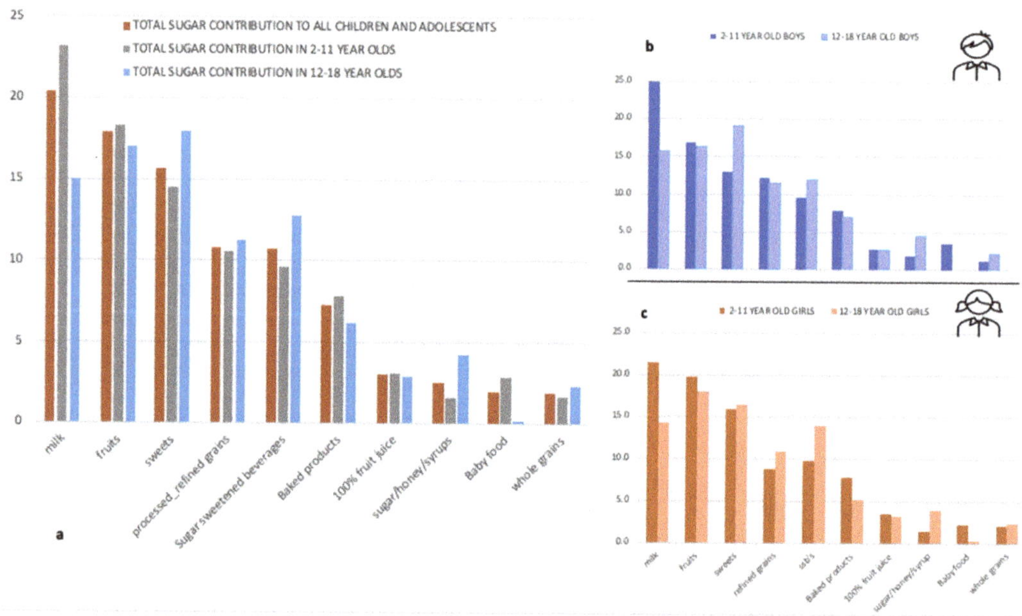

Figure 1. Total sugar contribution in children and adolescents, in total sample and by sex. (**a**). All children and adolescents; (**b**). boys; (**c**). girls.

Food groups contribution for total and added sugars are shown in Figures 1 and 2, respectively. For the total sample (Figure 1a), milk was the main contributor of total sugars (20.4%), followed by fruits (17.9%), sweets (15.7%), processed/refined grains (10.8%), and SSBs (10.7%). In total, these foods/food groups contributed 75.5% of total sugar intake. If baked products, that include cakes and cookies, are also accounted for and included, the contribution increased to almost 83.0%, 44.5% of which were from sources containing added sugars. When adolescents were differentiated from children, the main food group contributing to total sugars were sweets, especially among adolescent boys (Figure 1b), followed by fruits (17%), and the third main contributor was SSBs (12.8%); the latter being consistent in both boys and girls. The main food contributors in adolescent girls (Figure 1c) were fruits (18%) followed by sweets (16%), in comparison to boys which were sweets (18%), followed by fruits (16.2%). Overall, 100% juice contributed to 3% of total sugar intake and honey, added sugar, and syrups to 2.5%.

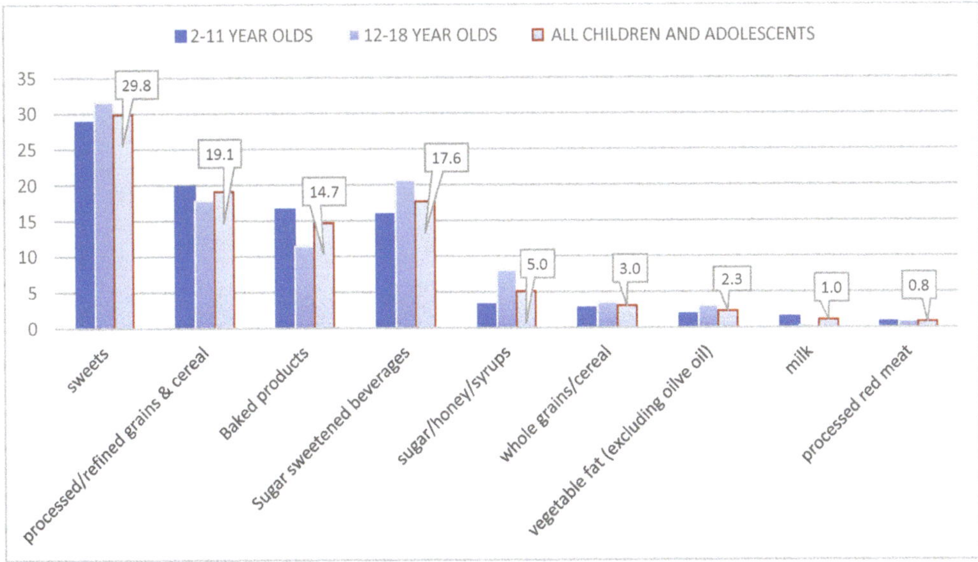

Figure 2. Main food groups contributors to added sugar intake in children and adolescents.

Figure 2 depicts the major food groups that contributed to added sugars, in total and by age group. The main two sources of added sugars were sweets (29.8%) and processed/refined grains and cereals (19.1%), in total and in both age groups. These were followed by SSBs (17.6%) in the total sample, due to high consumption among adolescents (20.6%), but was fourth in children (11.4%), following baked products (16.6%).

Table 2 summarizes the estimated mean energy, macronutrient, and selected recommended food group intakes, by level of added sugar consumption, according to current recommendations (≥10% compared to <10% of total calories). Total sugars were significantly lower, and total fat and animal protein were higher, in children and adolescents consuming <10% of energy from added sugars. Non-sugar carbohydrates were significantly higher whereas total fiber was lower, in children consuming <10% of energy from added sugars but did not differ in adolescents. No significant differences were found in vegetable protein, being low in both groups. Regarding specific recommended food groups, all were higher among children with greater added sugar intake, whereas in adolescents' fruit and animal protein intake was significantly higher for those consuming <10% of energy from added sugars. More specifically, 75% of children aged 2–11 years, with <10% of energy intake from added sugars, consumed no fruits or vegetables. Twenty five percent (25%) of the children consuming ≥10% added sugars also reported no fruit consumption and 75% consumed less than one vegetable portion per day (one fruit and one vegetable choice ~80 g in children). In comparison, half of the adolescents with <10% of energy from added sugars consumed one fruit portion and almost two vegetable portions, whereas those consuming 10% of added sugars had significantly lower fruit intake ($p < 0.029$).

Sensitivity analysis revealed that total sugar intake was 14.7% in under-reporters, 17.1% in plausible-reporters, and 14.3 % in over-reporters. Added sugar intake was 7.8%, 8.5%, and 8.3% respectively, with 31% of plausible- and 31.6% over-reporters consuming ≥10% of energy from added sugar. This was significantly lower among under-reporters with only 9.8% reporting added sugar intake ≥10% of total energy.

Table 2. Estimated mean energy, macronutrient, and selected food group intake in children and adolescents by added sugar intake status.

	Food Group Intake Per Day	Added Sugars Intake Status		Significance
		<10% Total Calories	≥10% Total Calories	*p* Value
Children (2–11 years)	**Food groups (median; IQR)**			
	Fruit, gr	0 (0, 116)	71.3 (0, 166)	<0.001
	Vegetables, gr	0 (0, 71)	55 (14, 96.2)	<0.001
	Animal protein, gr	53 [1]	54 (22, 86)	<0.001
	Non milk dairy, gr	30 [1]	20.1 (0, 46.5)	<0.001
	Energy, kcal	1378 (571)	1692 (777)	<0.001
	Macronutrients, % Energy			
	Non sugars carbohydrates	29.8 (6.3)	27.0 (8.9)	<0.001
	Total sugars	14.6 (4.3)	23.9 (8.1)	<0.001
	Animal protein	7.5 (7.5, 7.6)	7.1 (4.7, 10.3)	<0.001
	Vegetable protein	4.1 (1.2)	4.2 (1.8)	0.514
	Total Fat	40.1 (5.8)	36.7 (7.0)	<0.001
	Total Fiber	7.1 (7.1, 9.9)	10.7 (7.5, 16.5)	<0.001
Adolescents (12–18 years)	**Food groups (median; IQR)**			
	Fruit, gr	87 (0, 193)	4 (0, 147)	0.029
	Vegetables, gr	138 (59, 254)	110 (41, 199)	0.135
	Animal protein, gr	98 (52, 183)	59 (23, 106)	<0.001
	Non milk dairy, gr	30 (11, 73)	23 (7, 60)	0.180
	Energy, kcal	1862 (749)	1799 (780)	0.484
	Macronutrients, %Energy			
	Non sugars carbohydrates	29.4 (8.3)	29.3 (10.0)	0.889
	Total sugars	12.2 (6.5)	20.7 (6.0	<0.001
	Animal protein, %Energy	9.0 (5.5, 11.8)	6.4 (4.1, 10.5)	<0.001
	Vegetable protein, %Energy	4.0 (2.0)	4.2 (2.0)	0.607
	Total Fat, %Energy	42.1 (9.0)	37.7 (9.6)	<0.001
	Total Fiber, gr/day	12.9 (8.9, 18.3)	13.1 (9.4, 18.2)	0.726

[1] no IQR differences. Skewed variables are summarized using median and IQR, whereas normally distributed variables with mean (±sd). Group comparisons have been made by Student *t*-test vs Kruskal–Wallis test when skewed. Animal protein food group includes all meat, eggs, and seafood/fish. Vegetables include all starchy, non-starchy vegetables and potatoes. Fruit group includes fruit and 100% fruit juice. Non-milk dairy includes yogurt and cheese. Plant protein was not included in food groups, due to limited consumption. Non sugars carbohydrates include total carbohydrate intake after subtracting total sugar intake, and then expressed as %energy intake.

The likelihood of being overweight/obese was 2.33 times higher (95% CI: 1.298, 4.183, $p = 0.005$) for children and adolescents with total sugar consumption above the population median (13.6%) when specific food groups adjusted for energy were accounted for (Table 3; model 1), and 1.68 (95% CI: 1.008, 2.817, $p = 0.047$) when macronutrient intakes with respect to energy intake were accounted for (model 2). Children and adolescents consuming ≥10% of energy from added sugars were 2.57 times more likely (95% CI: 1.398, 4.717, $p = 0.002$) to be overweight/obese compared to those that consumed less <10% in model 1 and 1.77 time in model 2 (95% CI: 1.008, 3.096, $p = 0.047$). The results for the odds of overweight/obese when the median for added sugars was assessed, were very similar with those for ≥10%, as expected, as the median intake was 9.4%. In both cases the models were adjusted for level of physical activity, total screen time, sex, and parental educational level and professional status. The foods model was adjusted for total energy intake as well. Following the mixed effects logistic regressions, predictive probability of overweight/obesity, based on the models used, were derived.

Table 3. Odds of overweight/obesity in total sample using 2 models: (a) food groups and (b) macronutrients.

	Model 1			Model 2		
	Odds Ratio	**[95% Conf.Interval]**		**Odds Ratio**	**[95% Conf.Interval]**	
Total Sugar, % energy (above vs below median)	2.33	1.298	4.183	1.69	1.008	2.817
Added Sugars, % energy (above vs below median)	2.64	1.459	4.789	1.80	1.046	3.124
Added Sugars ≥10% energy	2.57	1.398	4.717	1.77	1.008	3.096

Results following mixed effects logistic regression; for total sugars population median was used; reference population were children <50th percentile. Model 1 includes: animal- and plant- protein, non-milk dairy, fruit, and vegetable intakes, adjusted for energy; Model 2 includes: macronutrient intakes (fiber, fat, protein, other non-sugar carbohydrates). Both models were adjusted for children's and adolescent's activity level (IPAQ), total screen time and for the primary guardian's educational level and professional status. Significance at $\alpha = 5\%$.

In Figure 3a,b, the predicted probabilities of being overweight/obese are depicted. In Figure 3a the probability increased significantly for higher total sugar and for ≥10% of added sugar intake, when accounting for specific food group intakes, as per Table 2. Specifically, consuming >10% of total energy from added sugar increases the probability of being overweight/obese by 8.3% in total sample ($p = 0.002$), by 7.4% in children ($p = 0.003$), and by 10.8% in adolescents ($p = 0.005$). The probability was lower for total sugar intake. In particular, it was 6.6% in children ($p = 0.005$), and 9.7% in adolescents ($p = 0.008$). When the model included macronutrients instead of food groups, the probability of being overweight/obese was lower in both cases compared to the first model (Figure 3b) but remained statistically significant.

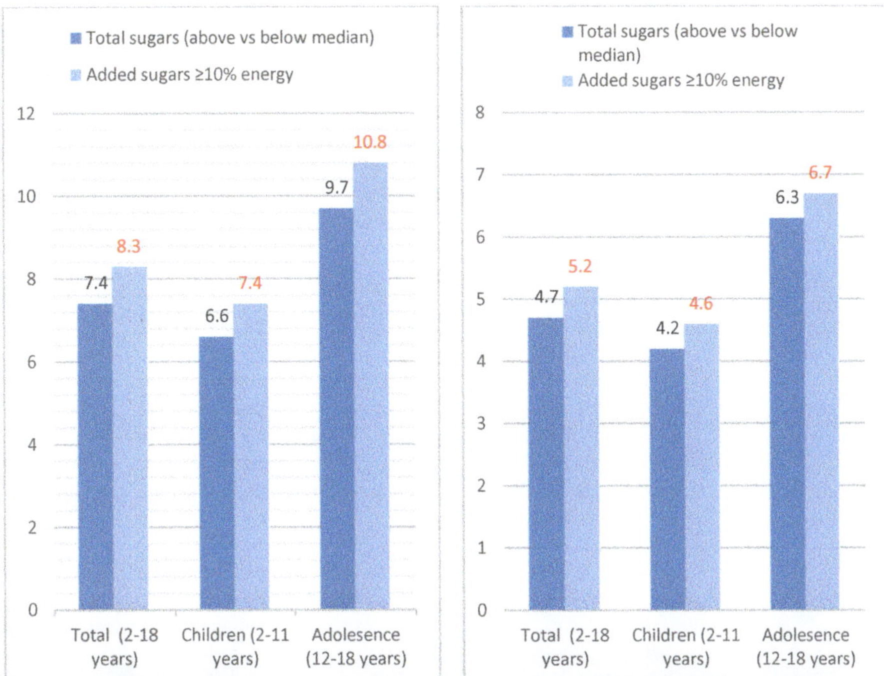

(a) Accounting for food groups **(b)** Accounting for macronutrient intake

Figure 3. Predicted probability of being overweight/obese for children and adolescents by total- and added-sugar intake. (**a**) Accounting for food groups. (**b**) Accounting for macronutrient intake.; values in red represent predicted probability when consuming added sugars ≥10% of energy.

4. Discussion

The present study examined total sugar and added sugar intakes, their major food contributors, and it assessed the relations between higher intake and likelihood of obesity in a representative sample of children and adolescents in Greece. The major finding of this study wasthat a high percentage of children and adolescents (20.6%) consumed added sugars above the recommended threshold of 10%, placing them at increased risk for overweight and obesity. It was also found that higher dietary total sugars intake was linked with higher added sugars, since approximately half of the amounts were derived from foods containing added sugars: sweets, processed grains, SSBs, and baked products. These findings included all children and adolescents assessed, excluding extreme over-reporters, to avoid biased errors in estimates. When under-reporters were excluded, which had a significant lower intake, this percentage reached 30%, further underlining the magnitude of added sugar intake with potential associated obesity risk. Specifically, the likelihood of being overweight or obese more than doubled for those with higher than recommended added sugar intakes and the predicted probability of becoming overweight/obese increased by 8.3%. If we were to extrapolate these findings to the total number of children and adolescents in Greece, it is estimated that 36 895 children and adolescents are likely to be or will probably become overweight and obese.

The definitions for sugars, including total and added, remains to be a challenge for researchers. The WHO On uses the term "free sugar" instead of "added sugar" in their sugar recommendations, which means that free sugar includes added and total sugars which are naturally present in fruit juices and fruit juice concentrates [37]. To overcome this problem, and account for all sugars but at the same time, distinguish the effects of added sugars, this study investigated total sugar as well as added sugar intakes. According to recent data, total sugar intake ranges from 17% to 34.8% in children and from 15.4% to 29.6% in adolescents worldwide [38]. Based on results from our study, total sugars were at the lower end of the worldwide ranges reported, but over half of total sugars were attained by addition. This was due to the fact that the majority of total sugars in this sample were derived from sweets, processed/refined grains and cereals, and SSBs, especially among adolescents, and fruit intake was very low. In addition, the median intake of added sugars among children and adolescents was close to the upper recommended intake, with one fifth of children and one fourth of adolescents exceeding it. These results are in agreement with an earlier study that reported high intake of added and naturally present simple sugars among school aged children [14]. In Canada the range of added sugars intake was estimated from 10% to 13% of total energy intake among different population subgroups, with higher intakes among adolescents [39], the latter being in agreement with the results of our study as well as results from a recent review of nationally representative surveys across the world [38]. This review reported that intakes of added sugars were higher among school aged children and adolescents, than in younger children.

It has been reported that higher added sugar intake is associated with lower diet quality [40,41]. In this study the majority of children and adolescents, irrespectively of added sugar intake, had poor dietary intakes as depicted by the median intakes of fruit, vegetables, fiber, and vegetable protein. This is in accordance with the results of other studies that found that even at lower deciles of added sugars intake, large percentages of children and adolescents had nutrient intakes below the average requirements [42], underlying poor dietary habits. Between studies controversies can be explained by differences in assessing the main foods that contribute to total and added sugar intake. In this population, other than sweets, processed/refined grains and cereals were the main contributors of total and added sugars, in both children and adolescents, foods that are usually enriched with many nutrients. Fruits including 100% fruit juices, although low in the majority of the population, also highly contributed, unlike drinks containing added sugars, that were at the lower end of food contributors.

The probability of overweight and obesity increased significantly with higher total sugar intake and higher added sugar intake above the recommendations. The potential effect of total sugar intake, irrespective of total energy consumption, was evaluated in relation to other food group intakes, granted that a recent systematic review suggested a diet with a lower percentage of obesogenic foods, including sugars, may reduce the risk of developing obesity among children [22]. Simple sugars overall have a high glycemic index, and lead to a faster and greater insulin secretion; a hormone that has an anabolic effect. This can be one major reason of the higher probability of overweight and obesity found in this study among children and adolescents with added sugar consumption above 10% of total energy in foods of low quality. The latter was reinforced by the fact that the likelihood of overweight and obesity was higher when the same model was adjusted for actual food intake, not only macronutrients, while maintaining total energy and other obesity confounders constant. To our knowledge, this is the first study that has compared the effects of food vs macronutrients on weight status, without seeking for specific dietary patterns but rather examining actual intakes of protective healthy foods and actual macronutrients. This information is important, as obesity is hard to treat [43], and although body weight can be modified by diet, treatments generally tend to have poor long term outcomes [44,45]. This is because obesity demands a more sophisticated approach than counting calories [44], including the understanding of pathophysiological and endocrine functions that occur post food consumption, reinforcing the results of this study. Furthermore, it has been reported that overconsumption of foods high in simple sugars replace more nutrient dense foods, and result in nutrient inadequacies [9,37]. This may be the case among children and adolescents in Greece based on a previous study reporting that most vitamins and essential minerals were derived from low-quality foods [46].

Study limitations include potential under-reporting of food-consumption, which could have led in an underestimation of energy, overall macronutrient intake, including total and added sugar intake. This said, based on the sensitivity analysis performed sugar intake can only be more but not less than the reported, since under-reporters had significantly lower consumption. Another limitation due to the cross-sectional nature of the study is that temporal relationship cannot be established; however, prediction probability analysis was used to estimate future trends, if levels of added sugars are not reduced and healthier dietary choices are not attained.

In conclusion children and adolescents need to develop healthy nutrition habits to maintain healthy body weight and prevent obesity development in adolescence and beyond. Interventions to prevent and control obesity focus on solutions which help acquire healthier behaviors [47]. Effort should be placed, not only to reduce added sugars, but to improve overall dietary intake in terms of foods consumed, emphasizing vegetables and plant-based protein. The results of this study confirm the strong recommendation of WHO for a reduced sugar intake, emphasizing on reducing intake to <10% of total energy intake for preventing and controlling excess body weight.

Supplementary Materials: The following are available online at https://www.mdpi.com/article/10.3390/children8080676/s1, Supplementary Table S1: Grouping of primary food groups for specific study purposes.

Author Contributions: Conceptualization, E.M. and A.Z.; Methodology, E.M., E.P. and N.C., Formal Analysis, E.M., Writing—Original Draft Preparation, E.M., N.C. and E.P., Writing—Review and Editing, A.Z., E.R., G.P.C., V.B., G.M., D.P. and R.M. Project Administration, A.Z., R.M. and G.M., Funding Acquisition, A.Z. All authors have read and agreed to the published version of the manuscript.

Funding: The study was co-funded by the European Union (European Social Fund) and Greece (Ministry of Health) under the Operational Program "Human Resources Development 2007–2013".

Institutional Review Board Statement: The study was conducted according to the guidelines of the Declaration of Helsinki, and approved by Ethics Committee of the Department of Food Science and Human Nutritional (Agricultural University of Athens) and the Hellenic Data Protection Authority (ethical approval received April 2013).

Acknowledgments: Special thanks to all the fieldworkers.

Conflicts of Interest: The authors declare no conflict of interest.

References

1. World Health Organization (WHO). Obesity and Overweight. 2020. Available online: https://www.who.int/news-room/fact-sheets/detail/obesity-and-overweight (accessed on 27 April 2021).
2. NCD Risk Factor Collaboration (NCD-RISC). Worldwide trends in body-mass index, underweight, overweight, and obesity from 1975 to 2016: A pooled analysis of 2416 population-based measurement studies in 128Â·9 million children, adolescents, and adults. *Lancet* **2017**, *390*, 2627–2642. [CrossRef]
3. Spinelli, A.; Buoncristiano, M.; Kovacs, V.A.; Yngve, A.; Spiroski, I.; Obreja, G.; Starc, G.; Pérez, N.; Rito, A.I.; Kunešová, M.; et al. Prevalence of Severe Obesity among Primary School Children in 21 European Countries. *Obes. Facts* **2019**, *12*, 244–258. [CrossRef]
4. Grundy, S.M.; Brewer, H.B., Jr.; Cleeman, J.I.; Smith, S.C., Jr.; Lenfant, C. Definition of metabolic syndrome: Report of the National Heart, Lung, and Blood Institute/American Heart Association conference on scientific issues related to definition. *Circulation* **2004**, *109*, 433–438. [CrossRef] [PubMed]
5. Powell, E.S.; Smith-Taillie, L.P.; Popkin, B.M. Added Sugars Intake Across the Distribution of US Children and Adult Consumers: 1977–2012. *J. Acad. Nutr. Diet.* **2016**, *116*, 1543–1550.e1541. [CrossRef]
6. Bray, G.A.; Popkin, B.M. Dietary sugar and body weight: Have we reached a crisis in the epidemic of obesity and diabetes? Health be damned! Pour on the sugar. *Diabetes Care* **2014**, *37*, 950–956. [CrossRef]
7. Kahn, R.; Sievenpiper, J.L. Response to Comment on Kahn and Sievenpiper. Dietary Sugar and Body Weight: Have We Reached a Crisis in the Epidemic of Obesity and Diabetes? We Have, but the Pox on Sugar Is Overwrought and Overworked. Diabetes Care 2014; 37: 957–962. *Diabetes Care* **2014**, *37*, e189. [CrossRef] [PubMed]
8. Ervin, R.B.; Ogden, C.L. Consumption of added sugars among U.S. adults, 2005–2010. *NCHS Data Brief.* **2013**, *122*, 1–8.
9. Erickson, J.; Slavin, J. Total, Added, and Free Sugars: Are Restrictive Guidelines Science-Based or Achievable? *Nutrients* **2015**, *7*, 2866–2878. [CrossRef]
10. Wang, H.; Jeong, H.; Kim, N.H.; Kang, Y.; Hwang, K.; Lee, H.; Hong, J.H.; Oh, K.S. Association between beverage intake and obesity in children: The Korea National Health and Nutrition Examination Survey (KNHANES) 2013–2015. *Nutr. Res. Pract.* **2018**, *12*, 307–314. [CrossRef] [PubMed]
11. Laverty, A.A.; Magee, L.; Monteiro, C.A.; Saxena, S.; Millett, C. Sugar and artificially sweetened beverage consumption and adiposity changes: National longitudinal study. *Int. J. Behav. Nutr. Phys. Act.* **2015**, *12*, 137. [CrossRef]
12. Millar, L.; Rowland, B.; Nichols, M.; Swinburn, B.; Bennett, C.; Skouteris, H.; Allender, S. Relationship between raised BMI and sugar sweetened beverage and high fat food consumption among children. *Obesity* **2014**, *22*, E96–E103. [CrossRef]
13. Yu, P.; Chen, Y.; Zhao, A.; Bai, Y.; Zheng, Y.; Zhao, W.; Zhang, Y. Consumption of sugar-sweetened beverages and its association with overweight among young children from China. *Public Health Nutr.* **2016**, *19*, 2336–2346. [CrossRef]
14. Farajian, P.; Risvas, G.; Panagiotakos, D.B.; Zampelas, A. Food sources of free sugars in children's diet and identification of lifestyle patterns associated with free sugars intake: The GRECO (Greek Childhood Obesity) study. *Public Health Nutr.* **2016**, *19*, 2326–2335. [CrossRef]
15. Nicklas, T.A.; O'Neil, C.E.; Liu, Y. Intake of added sugars is not associated with weight measures in children 6 to 18 years: National Health and Nutrition Examination Surveys 2003–2006. *Nutr. Res.* **2011**, *31*, 338–346. [CrossRef]
16. Lommi, S.; Figueiredo, R.A.O.; Tuorila, H.; Viljakainen, H. Frequent use of selected sugary products associates with thinness, but not overweight during preadolescence: A cross-sectional study. *Br. J. Nutr.* **2020**, *124*, 631–640. [CrossRef]
17. Lin, P.Y.; Lin, F.Y.; Chen, T.C.; Chen, W.L.; Doong, J.Y.; Shikanai, S.; Sarukura, N.; Yamamoto, S. Relationship between Sugar Intake and Obesity among School-Age Children in Kaohsiung, Taiwan. *J. Nutr. Sci. Vitaminol.* **2016**, *62*, 310–316. [CrossRef] [PubMed]
18. Shikanai, S.; Koung Ry, L.; Takeichi, H.; Emiko, S.; San, P.; Sarukura, N.; Kamoshita, S.; Yamamoto, S. Sugar intake and body weight in Cambodian and Japanese children. *J. Med. Investig.* **2014**, *61*, 72–78. [CrossRef] [PubMed]
19. Morales-Suarez-Varela, M.; Peraita-Costa, I.; Llopis-Morales, A.; Picó, Y.; Bes-Rastrollo, M.; Llopis-Gonzalez, A. Total Sugar Intake and Macro and Micronutrients in Children Aged 6–8 Years: The ANIVA Study. *Nutrients* **2020**, *12*, 349. [CrossRef] [PubMed]
20. Valente, H.; Teixeira, V.; Padrão, P.; Bessa, M.; Cordeiro, T.; Moreira, A.; Mitchell, V.; Lopes, C.; Mota, J.; Moreira, P. Sugar-sweetened beverage intake and overweight in children from a Mediterranean country. *Public Health Nutr.* **2011**, *14*, 127–132. [CrossRef]
21. Wang, J.; Shang, L.; Light, K.; O'Loughlin, J.; Paradis, G.; Gray-Donald, K. Associations between added sugar (solid vs. liquid) intakes, diet quality, and adiposity indicators in Canadian children. *Appl. Physiol. Nutr. Metab.* **2015**, *40*, 835–841. [CrossRef]
22. Liberali, R.; Kupek, E.; Assis, M.A.A.D. Dietary Patterns and Childhood Obesity Risk: A Systematic Review. *Child. Obes.* **2020**, *16*, 70–85. [CrossRef] [PubMed]
23. Magriplis, E.; Dimakopoulos, I.; Karageorgou, D.; Mitsopoulou, A.V.; Bakogianni, I.; Micha, R.; Michas, G.; Ntouroupi, T.; Tsaniklidou, S.M.; Argyri, K.; et al. Aims, design and preliminary findings of the Hellenic National Nutrition and Health Survey (HNNHS). *BMC Med. Res. Methodol.* **2019**, *19*, 37. [CrossRef] [PubMed]
24. Blanton, C.A.; Moshfegh, A.J.; Baer, D.J.; Kretsch, M.J. The USDA Automated Multiple-Pass Method accurately estimates group total energy and nutrient intake. *J. Nutr.* **2006**, *136*, 2594–2599. [CrossRef]

25. Moshfegh, A.J.; Rhodes, D.G.; Baer, D.J.; Murayi, T.; Clemens, J.C.; Rumpler, W.V.; Paul, D.R.; Sebastian, R.S.; Kuczynski, K.J.; Ingwersen, L.A.; et al. The US Department of Agriculture Automated Multiple-Pass Method reduces bias in the collection of energy intakes. *Am. J. Clin. Nutr.* **2008**, *88*, 324–332. [CrossRef] [PubMed]
26. Burrows, T.L.; Martin, R.J.; Collins, C.E. A systematic review of the validity of dietary assessment methods in children when compared with the method of doubly labeled water. *J. Am. Diet. Assoc.* **2010**, *110*, 1501–1510. [CrossRef]
27. Trichopoulou, A.; Georga, K. *Composition Tables of Foods and Greek Dishes*; Parisianou Publications: Athens, GA, USA, 2004.
28. Block, G.; Dresser, C.M.; Hartman, A.M.; Carroll, M.D. Nutrient sources in the American diet: Quantitative data from the NHANES II survey. I. Vitamins and minerals. *Am. J. Epidemiol.* **1985**, *122*, 13–26. [CrossRef]
29. U.S. Department of Health and Human Services; U.S. Department of Agriculture. *2015–2020 Dietary Guidelines for Americans*, 8th ed.; 2015. Available online: https://health.gov/our-work/food-nutrition/previous-dietary-guidelines/2015 (accessed on 9 May 2021).
30. Cole, T.J.; Lobstein, T. Extended international (IOTF) body mass index cut-offs for thinness, overweight and obesity. *Pediatric Obes.* **2012**, *7*, 284–294. [CrossRef] [PubMed]
31. Brophy, S.; Cooksey, R.; Gravenor, M.B.; Mistry, R.; Thomas, N.; Lyons, R.A.; Williams, R. Risk factors for childhood obesity at age 5: Analysis of the millennium cohort study. *BMC Public Health* **2009**, *9*, 467. [CrossRef]
32. Dwyer, G.M.; Hardy, L.L.; Peat, J.K.; Baur, L.A. The validity and reliability of a home environment preschool-age physical activity questionnaire (Pre-PAQ). *Int. J. Behav. Nutr. Phys. Act.* **2011**, *8*, 86. [CrossRef]
33. Hagströmer, M.; Bergman, P.; De Bourdeaudhuij, I.; Ortega, F.B.; Ruiz, J.R.; Manios, Y.; Rey-López, J.P.; Phillipp, K.; von Berlepsch, J.; Sjöström, M. Concurrent validity of a modified version of the International Physical Activity Questionnaire (IPAQ-A) in European adolescents: The HELENA Study. *Int. J. Obes.* **2008**, *32*, S42–S48. [CrossRef] [PubMed]
34. Arvidsson, D.; Slinde, F.; Hulthèn, L. Physical activity questionnaire for adolescents validated against doubly labelled water. *Eur. J. Clin. Nutr.* **2005**, *59*, 376–383. [CrossRef] [PubMed]
35. Börnhorst, C.; Huybrechts, I.; Hebestreit, A.; Krogh, V.; De Decker, A.; Barba, G.; Moreno, L.A.; Lissner, L.; Tornaritis, M.; Loit, H.M.; et al. Usual energy and macronutrient intakes in 2–9-year-old European children. *Int. J. Obes.* **2014**, *38*, S115–S123. [CrossRef] [PubMed]
36. Tooze, J.A.; Krebs-Smith, S.M.; Troiano, R.P.; Subar, A.F. The accuracy of the Goldberg method for classifying misreporters of energy intake on a food frequency questionnaire and 24-h recalls: Comparison with doubly labeled water. *Eur. J. Clin. Nutr.* **2012**, *66*, 569–576. [CrossRef]
37. Hess, J.; Latulippe, M.E.; Ayoob, K.; Slavin, J. The confusing world of dietary sugars: Definitions, intakes, food sources and international dietary recommendations. *Food Funct.* **2012**, *3*, 477–486. [CrossRef]
38. Newens, K.J.; Walton, J. A review of sugar consumption from nationally representative dietary surveys across the world. *J. Hum. Nutr. Diet.* **2016**, *29*, 225–240. [CrossRef]
39. Brisbois, T.D.; Marsden, S.L.; Anderson, G.H.; Sievenpiper, J.L. Estimated intakes and sources of total and added sugars in the Canadian diet. *Nutrients* **2014**, *6*, 1899–1912. [CrossRef] [PubMed]
40. Marriott, B.P.; Olsho, L.; Hadden, L.; Connor, P. Intake of added sugars and selected nutrients in the United States, National Health and Nutrition Examination Survey (NHANES) 2003–2006. *Crit. Rev. Food Sci. Nutr.* **2010**, *50*, 228–258. [CrossRef]
41. Joyce, T.; Gibney, M.J. The impact of added sugar consumption on overall dietary quality in Irish children and teenagers. *J. Hum. Nutr. Diet.* **2008**, *21*, 438–450. [CrossRef]
42. Fulgoni, V.L., III; Gaine, P.C.; Scott, M.O.; Ricciuto, L.; DiFrancesco, L. Association of Added Sugars Intake with Micronutrient Adequacy in US Children and Adolescents: NHANES 2009–2014. *Curr. Dev. Nutr.* **2019**, *3*, nzz126. [CrossRef]
43. Felitti, V.J.; Jakstis, K.; Pepper, V.; Ray, A. Obesity: Problem, solution, or both? *Perm. J.* **2010**, *14*, 24–30. [CrossRef]
44. Benton, D.; Young, H.A. Reducing Calorie Intake May Not Help You Lose Body Weight. *Perspect. Psychol. Sci.* **2017**, *12*, 703–714. [CrossRef]
45. McKelvey, L.M.; Saccente, J.E.; Swindle, T.M. Adverse Childhood Experiences in Infancy and Toddlerhood Predict Obesity and Health Outcomes in Middle Childhood. *Child. Obes.* **2019**, *15*, 206–215. [CrossRef] [PubMed]
46. Mitsopoulou, A.V.; Magriplis, E.; Michas, G.; Micha, R.; Chourdakis, M.; Chrousos, G.P.; Roma, E.; Panagiotakos, D.B.; Zampelas, A.; Karageorgou, D.; et al. Micronutrient dietary intakes and their food sources in adults: The Hellenic National Nutrition and Health Survey (HNNHS). *J. Hum. Nutr. Diet.* **2021**, *34*, 616–628. [CrossRef] [PubMed]
47. Akgul Gundogdu, N.; Sevig, E.U.; Guler, N. The effect of the solution-focused approach on nutrition-exercise attitudes and behaviours of overweight and obese adolescents: Randomised controlled trial. *J. Clin. Nurs.* **2018**, *27*, e1660–e1672. [CrossRef] [PubMed]

children

MDPI

Systematic Review

The Impact of Nutritional and Lifestyle Changes on Body Weight, Body Composition and Cardiometabolic Risk Factors in Children and Adolescents during the Pandemic of COVID-19: A Systematic Review

Kalliopi Karatzi †, Kalliopi-Anna Poulia †, Emilia Papakonstantinou and Antonis Zampelas *

Laboratory of Dietetics and Quality of Life, Department of Food Science and Human Nutrition, Agricultural University of Athens, Iera Odos 75, 11855 Athens, Greece; pkaratzi@aua.gr (K.K.); lpoulia@gmail.com (K.-A.P.); apapakonstantinou@gmail.com (E.P.)
* Correspondence: azampelas@aua.gr; Tel.: +30-210-5294701
† These authors contributed equally to this work.

Abstract: The coronavirus (COVID-19) pandemic and the measures taken by most countries to curb virus transmission, such as social distancing, distance learning, population, home confinement and disruption of all organized activities, has affected children and adolescents worldwide. The aim of this review was to assess the role of diet and lifestyle changes due to COVID-19 measures on body weight/composition and cardiometabolic risk factors in children and adolescents. An electronic search was conducted in PUBMED, COCHRANE, Google Scholar and SCOPUS databases up to 31 October 2021. 15 eligible studies were identified. According to the studies included in the analysis, COVID-19 measures seem to have had a negative impact on the diets and lifestyles of children and adolescents, with a consequent increase in body weight and central fat accumulation. On the other hand, the parental presence and control resulted in better glycaemic control in children with diabetes mellitus (DM) Type 1, but the effect of the pandemic in the glycaemic control of children with DM2 2 is controversial. Finally, diet and lifestyle changes had a differential impact on children's hypertension prevalence. These findings point to the need for public policy measures to prevent obesity and its complications, to and improve diet and lifestyle during the continuing and yet unresolved COVID-19 epidemic.

Keywords: diet; lifestyle; body composition; cardiometabolic risk; children; adolescents; COVID-19

check for updates

Citation: Karatzi, K.; Poulia, K.-A.; Papakonstantinou, E.; Zampelas, A. The Impact of Nutritional and Lifestyle Changes on Body Weight, Body Composition and Cardiometabolic Risk Factors in Children and Adolescents during the Pandemic of COVID-19: A Systematic Review. *Children* **2021**, *8*, 1130. https://doi.org/10.3390/children8121130

Received: 12 November 2021
Accepted: 29 November 2021
Published: 4 December 2021

1. Introduction

Coronavirus disease (COVID-19) caused by the severe acute respiratory syndrome coronavirus 2 (SARS-CoV-2) is one of the most serious pandemics worldwide that has infected and killed millions of people. To curb virus transmission, several preventive measures have been taken by most of the countries on all continents, such as the limitation of travelling, distance learning in all educational levels, tele-working, and total confinement of the general population at home. These measures, also known as lockdown, caused disruption of daily activities in all ages due to the need for social distancing to control the epidemiology of the disease [1].

Some of the major consequences of the pandemic measures, such as social distancing and the shutdown of live educational activities have affected lifestyle and dietary behaviors related to overall health in both children and their families. Interestingly it was suggested that the lockdown is responsible for the consumption of poor-quality food, such as ultra-processed, calorie-dense comfort food when compared to standard living conditions [2,3]. The abrupt interruption of organized physical activities such as sports are paralleled with increased sedentary habits and screen time due to the inability to perform social gathering, distance learning, cancellation of the function of recreation programs and playgrounds and confinement at home. These changes have also affected sleep quality and duration. More specifically, the cancelation of live participation in educational activities attenuated all unfavorable consequences, as school increases opportunities for structured and recreational physical activities, decreases screen time, reduces opportunities for snacking and enables earlier school-day bedtimes [4]. Therefore, children's social distancing enhanced feelings of boredom and stress, increased snacking of energy-dense foods and emotional eating, and at the same time reduced the participation in structured physical activities, and increased sedentary time and inadequate or low quality sleep time [5]. The consequent disproportion between energy intake and energy spent inevitably resulted in a positive energy balance, increased fat disposition and weight gain [2,6].

Childhood obesity is another epidemic and it is considered as the most important health problem of the 21st century concerning 38 million children under the age of 5, and over 340 million children and adolescents aged 5–19 [7,8]. It has a multi-factorial origin with genes, perinatal factors, social setting background, obesogenic environment, and lifestyle habits/dietary preferences, physical activity, sedentary time and sleep duration playing the most important roles [6].

The measures implemented due to the COVID-19 pandemic have created an obesogenic environment in children with less physical activity, longer sedentary time, unhealthy and energy dense food choices and sleep disturbances, which has probably led to even higher childhood obesity prevalence [9]. In turn, childhood obesity amplifies the risk of cardiometabolic conditions such as hypertension, dyslipidemia, diabetes and cardiovascular disease (CVD) [10–12]. At the same time, healthy food choices and regular physical activity are also considered to play a significant role in health in the early stages of life, which can also lower the risk of developing non-communicable (NCDs) diseases in adulthood [8].

These unprecedented living conditions, with confinement at home, increased food consumption and sedentary time during the day, may have led to significant alterations in body composition and body weight, as well as in deterioration of cardiometabolic health. As the COVID-19 pandemic remains a resisting and unresolved health problem, it continues to affect children's lives and overall health.

Therefore, the aim of this systematic review was to collect and critically evaluate all the available scientific work related to the interplay between diet and lifestyle changes and children's body composition and cardiometabolic risk during the COVID-19 pandemic, in order to enable public health initiatives for the prevention of the negative consequences of obesity and cardiometabolic risk early in life.

2. Materials and Methods

The present review was registered and published by PROSPERO, under the title "Systematic review of the role of lifestyle and diet on body composition changes and cardiometabolic risk factors in children during the pandemic of COVID-19" (registration number: CRD42021279972). The present work was prepared and presented in accordance with the guidelines of the PRISMA statements.

2.1. Search Strategy

A systematic search for eligible studies was performed through October 2021 by two separate reviewers on the PUBMED, COCHRANE, Google Scholar and SCOPUS databases using identical language, age, and date limits. The last day of searches in all mentioned databases was 31 October. Search keywords and medical subject headings (MESH terms) applied were: [("diet" or "dietary" or "lifestyle") and ("body composition" or "body weight" or "weight status" or "weight change" or "BMI" or "waist circumference" or "childhood obesity" or "children overweight/obese") or ("cardiometabolic" or "cardiovascular" or "hyperlipidemia" or "dyslipidemia" or "hypertension" or "diabetes" or "metabolic syndrome"]. Studies were limited to the English language and human studies in children aged 0–18 years. Reference lists and related articles of included studies were also examined for additional relevant articles (Figure 1).

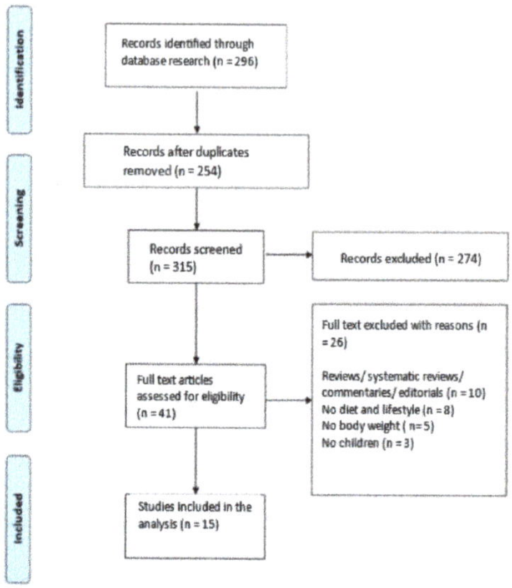

Figure 1. PRISMA flow chart of included studies.

2.2. Eligibility Criteria

The studies that were decided to be included in the review were examining the possible role of diet and lifestyle changes on children's body composition and cardiometabolic conditions. The inclusion criteria applied were peer-reviewed, epidemiological studies or clinical trials, English language, human studies, children 0–18 years of age, published from the year 2020 and onwards, with clear results regarding changes in body weight/body composition or parameters related to cardiometabolic health. The following exclusion criteria were applied: studies conducted in adults, animal studies, reviews, systematic reviews, meta-analyses, commentaries/letters and editorials.

2.3. Selection of Studies and Data Extraction

Two reviewers performed independent research in the databases mentioned before, for titles and abstracts of all the relative resources. The search results were then imported to EndNote Software for identification and removal of all duplicates. Afterwards the two reviewers removed articles according to eligibility criteria in order to determine the final number of studies to be included in the review. Discrepancies were resolved after discussion. After agreement, full text screening was carried out. Qualitative and quantitative data from all included articles were extracted by both reviewers. The extracted data included specific details for study design, population characteristics, assessment methods used and changes in diet, lifestyle, body composition or body weight and cardiometabolic conditions.

2.4. Risk of Bias

A risk of bias tool was conducted with the Revised Cochrane risk-of-bias tool for randomized trials (RoB 2) [13]. The risk of bias covered the six domains of bias as described in the tool: election bias, performance bias, detection bias, attrition bias, reporting bias, and other bias. Relative plots were created for all the studies included in the analysis (Figures 2 and 3).

3. Results

The initial search identified 296 references (Figure 1). After scanning for potential duplicates, 42 references were excluded. The remaining 254 references were screened according to title/abstract and reference lists, and relevant articles were screened. These procedures resulted in 315 references screened for relevance, out of which 274 were excluded. The remaining 41 articles underwent a full-text review based on inclusion and exclusion criteria. Further to this, 26 articles were excluded due to article type (editorials, commentaries or irrelevant reviews, $N = 10$) or due to inclusion/exclusion criteria, absence of diet and/or lifestyle outcomes ($N = 8$), absence of body composition/body weight outcomes ($N = 5$) adult population ($N = 3$). The final number of studies included in the present review is 15. These studies are summarized in Tables 1 and 2.

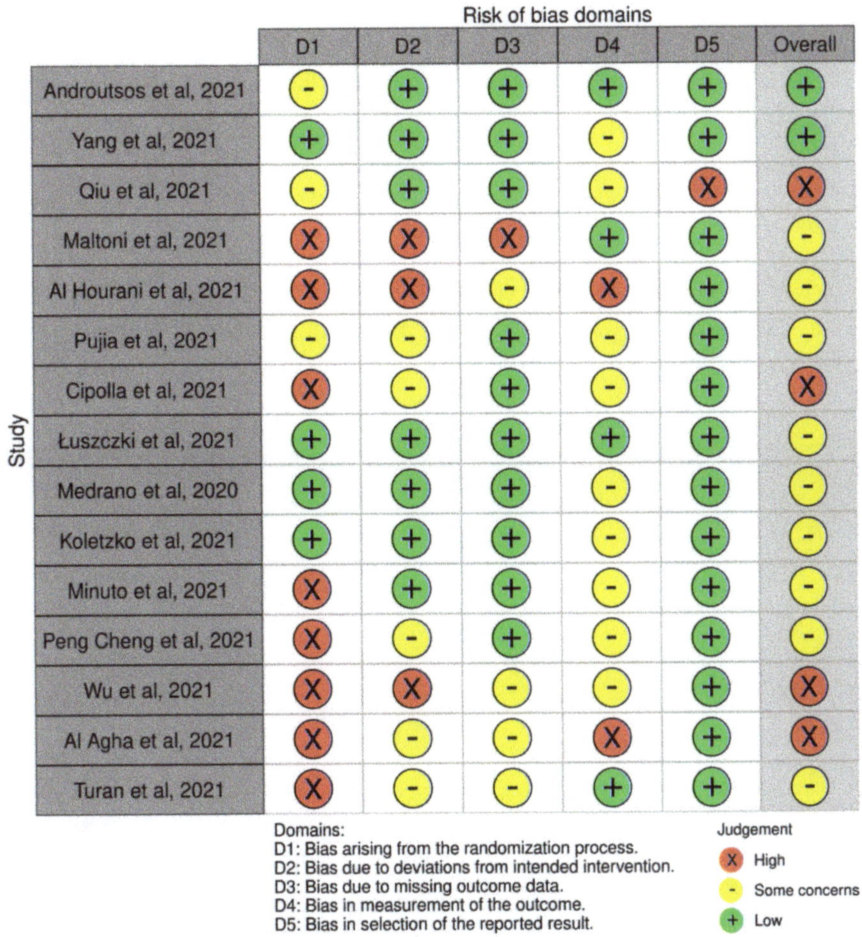

Figure 2. Studies with the representation of the Domains of risk of Bias.

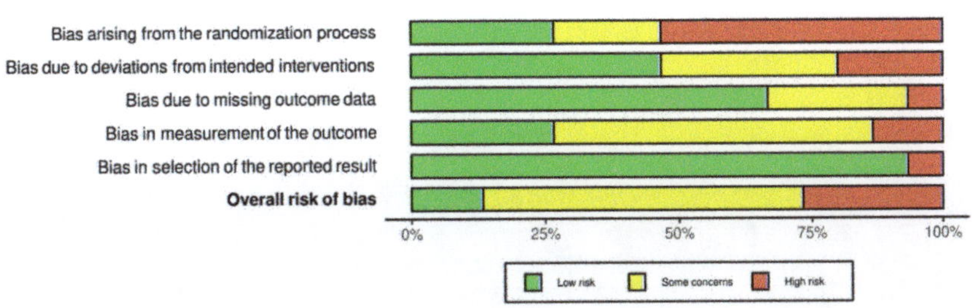

Figure 3. Overall presentation of the domains of risk among the studies.

Table 1. Characteristics of studies regarding the role of diet and lifestyle changes on body weight and body composition.

Identity	Country	Study Design	Participants (N)	Age Range (Years)	Diet Assessment	Lifestyle Assessment	Anthropometry and Body Composition Assessment	Outcome in Diet and Lifestyle	Outcome in Body Composition
Androutsos et al., 2021 [14]	Greece	Cross sectional Online survey	397	2–18	Questionnaire for the dietary habits of parents and children	Questionnaire on Sleep duration Screen time and Physical activity	Self-reported weight and height	↑ sleep duration ↑ screen time ↓ physical activity ↓ fast food ↑ fruit and vegetables ↑ breakfast	↑ Body weight in 35% of the participants
Yang et al., 2020 [15]	China	Retrospective study	10,082	16–18	N/A	Physical activity Questionnaire (IPAQ) long form	Self-reported weight and Height BMI	↓ physical activity frequency and intensity ↑ screen time	↑ BMI ↑ % Overweight/Obesity
Qiu et al., 2021 [16]	China	Cohort study	445	7–12	Questionnaire on diet	Questionnaires on lifestyle	Measured weight and Height	N/A	↑ % Overweight/Obesity
Maltoni et al., 2021 [17]	Italy	Cross sectional	51	10–18	Questionnaire on nutrition	Physical activity Questionnaire (IPAQ) sort form	Measured weight and Height Waist circumference	↑ Sedentary behavior ↓ Physical activity	↑ BMI ↑ Waist circumference ↑ Weight ↑ Weight/height ratio
Al Hourani et al., 2021 [18]	Jordan	Cross sectional	477	6–17	FFQ	Questionnaire on Screen time and Physical activity	Self-reported weight and height BMI for age Z scores	↑ Screen time ↑ Sedentary ↑Dietary intake	↑ BMI for age Z score ↑ Body Weight
Pujia et al., 2021 [19]	Italy	Cross sectional Online survey	439	5–14	Questionnaire on eating habits	Questionnaire on physical activity	Self-reported weight and height BMI	↑ Comfort food intake (chocolate, sweet snacks, and deserts ↑ pizza and bakery products ↓ sweetened beverages and candies ↑ sedentary lifestyle in adolescents	↑ Body Weight in 59.7% of the participants ↑ BMI
Cipolla et al., 2021 [20]	Italy	Cross sectional Telephone interview	64	8–18	Questionnaire on eating habits	Questionnaire on physical activity	Measured weight and height	↑ pizza and bakery products ↑ sedentary lifestyle ↑ Screen time	↑ BMI
Łuszczki et al., 2021 [21]	Poland	Cross sectional	1016	6–15	FFQ-6	Sleep quality and durationScreen timeSelf-reported physical activity	Measured weight and height BMI	↓ Sleep duration ↑ Sleep quality ↓ Physical activity ↑ Screen time ↓ fruit juices, carbonated sugar sweetened/diet drinks, meats ↓ canned food, fast f ood, snacks ↑ protein intake ↑ sweets	Non-significant changes

Table 1. *Cont.*

Identity	Country	Study Design	Participants (N)	Age Range (Years)	Diet Assessment	Lifestyle Assessment	Anthropometry and Body Composition Assessment	Outcome in Diet and Lifestyle	Outcome in Body Composition
Medrano et al., 2020 [22]	Spain	Longitudinal cohort study (MUGI project)	291	8–16	Adherence to Mediterranean diet (KIDMED)	Physical Activity and screen time during leisure were assessed by "The Youth Activity Profile" questionnaire (YAP)	Measured Height and Weight BMI Body Composition (BIA) Waist circumference	↓ Physical activity ↑ Screen time ↑ KIDMED score	Non-significant changes
Koletzko et al., 2021 [23]	Germany	Cross sectional Online survey	1000 parents with at least 1 child<14 y living in the same household	0–14	Questionnaire on diet habits	Questionnaire on Physical activity	Self-reported weight gain	↑ Fruit and Vegetables ↑ Carbohydrates ↑ Salty/sweet snacks ↑ Soft drinks ↓ Physical activity ↑ cooking at home	↑ Body weight in 9% of the children

BMI: Body Mass Index, FFQ: Food Frequency Questionnaire, N/A: non applicable, ↑ increase, ↓ decrease, ↑% increased percentage of patients, ↓% decreased percentage of patients.

Table 2. Study characteristics regarding the role of diet and lifestyle changes on cardiometabolic conditions.

Identity	Country	Study Design	Participants (N)	Age Range (Years)	Diet Assessment	Lifestyle Assessment	Cardiometabolic Risk Factors	Outcome in Diet and Lifestyle	Outcome in Cardiometabolic Risk Factors
Qiu et al., 2021 [16]	China	Cohort study	445	7–12	Questionnaire on diet	Questionnaires on lifestyle	Blood pressure measured	N/A	↑% with Elevated Blood pressure
Maltoni et al., 2021 [17]	Italy	Cross sectional	51	10–18	Questionnaire on nutrition	Physical activity Questionnaire (IPAQ) sort form	Blood pressure	↑ Sedentary behavior ↓ Physical activity	↓% with Elevated blood pressure
Minuto et al., 2021 [24]	Italy	Retrospective observational cohort study	202 101 (0–18) with T1DM	6–39	N/A	Self-reported physical activity	GCM for glucose monitoring HbA1c	↓ Physical activity	Improved glycemic control
Peng Cheng et al., 2021 [25]	Malaysia	Cross sectional	123 (93 patients with T1DM and 30 with T2DM)	0–18	Standardized questionnaire for diet and lifestyle	Standardized questionnaire for lifestyle Physical activity assessment with PAQ-C for children and PAQ-A for older children;	HbA1c	↓ meal frequency (skipping breakfast) ↓ physical activity ↑ screen time ↑ sleep duration	↑ HbA1c in T2DM and pubertal adolescents ↓ HbA1c in pre-pubertal T1DM children ↑ Weight and BMI SDS in T1DM ↓ Weight and BMI SDS in T2DM

Table 2. *Cont.*

Identity	Country	Study Design	Participants (N)	Age Range (Years)	Diet Assessment	Lifestyle Assessment	Cardiometabolic Risk Factors	Outcome in Diet and Lifestyle	Outcome in Cardiometabolic Risk Factors
Wu et al., 2021 [26]	China	Observational study Telephone interview	43 with T1DM	0–18	Questionnaire for dietary intake	Questionnaire for, physical exercise, sleep habits and emotions	GCM for glucose monitoring HbA1c	↑ number of snacks, ↑ sleep duration ↑ time for diabetes management ↓ physical activity	No significant changes
Al Agha et al., 2021 [27]	Saudi Arabia	Cross sectional study	150 with T1DM	2–18	Questionnaire for dietary habits	Questionnaire for physical activity and mood	Blood pressure (systolic and diastolic) HbA1c CGM	↓ physical activity ↑ consumption of carbohydrates and fast food ↓ mood	↑ HbA1c ↑ BMI↑ Blood pressure
Turan et al., 2021 [28]	Turkey	Cross sectional	100 with T1DM	3–18	Questionnaire on snack and meal frequency, CHO consumption	Physical Activity Questionnaire-A (PAQ-A) or Physical Activity Questionnaire-C (PAQ-C)	HbA1c BMI	↑ consumption of carbohydrates Delayed sleep times	↑ HbA1c ↓ No of hypoglycaemic events

GCM: Glucose Continuous Monitoring, BMI: Body Mass Index, T1DM: Type 1 Diabetes Mellitus, T2DM: Type 2 Diabetes Mellitus, HbA1: Glycosylated Hemoglobin, N/A: non applicable, CHO carbohydrates, ↑ increase, ↓ decrease. ↑% increased percentage of patients, ↓% decreased percentage of patients.

3.1. Study Characteristics

The studies included in the present systematic review regarding the role of diet and lifestyle on body composition/body weight (Table 1) originate from two continents, four from Asia [15,16,18,29] and seven from Europe [14,17,19–23].The included studies for the role of diet/lifestyle on cardiometabolic risk factors (Table 2) originate from three continents, one from the Middle East [27], four from Asia [16,25,26,28] and two from Europe [24,30]. Interestingly the countries included present with large variation in their socioeconomic status having both developed countries like Germany and developing countries like Malaysia. The age of the participants ranged from 0–18 years and the study sample varied from very small i.e., 43 volunteers, [26] to very large i.e., 10,000 participants [15]. Regarding the study methodology, there were 12 cross-sectional studies, two retrospective studies and one longitudinal study.

According to the risk assessment of the studies, it has been noted that most of them are rather small and are based on convenience sampling, with no actual randomization, and including self-reported measurements of height and weight. Therefore, the selection bias and the accuracy of the answers are questionable (Figure 1). Under this spectrum, the overall risk evaluation in Figure 3 presents to be in greater extent either moderate or high. There were also studies mentioning measurements without presenting the relative results. This fact can be attributed to the eagerness to publish results for a new and unknown situation and at the same time the difficulty of performing assessments in person due to the lockdown and the limited access to the healthcare facilities.

3.2. Main Exposures

Changes in body weight/body composition were evaluated in relation to measures of change in diet and/or lifestyle (Table 1). All studies used self-reported data either using known validated questionnaires such as the Food Frequency Questionnaire (FFQ) and the International Physical Activity Questionnaire (IPAQ), or other questionnaires collected either by phone, live interviews, or online surveys. The primary outcomes were changes in body weight, Body Mass Index (BMI) and waist circumference, which were either measured ($N = 5$) [16,17,20–22] or self-reported ($N = 5$) [14,15,18,19,23]. Regarding the association of diet and/or lifestyle changes on cardiometabolic conditions, the primary outcome was blood pressure ($N = 3$) or glycemic control ($N = 5$).

3.3. Studies on the Association of Diet and/or Lifestyle Changes on Body Weight/Body Composition

The majority of studies assessed changes of both diet and lifestyle, except for two (one retrospective and one cross-sectional) which solely assessed parameters of physical activity and sedentary behavior [15,17]. Apart from one study that reported no results, [16] all the other included studies reported increased consumption of fats and fast foods, as well as processed food, sweet and salty snacks and sugar sweetened beverages. Yet some studies also reported increased consumption of breakfast, fruit and vegetable intake and reduction in fruit juice and carbonated drinks consumption. Also, all studies shown in Table 1 reported a significant decrease in physical activity paralleled with a significant increase in sedentary and screen time, while two studies assessing sleep duration and sleep quality as part of the lifestyle assessment presented with controversial results [14,21]. Regarding the changes in body weight/body composition, eight out of the 10 studies included in the analysis and presented in Table 1 reported significant increases in body weight, BMI, waist circumference or body fat, which was evident even in the 59.7% of the study population [14–19,23]. According to Qiu et al., 28.1% of the children with normal BMI before the lockdown became overweight or obese, 42.4% of the overweight children became obese and 46.6% of the participants starting as normotensives presented abnormal BP levels at the end of the lockdown [16].

On the other hand, two studies reporting no changes in body weight/body composition measures during the pandemic also reported better adherence to the Mediterranean diet and reduced consumption of canned food, fast foods, snacks, fruit juices, sugar sweetened soft drinks and meat, resulting in an improved nutritional intake, an effect attributed to the increased family time and the time spent on cooking and meal preparation [21,22].

3.4. Studies on the Association of Diet and/or Lifestyle Changes on Cardiometabolic Risk Factors

Studies included in Table 2 presented outcomes regarding blood pressure and glycemic control, while no studies were retrieved for the association of diet and/or lifestyle changes during the pandemic with other cardiometabolic conditions. Regarding blood pressure, there are mixed results, as two studies reported increased blood pressure levels [16,27], while one study reported a relevant reduction [17], all paralleled with significant decreases in physical activity and increased time in sedentary behaviors. As regards glycaemic control studies, four studies were conducted solely in subjects with Type 1 diabetes mellitus (T1DM) [24,26–28] and one study included patients with both T1DM and type 2 diabetes mellitus (T2DM) [25]. Among them, two cross-sectional studies reported either increased plasma glucose levels by 18% [27] or increased glycosylated haemoglobin A1c (HbA1c) [28] levels by 0.32%, as a consequence of decreased physical activity, increased sedentary time and increased consumption of carbohydrates. Another cross-sectional study also reported elevated levels of sedentary behavior and decreased levels of physical activity together with altered meal frequency, mainly expressed as skipping breakfast. These dietary changes led to increased body weight and reduced HbA1c in prepubertal T1DM children due to closed parental control of the management of blood glucose levels. At the same time, it was connected with decreased body weight and elevated levels of HbA1c in children with T2DM and pubertal adolescents, signs of uncontrolled diabetes mellitus [25]. In another small cross-sectional study, the decrease of hypoglycemic episodes was reported, with no change observed in HbA1c, despite the decreased physical activity and increased number of daily snacks [26]. Finally, in a retrospective observational study, an improvement in glycemic control was also reported in children with T1DM despite the increase of sedentary activities and the limitation of physical activity, mainly due to the closer control of the food intake and the glucose measurements from their parents [24].

4. Discussion

The impact of the COVID-19 pandemic on children has yet to be fully examined. There are justified concerns, confirmed by international organizations, that the health crisis that has emerged over the past two years may have severely affected the nutrition and lifestyles of children, with many unfavorable consequences with regard to childhood obesity and other cardiometabolic risks. Unfortunately, there are concerns that the COVID-19 pandemic threatens to reverse the hard-won progress achieved in the previous years in pediatric health.

The present systematic review aimed at gathering all existing knowledge from recently published articles regarding changes in diet and lifestyle during the COVID-19 pandemic and their effect on body weight and body composition, as well as cardiometabolic risk factors in children. These data may enable the design and implementation of public health preventive measures during periods of lockdown, particularly aiming at the enhancement of physical activity and healthy nutritional choices, especially when their duration is uncertain.

One of the main outcomes of the present systematic review is the fact that around the globe, similar dietary and lifestyle patterns were recorded [31,32]. The deterioration of food quality and the increment of food and energy quantity, alongside the mandatory confinement and the lower levels of physical activity and energy expenditure, combined to create an obesogenic environment for children. These changes were more profound in families with disadvantaged socioeconomic conditions, reflecting the economic impact of COVID-19 on nutrition and public health [33]. One of the main concerns is that these changes, which might be difficult to be reversed, can have short term and long-term negative effects on the overall health of the children in their childhoods and later on when they reach adulthood.

The tele-education and the lockdown resulted in increased screen time spent on computers and mobile phones, not as recreation but also as an obligatory responsibility. As a result, physical activity levels were compromised, and sedentary time increased significantly in children and adolescents [14,16,18,19,21,23]. It has already been reported by previous studies that the amount of time spent engaged in sedentary activities during childhood and adolescence predicts overweight and obesity in adulthood [34], posing a reasonable concern of the possible effects of this pandemic on the obesity epidemics that we will have to face in the following years.

A significant modification to sleep quality and overall sleeping habits was reported in many of the included studies. Some studies reported increased sleeping hours [18,27,32], or decreased sleep duration but increased sleep quality [22], or delayed sleeping time at night [28], affecting several aspects of the health of children and adolescents. Insufficient or interrupted sleep relates to mood swings and attention problems and poor mental health, situations made worse during the COVID-19 pandemic [35]. Sufficient sleeping time is essential in childhood. The American Academy of Sleep Medicine recommends that toddlers (aged 3–5 years) should get 10–13 h of sleep per day, school-going children (ages 6–12 years) should get 9–12 h per night, and adolescents should get 8-10 h nightly [36]. During the confinement, most of the studies reported an increase in sleep duration, explained by the absence of the obligation to attend school and other activities, which was not always achieved during the confinement. Moreover, the flexibility in waking and bedtime hours was the reason for increased sleep duration [35]. However, as circadian disturbances relate to weight gain during school breaks, these changes may have an impact in weight gain in children during COVID-19 [4].

All reported alterations observed regarding diet and lifestyle habits seem to have worsened the already existing problem of bad lifestyle choices and pediatric obesity. It is noteworthy that in the pre-COVID 19 era, more than 80% of children around the world were inactive, more than 60% did not meet recommended screentime guidelines, and nearly a half have a scarce adherence to the Mediterranean diet [37–39]. If the assumption that this picture will worsen during the pandemic is correct, it will negatively affect children's current and future health, and many scientists express their fear that the COVID-19 pandemic might shift towards an obesity pandemic. It is essential that more studies should be carried out with the primary aim of measuring the impact of this pandemic on the nutritional habits and health status of children and adolescents in different contexts. One of the main goals of these studies should be to assess whether unhealthy nutritional habits are maintained in the long term or even improved, and to objectively assess children's nutritional parameters and compare them to the pre-pandemic era.

Another important aspect of the present review is the association of diet and lifestyle changes during the lockdown with cardiometabolic risk. Although the findings are controversial in some cases, there is limited evidence regarding worsening of glycemic control in children with T2DM and hypertension. On the contrary, the presence of parents and their active involvement in the management of T1DM resulted in a better management of children with T1DM. However, it is noteworthy that in most studies an increase in sweet and salty snacks, as well as sugar sweetened beverages, was observed, pointing out an increase in sugar and sodium intake, which makes us skeptical regarding possible underestimated unfavorable consequences due to COVID-19 on glycaemic control and hypertension. However, due to limited evidence, the heterogeneity of studies and the varied quality more well-designed studies examining the effects of COVID-19 on glycaemic control and hypertension are needed to draw credible conclusions.

Moreover, adolescence is known to be an overwhelming time, with physical, hormonal, and behavioral changes that were made even more profound during lockdown, increasing the prevalence of anxiety, boredom, and lack of motivation. Stress can negatively affect blood glucose and blood pressure management, alongside with the unbalanced nutrient and energy equilibrium, resulting in poor diabetes control and blood pressure in children with T2DM. Therefore, the lifestyle and stress management in younger patients with diabetes is undoubtedly of paramount importance to maintaining good diabetes self-care, lower cardiometabolic risk and overall better health.

5. Conclusions

The COVID-19 pandemic that emerged in the past two years is still uncontrolled, and measures to curb virus infection, such as confinement at home and social distancing, lead to a disruption of daily activities with yet unknown health consequences. The present systematic review is the first to our knowledge that gathered existent relevant studies highlighting the worsening of children and adolescents' diet and lifestyle with unhealthy food choices, increased sedentary time, and decreased physical activities, which led to increased body weight and abdominal fat accumulation. Also, obesity, alongside with unhealthy diet and lifestyle with increased consumption of fat, sugar and sodium and increased time spent on sedentary activities are significant risk factors for cardiometabolic unfavorable consequences, such as diabetes and hypertension that are poorly studied or such as dyslipidemia and metabolic syndrome which are not studied at all, posing a great threat for an outburst of the existent obesity epidemic and its serious health implications. It is highly possible that the real consequences of COVID-19 on obesity and cardiometabolic conditions are understudied and possibly neglected, and therefore future public health initiatives are needed to assess and control obesity and its cardiometabolic complications in an early and timely fashion during this unprecedented public health crisis.

Author Contributions: K.K., K.-A.P., E.P. and A.Z. have contributed substantially to the work reported. In specific all authors contributed to conceptualization and methodology, K.K. and K.-A.P. performed the research, assessment of bias and data curation, and all the authors (K.K., K.-A.P., E.P. and A.Z.) contributed to the original draft preparation, writing, review and editing. All authors have read and agreed to the published version of the manuscript.

Funding: This research received no external funding.

Institutional Review Board Statement: Not applicable.

Informed Consent Statement: Not applicable.

Data Availability Statement: Not applicable.

Conflicts of Interest: The authors declare no conflict of interest.

References

1. WHO: Coronavirus Disease (COVID-19) Pandemic. Available online: https://www.who.int/emergencies/diseases/novel-coronavirus-2019 (accessed on 9 November 2021).
2. Ammar, A.; Brach, M.; Trabelsi, K.; Chtourou, H.; Bukhari's, O.; Masmoudi, L.; Bouaziz, B.; Bentlage, E.; How, D.; Ahmed, M.; et al. Effects of COVID-19 Home Confinement on Eating Behaviour and Physical Activity: Results of the ECLB-COVID19 International Online Survey. *Nutrients* **2020**, *12*, 1583. [CrossRef] [PubMed]
3. Jia, P.; Luo, M.; Li, Y.; Zheng, J.S.; Xiao, Q.; Luo, J. Fast-food restaurant, unhealthy eating, and childhood obesity: A systematic review and meta-analysis. *Obes. Rev.* **2021**, *22*, e12944. [CrossRef] [PubMed]
4. Baranowski, T.; O'Connor, T.; Johnston, C.; Hughes, S.; Moreno, J.; Chen, T.A.; Meltzer, L.; Baranowski, J. School year versus summer differences in child weight gain: A narrative review. *Child. Obes.* **2014**, *10*, 18–24. [CrossRef]
5. Di Renzo, L.; Gualtieri, P.; Pivari, F.; Soldati, L.; Attinà, A.; Cinelli, G.; Leggeri, C.; Caparello, G.; Barrea, L.; Scerbo, F.; et al. Eating habits and lifestyle changes during COVID-19 lockdown: An Italian survey. *J. Transl. Med.* **2020**, *18*, 229. [CrossRef] [PubMed]
6. Browne, N.T.; Snethen, J.A.; Greenberg, C.S.; Frenn, M.; Kilanowski, J.F.; Gance-Cleveland, B.; Burke, P.J.; Lewandowski, L. When Pandemics Collide: The Impact of COVID-19 on Childhood Obesity. *J. Pediatr. Nurs.* **2021**, *56*, 90–98. [CrossRef]
7. OECD; WHO. Overweight and Obesity. In *Health at a Glance: Asia/Pacific 2020: Measuring Progress Towards Universal Health Coverage*; OECD Publishing: Paris, France, 2020. [CrossRef]
8. NCD Risk Factor Collaboration. Worldwide trends in body-mass index, underweight, overweight, and obesity from 1975 to 2016: A pooled analysis of 2416 population-based measurement studies in 128·9 million children, adolescents, and adults. *Lancet* **2017**, *390*, 2627–2642. [CrossRef]
9. Nogueira-de-Almeida, C.A.; Del Ciampo, L.A.; Ferraz, I.S.; Del Ciampo, I.R.L.; Contini, A.A.; Ued, F.D.V. COVID-19 and obesity in childhood and adolescence: A clinical review. *J. Pediatr.* **2020**, *96*, 546–558. [CrossRef]

10. Flynn, J.T.; Kaelber, D.C.; Baker-Smith, C.M.; Blowey, D.; Carroll, A.E.; Daniels, S.R.; de Ferranti, S.D.; Dionne, J.M.; Falkner, B.; Flinn, S.K.; et al. Clinical Practice Guideline for Screening and Management of High Blood Pressure in Children and Adolescents. *Pediatrics* **2017**, *140*, e20171904. [CrossRef] [PubMed]
11. Deeb, A.; Attia, S.; Mahmoud, S.; Elhaj, G.; Elfatih, A. Dyslipidemia and Fatty Liver Disease in Overweight and Obese Children. *J. Obes.* **2018**, *2018*, 8626818. [CrossRef]
12. Pulgaron, E.R.; Delamater, A.M. Obesity and type 2 diabetes in children: Epidemiology and treatment. *Curr. Diab. Rep.* **2014**, *14*, 508. [CrossRef]
13. Sterne, J.A.C.; Savović, J.; Page, M.J.; Elbers, R.G.; Blencowe, N.S.; Boutron, I.; Cates, C.J.; Cheng, H.Y.; Corbett, M.S.; Eldridge, S.M.; et al. RoB 2: A revised tool for assessing risk of bias in randomised trials. *BMJ* **2019**, *366*, l4898. [CrossRef] [PubMed]
14. Androutsos, O.; Perperidi, M.; Georgiou, C.; Chouliaras, G. Lifestyle Changes and Determinants of Children's and Adolescents' Body Weight Increase during the First COVID-19 Lockdown in Greece: The COV-EAT Study. *Nutrients* **2021**, *13*, 930. [CrossRef] [PubMed]
15. Yang, S.; Guo, B.; Ao, L.; Yang, C.; Zhang, L.; Zhou, J.; Jia, P. Obesity and activity patterns before and during COVID-19 lockdown among youths in China. *Clin. Obes.* **2020**, *10*, e12416. [CrossRef]
16. Qiu, N.; He, H.; Qiao, L.; Ding, Y.; Ji, S.; Guo, X.; Luo, J.; Luo, Z.; Li, Y.; Pang, H.; et al. Sex differences in changes in BMI and blood pressure in Chinese school-aged children during the COVID-19 quarantine. *Int. J. Obes.* **2021**, *45*, 2132–2136. [CrossRef] [PubMed]
17. Maltoni, G.; Zioutas, M.; Deiana, G.; Biserni, G.B.; Pession, A.; Zucchini, S. Gender differences in weight gain during lockdown due to COVID-19 pandemic in adolescents with obesity. *Nutr. Metab. Cardiovasc. Dis.* **2021**, *31*, 2181–2185. [CrossRef]
18. Al Hourani, H.; Alkhatib, B.; Abdullah, M. Impact of COVID-19 Lockdown on Body Weight, Eating Habits, and Physical Activity of Jordanian Children and Adolescents. *Disaster Med. Public Health Prep.* **2021**, *16*, 1–9. [CrossRef] [PubMed]
19. Pujia, R.; Ferro, Y.; Maurotti, S.; Khoory, J.; Gazzaruso, C.; Pujia, A.; Montalcini, T.; Mazza, E. The Effects of COVID-19 on the Eating Habits of Children and Adolescents in Italy: A Pilot Survey Study. *Nutrients* **2021**, *13*, 2641. [CrossRef] [PubMed]
20. Cipolla, C.; Curatola, A.; Ferretti, S.; Giugno, G.; Condemi, C.; Delogu, A.B.; Birritella, L.; Lazzareschi, I. Eating habits and lifestyle in children with obesity during the COVID19 lockdown: A survey in an Italian center. *Acta Biomed.* **2021**, *92*, e2021196. [PubMed]
21. Łuszczki, E.; Bartosiewicz, A.; Pezdan-Śliż, I.; Kuchciak, M.; Jagielski, P.; Oleksy, Ł.; Stolarczyk, A.; Dereń, K. Children's Eating Habits, Physical Activity, Sleep, and Media Usage before and during COVID-19 Pandemic in Poland. *Nutrients* **2021**, *13*, 2447. [CrossRef]
22. Medrano, M.; Cadenas-Sanchez, C.; Oses, M.; Arenaza, L.; Amasene, M.; Labayen, I. Changes in lifestyle behaviours during the COVID-19 confinement in Spanish children: A longitudinal analysis from the MUGI project. *Pediatr. Obes.* **2021**, *16*, e12731. [CrossRef]
23. Koletzko, B.; Holzapfel, C.; Schneider, U.; Hauner, H. Lifestyle and Body Weight Consequences of the COVID-19 Pandemic in Children: Increasing Disparity. *Ann. Nutr. Metab.* **2021**, *77*, 1–3. [CrossRef] [PubMed]
24. Minuto, N.; Bassi, M.; Montobbio, C.; Vinci, F.; Mercuri, C.; Perri, F.N.; Cabri, M.; Calevo, M.G.; d'Annunzio, G.; Maghnie, M. The Effect of Lockdown and Physical Activity on Glycemic Control in Italian Children and Young Patients with Type 1 Diabetes. *Front. Endocrinol.* **2021**, *12*, 690222. [CrossRef]
25. Cheng, H.P.; Wong, J.S.L.; Selveindran, N.M.; Hong, J.Y.H. Impact of COVID-19 lockdown on glycaemic control and lifestyle changes in children and adolescents with type 1 and type 2 diabetes mellitus. *Endocrine* **2021**, *73*, 499–506. [CrossRef] [PubMed]
26. Wu, X.; Luo, S.; Zheng, X.; Ding, Y.; Wang, S.; Ling, P.; Yue, T.; Xu, W.; Yan, J.; Weng, J. Glycemic control in children and teenagers with type 1 diabetes around lockdown for COVID-19: A continuous glucose monitoring-based observational study. *J. Diabetes Investig.* **2021**, *12*, 1708–1717. [CrossRef] [PubMed]
27. Al Agha, A.E.; Alharbi, R.S.; Almohammadi, O.A.; Yousef, S.Y.; Sulimani, A.E.; Alaama, R.A. Impact of COVID-19 lockdown on glycemic control in children and adolescents. *Saudi Med. J.* **2021**, *42*, 44–48. [CrossRef] [PubMed]
28. Turan, H.; Güneş Kaya, D.; Tarçın, G.; Evliyaoğlu, S.O. Effect of the COVID-19 quarantine on metabolic control in children and adolescents with type 1 diabetes. *Endocrinol. Diabetes Nutr.* **2021**. [CrossRef] [PubMed]
29. Vinker-Shuster, M.; Grossman, E.S.; Yeshayahu, Y. Increased Weight Gain of Children during the COVID-19 Lockdown. *Isr. Med. Assoc. J.* **2021**, *23*, 219–222.
30. Marigliano, M.; Maffeis, C. Glycemic control of children and adolescents with type 1 diabetes improved after COVID-19 lockdown in Italy. *Acta Diabetol.* **2021**, *58*, 661–664. [CrossRef]
31. Ruíz-Roso, M.B.; de Carvalho Padilha, P.; Matilla-Escalante, D.C.; Brun, P.; Ulloa, N.; Acevedo-Correa, D.; Arantes Ferreira Peres, W.; Martorell, M.; Rangel Bousquet Carrilho, T.; de Oliveira Cardoso, L.; et al. Changes of Physical Activity and Ultra-Processed Food Consumption in Adolescents from Different Countries during Covid-19 Pandemic: An Observational Study. *Nutrients* **2020**, *12*, 2289. [CrossRef]
32. Pietrobelli, A.; Pecoraro, L.; Ferruzzi, A.; Heo, M.; Faith, M.; Zoller, T.; Antoniazzi, F.; Piacentini, G.; Fearnbach, S.N.; Heymsfield, S.B. Effects of COVID-19 Lockdown on Lifestyle Behaviors in Children with Obesity Living in Verona, Italy: A Longitudinal Study. *Obesity* **2020**, *28*, 1382–1385. [CrossRef]
33. Aguilar-Martínez, A.; Bosque-Prous, M.; González-Casals, H.; Colillas-Malet, E.; Puigcorbé, S.; Esquius, L.; Espelt, A. Social Inequalities in Changes in Diet in Adolescents during Confinement Due to COVID-19 in Spain: The DESKcohort Project. *Nutrients* **2021**, *13*, 1577. [CrossRef] [PubMed]
34. Boone, J.E.; Gordon-Larsen, P.; Adair, L.S.; Popkin, B.M. Screen time and physical activity during adolescence: Longitudinal effects on obesity in young adulthood. *Int. J. Behav. Nutr. Phys. Act.* **2007**, *4*, 26. [CrossRef]

35. Bates, L.C.; Zieff, G.; Stanford, K.; Moore, J.B.; Kerr, Z.Y.; Hanson, E.D.; Barone Gibbs, B.; Kline, C.E.; Stoner, L. COVID-19 Impact on Behaviors across the 24-Hour Day in Children and Adolescents: Physical Activity, Sedentary Behavior, and Sleep. *Children* **2020**, *7*, 138. [CrossRef] [PubMed]
36. Paruthi, S.; Brooks, L.J.; D'Ambrosio, C.; Hall, W.A.; Kotagal, S.; Lloyd, R.M.; Malow, B.A.; Maski, K.; Nichols, C.; Quan, S.F.; et al. Recommended Amount of Sleep for Pediatric Populations: A Consensus Statement of the American Academy of Sleep Medicine. *J. Clin. Sleep Med.* **2016**, *12*, 785–786. [CrossRef]
37. WHO: Prevalence of Insufficient Physical Activity. Global Health Observatory (GHO) Data. Available online: https://www.who.int/gho/ncd/risk_factors/physical_activity_text/en/ (accessed on 9 November 2021).
38. Atkin, A.J.; Sharp, S.J.; Corder, K.; van Sluijs, E.M. Prevalence and correlates of screen time in youth: An international perspective. *Am. J. Prev. Med.* **2014**, *47*, 803–807. [CrossRef]
39. Grosso, G.; Galvano, F. Mediterranean diet adherence in children and adolescents in southern European countries. *NFS J.* **2016**, *3*, 13–19. [CrossRef]

children

MDPI

Article

Moderate-to-Vigorous Physical Activity and Body Composition in Children from the Spanish Region of Aragon

Pilar Ferrer-Santos [1], Iris Iglesia [1,2,3,*], Borja Muñiz-Pardos [1,2,4], María Luisa Miguel-Berges [1,2,5], Paloma Flores-Barrantes [1,2,5], Luis A. Moreno [1,2,5] and Gerardo Rodríguez-Martínez [1,2,3,6]

1 Instituto de Investigación Sanitaria Aragón (IIS Aragón), 50009 Zaragoza, Spain; pilarferrersantos@gmail.com (P.F.-S.); bmuniz@unizar.es (B.M.-P.); mlmiguel@unizar.es (M.L.M.-B.); pfloba@unizar.es (P.F.-B.); lmoreno@unizar.es (L.A.M.); gerard@unizar.es (G.R.-M.)
2 Growth, Exercise, Nutrition and Development (GENUD) Research Group, Instituto Agroalimentario de Aragón (IA2), University of Zaragoza, 50009 Zaragoza, Spain
3 Red de Salud Materno Infantil y del Desarrollo (SAMID), Instituto de Salud Carlos III, 28029 Madrid, Spain
4 Facultad de Ciencias de la Salud y del Deporte, Departamento de Fisiatría y Enfermería, Universidad de Zaragoza, 50009 Zaragoza, Spain
5 CIBER Fisiopatología de la Obesidad y Nutrición, Instituto de Salud Carlos III, 28040 Madrid, Spain
6 Departament of Pediatry, Faculty of Medicine, University of Zaragoza, 50009 Zaragoza, Spain
* Correspondence: iglesia@unizar.es; Tel.: +34-876843756

Abstract: Most of the studies analyzing the effect of moderate to vigorous physical activity (MVPA) on children's health do not contain information on early stages or do not use accurate methods. We investigated the association between PA and body composition using objective methods, perinatal data, lifestyle behaviors, and World Health Organization (WHO) physical activity (PA) recommendations. The CALINA study is a longitudinal observational cohort study of children born in Aragon (Spain) in 2009. A total of 308 7-year-old children (52.3% boys) were assessed. We used dual-energy X-ray absorptiometry (DXA) and accelerometry. Rapid weight gain until 12 months and lifestyle behaviors were considered as covariates both in the ANCOVA and linear regression models. A higher percentage of boys met the WHO PA recommendations compared to girls (69.6% vs. 40.9%, respectively; $p < 0.001$). There was a negative association between MVPA and subtotal fat and abdominal fat in both girls and boys. After adjusting for perinatal and lifestyle variables, we found that subtotal body fat, abdominal fat, and fat mass index (FMI) were significantly lower in those classified as active. MVPA was associated with body fat both in boys and girls. More research is needed to identify the cutoffs points of MVPA that generate benefit to boys and girls in all body composition components.

Keywords: physical activity; children; body composition; accelerometry; dual-energy X-ray absorptiometry

Citation: Ferrer-Santos, P.; Iglesia, I.; Muñiz-Pardos, B.; Miguel-Berges, M.L.; Flores-Barrantes, P.; Moreno, L.A.; Rodríguez-Martínez, G. Moderate-to-Vigorous Physical Activity and Body Composition in Children from the Spanish Region of Aragon. *Children* **2021**, *8*, 341. https://doi.org/10.3390/children8050341

Academic Editor: Zoe Knowles

Received: 16 March 2021
Accepted: 23 April 2021
Published: 26 April 2021

1. Introduction

Obesity is a condition characterized by the excess of body fat, affecting both adults and children, and there has been an alarming increase of childhood obesity in the last decades [1]. This is of relevance as both excess body weight and adiposity are associated with a number of comorbidities affecting almost every system in the body, including psychological problems [2]. These comorbidities are in the top leading causes of death in the world [3].

Overweight and obesity prevalence in Spanish school-age children in 2019 were 23.3% and 17.9%, respectively [4]. Maternal smoking, rapid infant growth, or short breastfeeding have shown to be amongst the earliest factors contributing to later excess of body weight or adiposity [2,5]. However, still there is scarce evidence on large cohorts followed since birth to really understand until what extent such factors influence body composition or future health.

Moreover, already during childhood there are several risk factors for overweight and obesity such as those related to lifestyle behaviors including eating habits, low levels of physical activity (PA), or high levels of sedentarism [6,7]. Specifically, previous studies have shown that in school-aged children, high levels of PA are associated with low body fat and high muscle mass content [8], as well as low abdominal fat or low risk of cardiovascular disease risk factors and metabolic syndrome [9].

Due to the important benefits of practicing PA, the World Health Organization (WHO) recommended, for children aged 5–17, a minimum of one hour per day of moderate-to-vigorous physical activity (MVPA). Although this recommendation has shown several benefits [10], only 5% of the official PA recommendations in Spain are based on the WHO's advice mentioned [11].

Most studies analyzing the influence of PA in children's body composition do not account for perinatal data, and many of them do not use accurate methods to assess either PA (e.g., accelerometery) or body composition (e.g., DXA). In this, caution should be taken when considering the tool to assess PA levels, as qualitative methods (i.e., questionnaires) have been shown to differ from objective methods (e.g., accelerometery). For example, Rääsk et al. [12] showed that MVPA as assessed with questionnaires was overestimated in less active boys, but underestimated in more active boys, when compared to accelerometery. Nevertheless, and despite being an objective method, accelerometery also has its own pitfalls. A previous study [13] highlighted the functional limitations of these wearable devices depending on the population, stating that comparison between studies are challenging given the different cut-points used. This issue limits the PA interpretation and comparison between studies and need to be considered in studies assessing PA levels. Nevertheless, the use of accelerometery with infant populations has shown great accuracy in previous studies [14,15]. An additional limitation of the existing studies is that they do not appropriately analyze the influence of current WHO PA recommendations in terms of compliance in school-age children with regard to body composition indicators [16,17]. Our aim was to investigate such association objectively in a pediatric population, using accelerometery to assess PA and current WHO recommendations and dual X-ray absorptiometry to assess body composition.

2. Methodology

2.1. Study Design

This study included data from an observational study named "Growth and Feeding during Early Childhood in Children from Aragón (CALINA)", which is based on a cohort of children born in the region of Aragon (Spain) in 2009. The initial sample recruited for this project included 1602 subjects [18]. These children were followed every month during the first year of life and every year since then until they reached the age of 7. From September 2016 to September 2017, we contacted all the families recruited in Zaragoza back in 2009 to be re-assessed, requesting them to attend the laboratory located at the University of Zaragoza. A total of 415 families out of 952 agreed to participate, and 339 were finally included in this analysis in terms of the variables required for it (176 boys and 163 girls).

This study was approved by the Ethics Committee in Clinical Research of the Government of Aragon (ref. CP PI13/00105, Spain). The project adhered to the Declaration of Helsinki [19], and all the parents agreed and signed an informed consent form.

2.2. Data Collection

In 2009, the pediatricians of the selected primary care centers collected demographic data including immigrant background, obstetric history including if the mother smoked during pregnancy or not, perinatal history, and BMI of parents at child-birth, among other data [18]. Moreover, as a growth marker of the early postnatal period, we calculated and classified the children between those who were rapid weight gainers and those who were not on the basis of the definition of rapid weight gain (RWG), when a positive change in weight-for-age z-score > 0.67 between birth and 12 months [20]. Weight and recumbent height during the first year of life was measured by well-trained health professionals.

At this stage, demographic factors such as migrant background were also obtained, as well as maternal lifestyle behavior such as smoking habit during pregnancy, which were subsequently used in this analyses as covariates. The election of these variables as covariates were based on the results obtained by a colleague who observed in a recent publication that in our sample, migrant background of the mother, smoking habit during pregnancy, RWG, and maternal and paternal BMI were associated with body composition of children belonging to our study but in a previous follow-up (when they were around 6 years). These variables then, including parental BMI, were used as covariates in our study by assuming that will be also related with their children body composition at the age of approximately 7 years old.

Maternal and paternal BMI were self-reported in the follow-up performed between 2016 and 2017 and also used as covariates in this analysis.

In the 2016–2017 assessment, we collected the following data from the children when they were approximately 7 years-old:

Weekly screen time (WST): Using a previously validated questionnaire, parents reported the number of hours of TV/DVD/video viewing and computer/game console use of their child both for a typical day on weekdays and on weekend days. We summed the reported hours per day on weekdays and weekend days to obtain the total WST (hours in week days + hours in weekend days/7 days per week) [21].

Physical activity (PA): PA was objectively assessed with an Actigraph accelerometer (Actigraph GT3X; Manufacturing Technology Inc. Pensacola, FL, USA). Subjects were asked to wear the belt-like accelerometer on the hip all day during a complete week or at least 3 days per week and 1 day per weekend, recording a minimum of 5 h per day. Children and their parents were instructed to remove the accelerometer only during water-based activities, sleeping, and impact sports, registering duration and reason for removal in a formulary. PA was expressed as average in counts per minute (cpm) and minutes per day of light, moderate, and moderate-to-vigorous PA according to Evenson cut-points [22], with light/moderate PA (101–2295 cpm) and MVPA ($\geq$ 2296 cpm). According to WHO recommendations for PA [23], we classified the sample into two groups: active children, including those who did 60 or more minutes per day of MVPA, and inactive children, those performing less than 60 min of MVPA per day.

Diet: Dietary intake was self-reported by parents through a semi-quantitative food frequency questionnaire (FFQ) [24,25] that had been previously used and validated in the multifactorial evidence-based approach using behavioral models in understanding and promoting fun, healthy food, play, and policy for the prevention of obesity in early childhood study (ToyBox-study). In short, the FFQ consists of a list of foods and beverages with response categories to indicate usual frequency of consumption over the selected time period. We calculated the Diet Quality Index (DQI), which is a largely used index, in cohorts with similar characteristics in order to assess diet in terms of three subcomponents: dietary diversity, quality, and equilibrium [26].

Body composition:

1. Height and weight measurements were measured using a stadiometer with a precision of 1 mm (SECA 225, Germany) and Bioimpedance (BI) (Tanita BC-418, Japan) scale. Determination of z-score values of BMI for age (z-BMI) for girls and for boys was performed using the WHO Anthro Software, according to the WHO growth standards of 2006–2007 [27].
2. Dual-energy X-ray absorptiometry (DXA). DXA scans were performed in a supine position, wearing light clothing with no metal and no shoes or jewelry (21). All DXA scan tests were analyzed by the same researcher using Hologic Explorer scanner and a pediatric version of the software QDR-Explorer, Hologic Corp., software version 12.4 (Bedford, MA, USA). Lean mass (body mass– (FM + bone mass)), percentage body fat mass (percentage of fat grams/total mass). Fat mass index (FMI) was a continuous

variable calculated for each participant from data obtained from DXA as fat mass in kilograms/height in square meters. Additionally, fat-free mass index (FFMI) kg/height in meters^2 was also used in this study [28]. Abdominal adiposity was assessed at a delimited region that was drawn on the digital scan image, delimiting the lower horizontal border on the top of the iliac crest and the upper border parallel to the end of the lowest rib [29].

2.3. Statistical Analysis

All analyses were conducted with the SPSS program v22. First, we studied the descriptive characteristics of the sample. Chi squared tests were used to contrast differences between groups in case of categorical variables, and *t*-test or Mann–Whitney test were used for continuous variables depending on the assumption of normality. Statistical significance was set at $p < 0.05$. We studied the differences in body composition variables between two groups: those who met the PA recommendations and those who did not, using *t*-test or Mann–Whitney test.

Linear regression models were performed to check the associations between PA (in minutes of MVPA per week) and different body composition items. Three models were created: the first one without any adjustment, the second model was adjusted by RWG, BMI of the parents, and smoking and migrant status of the mother at birth of the child, as well as the z-score BMI of the children at 7 years. The third model was also adjusted by DQI and WST.

Finally, both in active and inactive children (when they meet or not PA recommendations), we performed an estimation of body composition parameters using analysis of covariance (ANCOVA) models adjusted by z-BMI if they had or did not have RWG, BMI of the parents, smoking habit of the mother during pregnancy, education of the mother, origin of the family, DQI, and WST.

3. Results

The characteristics of the sample are shown in Table 1. From the 308 subjects, 161 were boys and 147 were girls. When comparing boys to girls, boys had significantly lower subtotal fat mass and FMI, and significantly higher subtotal lean mass and FFMI (Table 1). Moreover, boys had significantly lower abdominal fat than girls.

According to lifestyle behaviors, boys exhibited a significantly higher screen time use per week than girls of 825 min (CI 763 min–887 min) vs. 728 min (CI 671 min–786 min), respectively ($p = 0.018$), while there were no significant differences for DQI. Moreover, a significantly higher percentage of boys met WHO PA recommendations (69.6% vs. 40.9%, respectively; $p < 0.001$) and achieved higher amount of MVPA per day 73 min/day (CI 69–76) vs. 57.61 min/day (CI 55–61), respectively; $p < 0.001$) than girls. We did not find statistical differences for family origin, RWG during the first 12 months of life, or maternal BMI before pregnancy between boys and girls.

After adjusting by RWG, parental BMI, maternal smoking during pregnancy, origin of the mother, child BMI z-score at 7 years, DQI, and WST, we found that body subtotal fat, abdominal fat, and FMI were all significantly lower both in boys and girls meeting PA recommendations (Table 2). Active boys had significantly lower subtotal fat mass, lower FMI, higher FFMI, less abdominal fat mass, and less abdominal percentage of fat than inactive boys. Active boys also had significantly lower total body percentage of fat. In girls, active individuals had less subtotal fat mass, lower FMI, higher FFMI, less abdominal fat mass, less abdominal percentage of fat, and less total body percentage of fat (Table 2).

Table 1. Main characteristics of the sample.

	Total N = 308	Boys N = 161	Girls N = 147	p-Value
Body Composition				
	Mean (95% CI)	**Mean (95% CI)**	**Mean (95% CI)**	
Height (m) [a]	1.26 (1.25–1.27)	1.27 (1.26–1.28)	1.25 (1.24–1.26)	**0.04**
Weight (kg) [a]	27.40 (26.82–27.96)	27.81 (26.98–28.63)	26.93 (26.15–27.72)	0.34
Subtotal fat (g) [a]	7.00 (6.68–7.32)	6.55 (6.11–6.99)	7.49 (7.04–7.95)	**<0.001**
Subtotal fat (%) [a]	29.19 (28.44–29.94)	26.74 (25.79–27.69)	31.87 (30.85–32.89)	**<0.001**
Subtotal lean (g) [a]	16.29 (15.99–16.59)	17.10 (16.67–17.52)	15.41 (15.03 –15.35)	**<0.001**
BMI (kg/m^2) [a]	17.14 (16.88–17.41)	17.19 (16.81–17.56)	17.09 (16.71–17.47)	0.99
z-BMI (kg/m^2) [b]	0.71 (1.15)	0.76 (1.26)	0.64 (1.02)	0.52
FMI (kg/m^2) [a]	4.36 (4.18–4.56)	4.02 (3.78–4.27)	4.74 (4.48–5.00)	**<0.001**
FFMI (kg/m^2) [a]	10.20 (10.08–10.32)	10.57 (10.40–10.74)	9.79 (9.64–9.95)	**<0.001**
Abdominal fat (g) [a]	355.25 (329.06–381.86)	327.68 (292.28–363.07)	385.46 (346.76–424.16)	**0.001**
Abdominal lean (g) [a]	1.01 (0.975–1.042)	1.039 (989–1.089)	975.16 (931.30–1019.02)	0.26
Abdominal fat (%) [a]	24.39 (23.51–25.26)	22.34 (21.28–23.4)	26.63 (25.29–27.96)	**<0.001**
Lifestyle Behaviors				
	Total N = 308 Mean (95% CI)	**Boys N = 161 Mean (95% CI)**	**Girls N = 147 Mean (95% CI)**	**p-Value**
DQI (%) [a]	81 (80–82)	82 (80–83)	80 (78–82)	0.33
WST (mins) [a]	779 (736–821)	825 (763–887)	728 (671–786)	**0.018**
MVPA (min/day) [a]	65 (63–68)	73 (69–76)	58 (55–61)	**<0.001**
Meeting WHO MVPA recommendations	**% (*n*)**	**% (*n*)**	**% (*n*)**	
Yes (active)	55.8 (184)	69.6 (119)	40.9 (65)	**<0.001**
No (inactive)	44.2 (146)	30.4 (52)	59.1 (94)	
Family and Perinatal Factors				
	Total N = 308 % (*n*)	**Boys N = 161 % (*n*)**	**Girls N = 147 % (*n*)**	**p-Value**
Family origin [a]				
Immigrant	11.4 (35)	8.1 (13)	15 (22)	0.072
Spanish	88.6 (273)	91.9 (148)	85 (125)	
RWG at 12 months [a]				
Yes	35.9 (106)	38.1 (59)	33.6 (47)	0.47
No	64.1 (189)	61.9 (95)	66.4 (93)	
Smoking during pregnancy [a]				
Yes	15.3 (47)	15.5 (25)	15 (22)	1
No	84.7 (261)	84.5 (136)	85 (125)	
	Mean (CI)	**Mean (CI)**	**Mean (CI)**	
Maternal BMI (kg/m^2) [a]	23.69 (23.20–24.19)	24.04 (23.33–24.75)	23.32 (22.62–24.01)	0.118
Parental BMI (kg/m^2) [a]	25.85 (25.49–26.20)	26.04 (25.54–26.55)	25.64 (25.13–26.14)	0.278

Statistically significant differences (p < 0.05) are highlighted in bold. Mann–Whitney test was used for studying differences between boys and girls for non-parametric variables [a] (where CI are presented) and t-test for parametric variables [b] (where SD is presented). The sample for the variable RWG was 295 due to the loss of follow-up of some children in the first year. BMI: body mass index; FMI: fat mass index; FFMI: fat-free mass index; DQI: dietary quality index; WST: weekly screen time; WHO: World Health Organization; MVPA: moderate-to-vigorous PA.

Associations between PA and body composition accounting for family, perinatal, postnatal, and lifestyle behavior variables are shown in Table 3 for boys and girls. In all regression models, irrespective of further adjustments, there was a significant negative association between MVPA and subtotal fat mass, abdominal fat percentage, and FMI in both girls and boys. There was also a significant positive association between MVPA and FFMI, and there were no significant association between subtotal lean mass and PA neither in boys nor girls. In models 2 and 3, these associations remained significant after adjusting by RWG, parental BMI, maternal smoking during pregnancy, origin of the family, child BMI z-score at 7 years, DQI, and WST.

Table 2. Adjusted means of body composition parameters between being active or not by sex. Data are adjusted by RWG, BMI of the parents, smoking during pregnancy, migrant origin of the mother, child BMI z-score at 7 years, DQI, and WST.

	Subtotal Fat Mass (kg)		**Subtotal Lean Mass (kg)**	
	Mean	**SE**	**Mean**	**SE**
		Boys		
Active	6.322	0.13	17.228	0.188
Inactive	7.149	0.20	16.748	0.283
p-value	**0.001**		0.16	
		Girls		
Active	7.02	0.182	15.631	0.238
Inactive	7.74	0.147	15.138	0.192
p-value	<0.01		0.12	
	FMI		**FFMI**	
	Mean	**SE**	**Mean**	**SE**
		Boys		
Active	3.87	0.069	10.63	0.057
Inactive	4.40	0.105	10.36	0.085
p-value	**<0.001**		**<0.001**	
		Girls		
Active	4.48	0.09	9.98	0.083
Inactive	4.88	0.08	9.61	0.067
p-value	**<0.001**		**0.001**	
	Abdominal Fat Mass (g)		**Abdominal Lean Mass (g)**	
	Mean	**SE**	**Mean**	**SE**
		Boys		
Active	309.68	13.50	1048.98	24.15
Inactive	371.75	20.39	1032.74	26.45
p-value	**0.013**		0.71	
		Girls		
Active	347.23	17.99	984.61	30.75
Inactive	400.32	14.54	945.17	24.87
p-value	0.025		0.33	
	Abdominal Fat (%)		**Total Body Fat (%)**	
	Mean	**SE**	**Mean**	**SE**
		Boys		
Active	21.39	0.45	25.90	0.34
Inactive	24.48	0.68	28.88	0.52
p-value	**<0.001**		**<0.01**	
		Girls		
Active	24.50	0.69	30.21	0.54
Inactive	27.97	0.56	32.94	0.44
p-value	**<0.001**		**<0.01**	

Statistically significant differences ($p < 0.05$) are highlighted in bold. SE: standard error; FMI: fat mass index; FFMI: fat-free mass index; DQI: diet quality index; WST: weekly screen time.

Table 3. Associations between MVPA and body composition parameters. Each model was adjusted by a number of perinatal and lifestyle behavior variables, as pointed out below.

	Model 1				Model 2				Model 3			
	Boys		Girls		Boys		Girls		Boys		Girls	
	β	*p*-Value	β	*p*-Value	β	*p*-Value	β	*p*-Value	β	*p*-Value	β	*p*-Value
Subtotal fat mass (g)	−0.074	**0.05**	−0.083	**0.037**	−0.092	**0.022**	−0.100	**0.016**	−0.076	**0.061**	−0.104	**0.013**
Subtotal lean mass (g)	0.079	0.14	0.033	0.59	0.090	0.111	0.094	0.15	0.109	0.057	0.10	0.127
Abdominal fat mass (g)	−0.027	0.57	−0.024	0.61	−0.024	0.63	−0.052	0.28	−0.007	0.89	−0.063	0.20
Abdominal lean (g)	0.099	0.10	0.092	0.17	0.108	0.08	0.124	0.09	0.125	**0.05**	0.111	0.13
Abdominal fat (%)	−0.128	**0.01**	−0.129	**0.01**	−0.150	**0.008**	−0.170	**0.001**	−0.143	**0.013**	−0.173	**0.001**
FMI (Kg/m^2)	−0.108	**0.002**	−0.081	**0.03**	−0.116	**0.002**	−0.109	**0.004**	−0.109	**0.005**	−0.113	**0.004**
FFMI (Kg/m^2)	0.084	0.06	0.102	**0.05**	0.123	**0.004**	−0.108	**0.005**	0.125	**0.004**	0.160	**0.004**

Model 2: Adjusted by RWG, BMI of the parents, and smoking and migrant status of the mother at birth of the child, BMI z-score of the children at 7 years. Model 3: adjustments of model 2 + DQI and WST. Associations were analyzed by lineal regression in 3 models adjusted by different variables. Statistically significant associations ($p < 0.05$) are highlighted in bold. β corresponds to standardized coefficients. RWG: rapid weight gain; BMI: body mass index; DQI: diet quality index; WST: weekly screen time; FMI: fat mass index; FFMI: fat-free mass index.

4. Discussion

As far as we are concerned, this is the first study to assess body composition parameters using DXA in such a high number of 7-year-old children followed up since they were born. The main outcome of this study was that active boys and girls (i.e., the group who met WHO PA recommendations) had lower percentage of fat mass and higher percentage of fat-free mass than their non-active counterparts.

Previous studies showed that higher levels of MVPA are associated with a better body composition in children and a lower cardiovascular risk [10]. These are the main reasons why WHO recommendations include performing at least 60 min of MVPA per day in school-age children [23]. However, there is a limited number of studies examining the benefits of meeting these recommendations in terms of body composition in children or later outcomes.

After adjusting by BMI and perinatal factors, both boys and girls had a significantly negative association between PA and fat mass. This association has been witnessed before in studies using samples with 9–10-year-old children [30]. Moreover, in a 3-year longitudinal study [31], Ara et al. showed that children who regularly participated in at least 3 h per week of sports activities were more prone to avoid total and regional fat mass accumulation. Furthermore, in adolescents participating in the HELENA study [29], it was revealed that MVPA was associated with total and central body fat in adolescents from several countries of Europe. However, this was performed in adolescents not in children.

Our final outcome of interest was to observe differences in body composition on the basis of the fulfillment of MVPA recommendations. We observed both boys and girls in terms of fat mass and lean mass (in girls for lean component only for FFMI). However, other authors, on the basis of a large cohort of European children, observed that current PA recommendations were appropriate for girls but not for boys in terms of reducing cardiovascular risk factors [9]. In this sense, Andersen et al. [32] showed that 85 min of daily PA (rather than 60 min) is likely to be a more appropriate threshold to try reducing cardiovascular risk in boys. Moreover, recently, a very similar publication [33] based on a sample of children from the south of Spain also elucidated that while a low proportion of school-aged children met PA WHO recommendations, a higher proportion of them showed normal weight, no abdominal obesity, and low adiposity in comparison with those who did not meet the recommendations. This is evidence that future studies should elucidate on the effects of lifestyle behaviors including PA, already at early stages, in order to reduce adiposity but also to improve the indices of early cardiovascular risk.

In Canada and the United States, there are no studies analyzing the benefits of accomplishing the PA WHO recommendations in body composition in young children, but there is a further 24-h movement guideline that includes three main recommendations to avoid body composition impairments: 9 to 11 h/night of sleep, $\leq$2 h/day of screen time, and at least 60 min/day of MVPA. On the basis of the results on a recent systematic review considering these three factors, PA specifically, MVPA was most consistently associated with desirable health indicators, including adiposity, in comparison with the other two [34]. In a study from Román-Viñas [35], the main conclusion was that meeting the 24-h PA recommendation elicited a reduced z-BMI. The same results were observed in adolescents [36], with those who met the WHO recommendations having a lower BMI and lower levels of total and central body fat.

According to the fat-free mass variable, boys showed significantly higher fat-free mass than girls. This difference in body composition between boys and girls might be explained due to a different biological distribution of fat and also because boys usually perform higher levels of PA than girls. In the IDEFICS study [8,21], boys also showed slightly higher mean values in all PA variables except for light PA and inactive time. Moreover, in the same study, the proportion of children who watch TV more than 1 h/day was 29% (33% of males and 25% of females), more evident during weekend days. In previous studies [37], boys had less inactive time but more screen time than girls.

One of the major strengths of this study is that it is based on a longitudinal cohort of children, which is the largest recruited in the region of Aragon in Spain. We had primarily perinatal and family collected data from these children from birth, which strengthens this kind of analysis in children between lifestyle behaviors and body composition. Body composition variables were assessed through anthropometry and also with DXA, which is considered an accurate and precise method especially for body composition measurements. PA data were collected using accelerometers, instead of questionnaires, which is also a more reliable method [22]. Furthermore, the measurements were performed by a trained group of health professionals. Lastly, the fact that we took into account other lifestyle behaviors such as diet or sedentarism to adjust our analyses allowed us to decrease potential bias in our results.

However, this study is not without limitations. The main limitation is that the children included in our sample were selected by convenience and those who accepted at 7 years were perhaps those from families that care about health and give more importance to monitor health-related habits; however, we cannot confirm if there is a sample bias in our results. Moreover, the fact that we restricted the validity of the accelerometer measurements to 5 h per day instead of 8 h as recommended, which was done in order to increase the sample size, might suppose an underestimation of total MVPA, but this bias probably underestimates the power of our results too, but going in the same direction in view of the available literature. However, there might have been cases wherein the same children recorded 5 h as a minimum for some of the days, but others more than that, and previous authors have described the validity of 4 to 5 days as reliable (0.80) as children of these ages varied their day-to-day MVPA less than older children [38]. In this sense, it is also worth mentioning that it would have been of interest to also assess the PA at light levels as it might be more accurate and representative at these ages. However, evidence on the impact of PA of light intensity on adiposity and cardiometabolic risk markers is still not sufficiently supported and controversial [39]. Finally, it is worth mentioning that DXA assessments were not performed in fasting status but on early afternoon during weekdays. There is no evidence in terms of children, but in adults [40], similarly for males and females, it seems that after feeding, independently of macronutrient composition, there is an over estimation of lean mass in detriment of fat mass. However, in this study, assessment was performed just after meals (but not clearly specified what time after), while in our study, a minimum of 2 h was passed.

5. Conclusions

Despite the fact that the results are applicable only to a sample of children from a region of northern Spain, these children belong to a cohort that was initially representative (at birth) of this region, and therefore valid in terms of sociodemographic representation. Moreover, the standardized procedures and the well-trained health professionals in charge of the measurements, together with the fact that the obtained results are in line with those obtained in bigger studies, allow us to conclude that, since early ages, it seems that there is a large difference between PA levels between boys and girls, although there is also an important difference in screen time in favor of girls in this case. As early as during childhood, these differences are already remarkable in terms of body composition, specifically in terms of fat mass being higher in girls and lean mass being higher in boys. Moreover, apart from gender differences, it seems that those meeting WHO PA recommendations have a benefit in terms of body composition, especially in total body fat and abdominal fat. Future research questions should address what the barriers for children are in order to meet PA recommendations, especially for girls, so that we can develop intervention studies focusing on propitiating a friendly environment for the practice of MVPA. Finally, it would be ideal to follow these kinds of longitudinal studies until later stages in life in order to better understand the weight of these factors in the development of future chronic diseases.

Author Contributions: Conceptualization, P.F.-S., G.R.-M., and I.I.; methodology, I.I., B.M.-P., L.A.M., and G.R.-M.; formal analysis, P.F.-S., I.I., and G.R.-M.; investigation, P.F.-S., I.I., B.M.-P., M.L.M.-B., P.F.-B., and G.R.-M.; resources, G.R.-M.; data curation, I.I.; writing—original draft preparation, P.F.-S.; writing—review and editing, P.F.-S., I.I., B.M.-P., M.L.M.-B., P.F.-B., L.A.M., and G.R.-M.; supervision, L.A.M. and G.R.-M.; project administration, G.R.-M. funding acquisition, G.R.-M. All authors have read and agreed to the published version of the manuscript.

Funding: This study was supported by three grants from the Carlos III Health Institute: (i) PI08/0559: Aragon Health Sciences Institute for the project Growth and Feeding in Infants from Aragon (CALINA); (ii) PI13/02359 environmental factors influencing early development of obesity during childhood and body composition programming; and (iii) RD12/0026: Maternal, Child Health and Development Network (Retic SAMID) RETICS funded by the PN I + D+I 2008–2011 (Spain), ISCIII-Sub-Directorate General for Research Assessment and Promotion and the European Regional Development Fund (ERDF).

Institutional Review Board Statement: The study was conducted according to the guidelines of the Declaration of Helsinki, and approved by the Ethics Committee of CEIC Aragón (CEICA) (protocol code PI13/00105, date of approval 5 October 2016).

Informed Consent Statement: Informed consent was obtained from all the parents of the children involved in the study.

Data Availability Statement: The data presented in this study are available on request from the corresponding author. The data are not publicly available due to data protection issues.

Acknowledgments: Thanks to the CALINA children and their parents who generously volunteered and participated in this project.

Conflicts of Interest: The authors declare no conflict of interest.

References

1. Ogden, C.L.; Carroll, M.D.; Lawman, H.G.; Fryar, C.D.; Kruszon-Moran, D.; Kit, B.K.; Flegal, K.M. Trends in Obesity Prevalence Among Children and Adolescents in the United States, 1988–1994 Through 2013–2014. *J. Am. Med. Assoc.* **2016**, *315*, 2292–2299. [CrossRef]
2. Kumar, S.; Kelly, A.S. Review of Childhood Obesity: From Epidemiology, Etiology, and Comorbidities to Clinical Assessment and Treatment. *Mayo Clin. Proc.* **2017**, *92*, 251–265. [CrossRef] [PubMed]
3. World Health Organization. Noncommunicable Diseases. 2018. Available online: https://www.who.int/news-room/fact-sheets/detail/noncommunicable-diseases (accessed on 7 April 2021).
4. Evaluación y Seguimiento de la Estrategia NAOS: Conjunto Mínimo de Indicadores. 2010. Available online: http://www.aecosan.msssi.gob.es/AECOSAN/web/nutricion/subseccion/indicadores.htm (accessed on 15 March 2021).
5. Monasta, L.; Batty, G.D.; Cattaneo, A.; Lutje, V.; Ronfani, L.; Van Lenthe, F.J.; Brug, J. Early-life determinants of overweight and obesity: A review of systematic reviews. *Obes. Rev.* **2010**, *11*, 695–708. [CrossRef]
6. Duncanson, K.; Shrewsbury, V.; Burrows, T.; Chai, L.K.; Ashton, L.; Taylor, R.; Gow, M.; Ho, M.; Ells, L.; Stewart, L.; et al. Impact of weight management nutrition interventions on dietary outcomes in children and adolescents with overweight or obesity: A systematic review with meta-analysis. *J. Hum. Nutr. Diet.* **2021**, *34*, 147–177. [CrossRef] [PubMed]
7. Wu, X.Y.; Han, L.H.; Zhang, J.H.; Luo, S.; Hu, J.W.; Sun, K. The influence of physical activity, sedentary behavior on health-related quality of life among the general population of children and adolescents: A systematic review. *PLoS ONE* **2017**, *12*, e0187668. [CrossRef] [PubMed]
8. Herrmann, D.; Buck, C.; Sioen, I.; Kouride, Y.; Mårild, S.; Molnar, D.; Mouratidou, T.; Pitsiladis, Y.; Russo, P.; Veidebaum, T.; et al. Impact of physical activity, sedentary behaviour and muscle strength on bone stiffness in 2-10-year-old children-cross-sectional results from the IDEFICS study. *Int. J. Behav. Nutr. Phys. Act.* **2015**, *12*, 112. [CrossRef]
9. Jiménez-Pavón, D.; Konstabel, K.; Bergman, P.; Ahrens, W.; Pohlabeln, H.; Hadjigeorgiou, C.; Siani, A.; Iacoviello, L.; Molnár, D.; De Henauw, S.; et al. Physical activity and clustered cardiovascular disease risk factors in young children: A cross-sectional study (the IDEFICS study). *BMC Med.* **2013**, *11*, 172. [CrossRef]
10. Poitras, V.J.; Gray, C.E.; Borghese, M.M.; Carson, V.; Chaput, J.-P.; Janssen, I.; Katzmarzyk, P.T.; Pate, R.R.; Gorber, S.C.; Kho, M.E.; et al. Systematic review of the relationships between objectively measured physical activity and health indicators in school-aged children and youth. *Appl. Physiol. Nutr. Metab.* **2016**, *41 (Suppl. 3)* (Suppl. 3), S197–S239. [CrossRef]
11. Chodzko-Zajko, W.J.; Schwingel, A.; Romo-Pérez, V. A critical analysis of physical activity recommendations in Spain. *Gac. Sanit.* **2012**, *26*, 525–533. [CrossRef]
12. Rääsk, T.; Mäestu, J.; Lätt, E.; Jürimäe, J.; Jürimäe, T.; Vainik, U.; Konstabel, K. Comparison of IPAQ-SF and Two Other Physical Activity Questionnaires with Accelerometer in Adolescent Boys. *PLoS ONE* **2017**, *12*, e0169527. [CrossRef]

13. Strath, S.J.; Pfeiffer, K.A.; Whitt-Glover, M.C. Accelerometer use with children, older adults, and adults with functional limitations. *Med. Sci. Sports Exerc.* **2012**, *44 (Suppl. 1)* (Suppl. 1), S77–S85. [CrossRef]
14. Westerterp, K.R. Physical activity assessment with accelerometers. *Int. J. Obes. Relat. Metab. Disord.* **1999**, *23* (Suppl. 3), S45–S49. [CrossRef]
15. Migueles, J.H.; Cadenas-Sanchez, C.; Ekelund, U.; Nyström, C.D.; Mora-Gonzalez, J.; Löf, M.; Labayen, I.; Ruiz, J.R.; Ortega, F.B. Accelerometer Data Collection and Processing Criteria to Assess Physical Activity and Other Outcomes: A Systematic Review and Practical Considerations. *Sports Med.* **2017**, *47*, 1821–1845. [CrossRef]
16. Steene-Johannessen, J.; Anderssen, S.A.; van der Ploeg, H.P.; Hendriksen, I.J.M.; Donnelly, A.E.; Brage, S.; Ekelund, U. Are Self-report Measures Able to Define Individuals as Physically Active or Inactive? *Med. Sci. Sports Exerc.* **2016**, *48*, 235–244. [CrossRef]
17. Guinhouya, B.; Samouda, H.; de Beaufort, C. Level of physical activity among children and adolescents in Europe: A review of physical activity assessed objectively by accelerometry. *Public Health* **2013**, *127*, 301–311. [CrossRef]
18. Cuadrón Andrés, L.; Samper Villagrasa, M.P.; Álvarez Sauras, M.L.; Lasarte Velillas, J.J.; Rodríguez Martínez, G.; Calina, G.C. Breastfeeding prevalence during the first year of life in Aragon. CALINA study. *An. Pediatr.* **2013**, *79*, 312–318. [CrossRef]
19. Human, D. Declaration of Helsinki. *Lancet* **2001**, *357*, 236. [CrossRef]
20. Nanri, H.; Shirasawa, T.; Ochiai, H.; Nomoto, S.; Hoshino, H.; Kokaze, A. Rapid weight gain during infancy and early childhood is related to higher anthropometric measurements in preadolescence. *Child Care Health Dev.* **2017**, *43*, 435–440. [CrossRef] [PubMed]
21. Santaliestra-Pasías, A.M.; Mouratidou, T.; Verbestel, V.; Bammann, K.; Molnar, D.; Sieri, S.; Siani, A.; Veidebaum, T.; Mårild, S.; Lissner, L.; et al. Physical activity and sedentary behaviour in European children: The IDEFICS study. *Public Health Nutr.* **2014**, *17*, 2295–2306. [CrossRef] [PubMed]
22. Cooper, A.R.; Goodman, A.; Page, A.S.; Sherar, L.B.; Esliger, D.W.; Van Sluijs, E.M.; Andersen, L.B.; Anderssen, S.; Cardon, G.; Davey, R.; et al. Objectively measured physical activity and sedentary time in youth: The International children's accelerometry database (ICAD). *Int. J. Behav. Nutr. Phys. Act.* **2015**, *12*, 113. [CrossRef]
23. WHO. About WHO Guidelines. 2017. Available online: https://www.who.int/maternal_child_adolescent/guidelines/about-guidelines/en/ (accessed on 15 March 2021).
24. Manios, Y.; Androutsos, O.; Katsarou, C.; Iotova, V.; Socha, P.; Geyer, C.; Moreno, L.; Koletzko, B.; De Bourdeaudhuij, I. Designing and implementing a kindergarten-based, family-involved intervention to prevent obesity in early childhood: The ToyBox-study. *Obes. Rev.* **2014**, *15* (Suppl. 3), 5–13. [CrossRef]
25. Mouratidou, T.; Miguel, M.L.; Androutsos, O.; Manios, Y.; De Bourdeaudhuij, I.; Cardon, G.; Kulaga, Z.; Socha, P.; Galcheva, S.; Iotova, V.; et al. Tools, harmonization and standardization procedures of the impact and outcome evaluation indices obtained during a kindergarten-based, family-involved intervention to prevent obesity in early childhood: The ToyBox-study. *Obes. Rev.* **2014**, *15* (Suppl. 3), 53–60. [CrossRef]
26. Huybrechts, I.; Vereecken, C.; De Bacquer, D.; Vandevijvere, S.; Van Oyen, H.; Maes, L.; Vanhauwaert, E.; Temme, L.; De Backer, G.; De Henauw, S. Reproducibility and validity of a diet quality index for children assessed using a FFQ. *Br. J. Nutr.* **2010**, *104*, 135–144. [CrossRef]
27. Group WMGRS. WHO Child Growth Standards based on length/height, weight and age. *Acta Paediatr Suppl.* **2006**, *450*, 76–85.
28. VanItallie, T.B.; Yang, M.U.; Heymsfield, S.B.; Funk, R.C.; Boileau, R.A. Height-normalized indices of the body's fat-free mass and fat mass: Potentially useful indicators of nutritional status. *Am. J. Clin. Nutr.* **1990**, *52*, 953–959. [CrossRef] [PubMed]
29. Moliner-Urdiales, D.; Ruiz, J.R.; Ortega, F.B.; Rey-Lopez, J.P.; Vicente-Rodriguez, G.; Espana-Romero, V.; Munguía-Izquierdo, D.; Castillo, M.J.; Sjöström, M.; Moreno, L.A. Association of objectively assessed physical activity with total and central body fat in Spanish adolescents; the HELENA Study. *Int. J. Obes.* **2009**, *33*, 1126–1135. [CrossRef] [PubMed]
30. Ekelund, U.; Sardinha, L.B.; Anderssen, S.A.; Harro, M.; Franks, P.W.; Brage, S.; Cooper, A.R.; Andersen, L.B.; Riddoch, C.; Froberg, K. Associations between objectively assessed physical activity and indicators of body fatness in 9- to 10-y-old European children: A population-based study from 4 distinct regions in Europe (the European Youth Heart Study). *Am. J. Clin. Nutr.* **2004**, *80*, 584–590. [CrossRef]
31. Ara, I.; Vicente-Rodriguez, G.; Perez-Gomez, J.; Jimenez-Ramirez, J.; Serrano-Sanchez, J.A.; Dorado, C.; Calbet, J.A.L. Influence of extracurricular sport activities on body composition and physical fitness in boys: A 3-year longitudinal study. *Int. J. Obes.* **2006**, *30*, 1062–1071. [CrossRef]
32. Andersen, L.B.; Riddoch, C.; Kriemler, S.; Hills, A.P. Physical activity and cardiovascular risk factors in children. *Br. J. Sports Med.* **2011**, *45*, 871–876. [CrossRef]
33. López-Gil, J.F.; Brazo-Sayavera, J.; De Campos, W.; Lucas, J.L.Y. Meeting the Physical Activity Recommendations and Its Relationship with Obesity-Related Parameters, Physical Fitness, Screen Time, and Mediterranean Diet in Schoolchildren. *Children* **2020**, *7*, 263. [CrossRef]
34. Saunders, T.J.; Gray, C.E.; Poitras, V.J.; Chaput, J.-P.; Janssen, I.; Katzmarzyk, P.T.; Olds, T.; Gorber, S.C.; Kho, M.E.; Sampson, M.; et al. Combinations of physical activity, sedentary behaviour and sleep: Relationships with health indicators in school-aged children and youth. *Appl. Physiol. Nutr. Metab.* **2016**, *41 (Suppl. 3)* (Suppl. 3), S283–S293. [CrossRef]
35. Roman-Viñas, B.; Chaput, J.P.; Katzmarzyk, P.T.; Fogelholm, M.; Lambert, E.V.; Maher, C.; Maia, J.; Olds, T.; Onywera, V.; Sarmiento, O.L.; et al. Proportion of children meeting recommendations for 24-hour movement guidelines and associations with adiposity in a 12-country study. *Int. J. Behav. Nutr. Phys. Act.* **2016**, *13*, 123. [CrossRef] [PubMed]

36. Martinez-Gomez, D.; Ruiz, J.R.; Ortega, F.B.; Veiga, O.L.; Moliner-Urdiales, D.; Mauro, B.; Galfo, M.; Manios, Y.; Widhalm, K.; Béghin, L.; et al. Recommended levels of physical activity to avoid an excess of body fat in European adolescents: The HELENA Study. *Am. J. Prev. Med.* **2010**, *39*, 203–211. [CrossRef] [PubMed]
37. Leblanc, A.G.; Katzmarzyk, P.T.; Barreira, T.V.; Broyles, S.T.; Chaput, J.-P.; Church, T.S.; Fogelholm, M.; Harrington, D.M.; Hu, G.; Kuriyan, R.; et al. Correlates of Total Sedentary Time and Screen Time in 9-11 Year-Old Children around the World: The International Study of Childhood Obesity, Lifestyle and the Environment. *PLoS ONE* **2015**, *10*, e0129622. [CrossRef] [PubMed]
38. Trost, S.G.; Pate, R.R.; Freedson, P.S.; Sallis, J.F.; Taylor, W.C. Using objective physical activity measures with youth: How many days of monitoring are needed? *Med. Sci. Sports Exerc.* **2000**, *32*, 426–431. [CrossRef] [PubMed]
39. Ayala-Marín, A.M.; Iguacel, I.; De Miguel-Etayo, P.; Moreno, L.A. Consideration of Social Disadvantages for Understanding and Preventing Obesity in Children. *Front. Public Health* **2020**, *8*, 423. [CrossRef] [PubMed]
40. Tinsley, G.M.; Morales, E.; Forsse, J.S.; Grandjean, P.W. Impact of Acute Dietary Manipulations on DXA and BIA Body Composition Estimates. *Med. Sci. Sports Exerc.* **2017**, *49*, 823–832. [CrossRef]

children

MDPI

Article

Changes in Children's Body Composition and Posture during Puberty Growth

Wojciech Rusek [1], Joanna Baran [2,3,*], Justyna Leszczak [2,3], Marzena Adamczyk [1,4], Rafał Baran [5], Aneta Weres [2,3], Grzegorz Inglot [6,7], Ewelina Czenczek-Lewandowska [2,3] and Teresa Pop [2,3]

1. Rehabilitation Centre Rehamed-Center Sp z o.o. Tajecina 66A, 36-002 Tajęcina, Poland; rusekaw@interia.pl (W.R.); m.adamczyk@rehamed-center.pl (M.A.)
2. Institute of Health Sciences, Medical College, University of Rzeszów, Al. mjr. W. Kopisto 2a, 35-310 Rzeszów, Poland; leszczakjustyna.ur@gmail.com (J.L.); anetaweres.ur@gmail.com (A.W.); e.czenczek@univ.rzeszow.pl (E.C.-L.); popter@interia.pl (T.P.)
3. Natural and Medical Center for Innovative Research, ul. Litawora 2, 35-310 Rzeszów, Poland
4. RehaKlinika Sp. z o.o., ul. Pogodne Wzgórze 12, 35-317 Rzeszów, Poland
5. Solution-Statistical Analysis, ul. Stojałowskiego 4/73, 35-120 Rzeszów, Poland; rafal.barann@gmail.com
6. Institute of Medical Sciences, Medical College, University of Rzeszów, Al. mjr. W. Kopisto 2a, 35-310 Rzeszów, Poland; iglooo@interia.pl
7. Individual Medical Practice Grzegorz Inglot, ul. Myśliwska 9, 36-060 Glogow Malopolski, Poland

* Correspondence: joannabaran.ur@gmail.com

Abstract: The main goal of our study was to determine how the age of children, puberty and anthropometric parameters affect the formation of body composition and faulty body posture development in children. The secondary goal was to determine in which body segments abnormalities most often occur and how gender differentiates the occurrence of adverse changes in children's body posture and body composition during puberty. The study group consisted of 464 schoolchildren aged from 6–16. Body posture was assessed with the Zebris system. The composition of the body mass was tested with Tanita MC 780 MA body mass analyzer and the body height was measured using a portable stadiometer PORTSTAND 210. The participants were further divided due to the age of puberty. Tanner division was adopted. The cut-off age for girls is ≥10 years and for boys it is ≥12 years. The analyses applied descriptive statistics, the Pearson correlation, stepwise regression analysis and the t-test. The accepted level of significance was $p < 0.05$. The pelvic obliquity was lower in older children (beta = −0.15). We also see that age played a significant role in the difference in the height of the right pelvis (beta = −0.28), and the difference in the height of the right shoulder (beta = 0.23). Regression analysis showed that the content of adipose tissue (FAT%) increased with body mass index (BMI) and decreased with increasing weight, age, and height. Moreover, the FAT% was lower in boys than in girls (beta negative equal to −0.39). It turned out that older children (puberty), had greater asymmetry in the right shoulder blade ($p < 0.001$) and right shoulder ($p = 0.003$). On the other hand, younger children (who were still before puberty) had greater anomalies in the left trunk inclination ($p = 0.048$) as well as in the pelvic obliquity ($p = 0.008$). Girls in puberty were characterized by greater asymmetry on the right side, including the shoulders ($p = 0.001$), the scapula ($p = 0.001$) and the pelvis ($p < 0.001$). In boys, the problem related only to the asymmetry of the shoulder blades ($p < 0.001$). Girls were characterized by a greater increase in adipose tissue and boys by muscle tissue. Significant differences also appeared in the body posture of the examined children. Greater asymmetry within scapulas and shoulders were seen in children during puberty. Therefore, a growing child should be closely monitored to protect them from the adverse consequences of poor posture or excessive accumulation of adipose tissue in the body.

Keywords: asymmetry; body composition; body posture; puberty

Citation: Rusek, W.; Baran, J.; Leszczak, J.; Adamczyk, M.; Baran, R.; Weres, A.; Inglot, G.; Czenczek-Lewandowska, E.; Pop, T. Changes in Children's Body Composition and Posture during Puberty Growth. *Children* **2021**, *8*, 288. https://doi.org/10.3390/children8040288

Academic Editor: Odysseas An-droutsos

Received: 23 February 2021
Accepted: 7 April 2021
Published: 8 April 2021

1. Introduction

Postural defects as well as excessive weight and obesity are among the most common issues faced by children in developmental age [1]. It is a well-known fact that incorrect positioning in the womb, asymmetry of muscle tone, premature birth [2] and adoption of incorrect posture while sitting or playing are just some of the factors contributing to the development of faulty body posture in children. This is true regardless of the issue: knock knees, flat feet, scoliosis, etc. The problem at hand affects the entire body of a child, because incorrect alignment of structures in one area brings about more misalignment in another. According to current data, 34–50% children and adolescents have different degrees of incorrect posture [3,4]. Professional literature specifies two critical periods for postural development: the first one around the age of 7 is referred to as "school years", and the second as "puberty" [5]. Incorrect body posture in childhood has consequences in adulthood, including reduction of circulatory and respiratory efficiency and vital lung capacity, pain in the spine and related structures, as well as displacement of internal organs [6–8].

The situation is similar with the composition of body mass, in particular with the content of adipose tissue and the development of excessive weight and obesity in children [9]. The origins of the issue may be traced back to fetal age, when the following factors contributing to the development of obesity in later age are pointed out: excessive gestational weight gain (GWG) [10–12], poorly balanced diet [13], and smoking [14]. All these factors may contribute to the abnormal birth weight of a child and excessive body mass index (BMI) in the subsequent years of life. Another study suggests that moderate to vigorous physical activity (MVPA) and vigorous physical activity (VPA) may attenuate the increased risk of an unfavorable body composition and BMI due to high maternal pre-pregnancy BMI and rapid infant weight gain in boys, but not in girls [15].

Fat content changes in the human body with age and is also gender-specific. Girls and women have a higher body fat content than boys and men. This has to do with the reproductive and endocrine function of the body. Moreover, the distribution of adipose tissue also varies with gender and age [16]. During puberty, dramatic hormonal fluctuations as well as a rapid growth in body size occur and are accompanied by marked changes in body composition [17]. The main goal of our study was to determine how the age of children, puberty and anthropometric parameters affect the formation of body composition and faulty body posture development in children. The secondary goal was to determine in which body segments abnormalities most often occur and how gender differentiates the occurrence of adverse changes in children's body posture and body composition during puberty.

2. Materials and Methods

2.1. Participants

Before the beginning of the study sample size was calculated with reference to the total number of children (n = 3790) living in the Trzebownisko Municipality, a rural area in south-eastern Poland. We used a 95% confidence level and a confidence interval (CI) of 0.05. It was calculated that the minimum sample size should be 349. The study took into account 464 school-age children, from 6 to 16 years of age (234 boys and 230 girls). 464 schoolchildren aged from 6–16 participated in the research, 234 of whom were boys and 230 of girls. This study was conducted in eight randomly selected schools (five primary and three secondary schools) in Trzebownisko commune. To ensure the reliability of the measurements, all students were tested in the morning on an empty stomach by the same members of a qualified team. The following inclusion criteria were used in the study: age from 6 to 16 years, consent of parents and children to the study, no certificate of intellectual or motor disability, students in fasting state. Exclusion criteria: metal implants, electronic implants, menstruation, a certificate of mental or motor disability, epilepsy, and failure to refrain from eating on the morning of the examination. Children with neonatal hypertonia or muscle hypotension, chronic neurological diseases or previous injuries and surgery in the last 6 months prior to the study were excluded from the study group.

The participants were informed about the course of the study. The tests were conducted after written consent was received from the headmasters of the schools, parents of the children participating in the project and the children themselves.

The participants were further divided according to the age of puberty. Tanner division was adopted. The cut-off age for girls is ≥10 years and for boys it is ≥12 years [18].

2.2. Body Posture

Body posture was assessed with the Zebris system. The Zebris system is a very extensive diagnostic system with a wide range of applications. It includes accessories for assessing body posture, gait, range of motion, balance and ground forces reaction. It is used in medicine, dentistry and sports, etc. This system shows high sensitivity, precision and measurement reliability [19,20]. Measurement error of the system is estimated at about 5.5% [21]. The Zebris system consists of a measuring unit, a system of micro transmitters and an ultrasonic pointer with which topographic points from the skeleton are scanned, and then compiled in a computer system in the form of a report containing numerical data specifying lengths, angles and degrees for individual parameters, as well as graphs and figures of spine lines and other anthropometric parameters.

Based on the defined topographic points, the software computes the values of selected body posture parameters [22–24]. The following parameters were taken into account in the assessment of posture: total length of spine (mm), thoracic length (mm), lumbar length (mm), pelvic torsion (degree), pelvic obliquity (degree), pelvic/shoulder obliquity (degree), scapula distance right (mm), scapula distance left (mm), scapula distance difference (mm), pelvic height difference right (mm), pelvic height difference left (mm), shoulder height difference right (mm), shoulder height difference left (mm), lateral inclination left (degree), lateral inclination right (degree). For parameters such as the shoulders or pelvis, the right or left side means the upper arm. In the case of lateral inclination, the right or left side indicates to which side the torso is inclined.

For a detailed comparison of children's posture parameters and body weight composition by gender, as well as relation between body composition and body posture parameters, please see another publication from our project [25].

2.3. Anthropometric Measurements and Bioelectrical Impedance

The composition of the body mass was tested with Tanita MC 780 MA (Tanita, Tokyo, Japan) body mass analyzer, which operates on the basis of the phenomenon of bioelectrical impedance (BIA). The bioelectrical impedance method is recognized in the medical and scientific world. Many scientists and practitioners use it. Numerous studies have been carried out on the accuracy and reliability of measurements in various age and clinical groups [26–31]. By analyzing the composition of body weight, the content of adipose tissue (FAT% and kg), lean tissue (fat free mass (FFM)kg), muscle tissue (PMM kg), total water content (TBW%), basic metabolism (BMR) and impedance (IMP) were assessed. Additionally, the parameters of the body mass component analysis segments were measured in the trunk (TR), lower right and left limb (RLP, LLL), and upper right and left limb (RAP, LAL).

A PORTSTAND 210 (Charder, Taichung, Taiwan) portable stadiometer was used to measure the height of the subjects. An accuracy level of 0.1 cm was adopted. Body weight was measured with an accuracy of 0.1 kg using a Tanita analyzer.

With data on weight and height, the child's BMI was calculated.

2.4. Statistical Analysis

The analyses applied descriptive statistics (mean, median, number of subjects, first quartile, third quartile, standard deviation). Statistical analysis was performed on the Statistica 10.0 software developed by StatSoft (Tulsa, OK, USA), and Microsoft Excel. In order to check the correlation between the age and the parameters of posture and body weight, the Pearson correlation was used. Then, the analysis was extended to assess the impact of selected factors on changes in the parameters of body posture and body weight using stepwise regression analysis. The final stage was to evaluate the differences in posture parameters and body weight composition between pre-puberty and during puberty. These differences were assessed by the t-test. The accepted level of significance was $p < 0.05$.

2.5. Ethical Approval

The approval of the Bioethics Committee of the University of Rzeszów No. 2016/06/28 on 28 June 2016 was obtained for conduct of the research.

3. Results

The study group of 464 children in total included a similar number of girls (230) and boys (234). The mean age of the examined girls and boys (11.55 vs. 11.49) and BMI (18.64 vs. 18.94) were also similar (Table 1).

Table 1. Characteristics of the study group.

Variables	*n*	Mean (SD)	Q_1	Median	Q_3
		All			
Age (years)	464	11.52 (2.99)	9.00	12.00	14.00
Height (cm)	464	152.48 (17.77)	137.00	155.50	167.00
Weight (kg)	464	45.39 (16.81)	30.40	44.75	56.45
Body mass index (BMIkg/m^2)	464	18.80 (3.75)	16.10	18.24	20.97
Total length of spine (mm)	464	430.38 (52.62)	388.50	433.00	469.50
Thoracic Length (mm)	464	307.89 (35.87)	281.00	307.00	332.00
Lumbar Length (mm)	464	90.10 (20.04)	76.00	91.00	105.00
Pelvic torsion (degree)	464	3.66 (3.43)	1.30	2.90	4.80
Pelvic obliquity (degree)	464	2.11 (1.77)	0.70	1.70	2.90
Pelvic/shoulder obliquity (degree)	464	2.97 (7.83)	1.10	2.15	3.70
Scapula distance right (mm)	464	46.21(14.33	37.00	46.00	56.00
Scapula distance left (mm)	464	47.06 (32.82)	37.00	45.00	54.00
Scapula distance difference (mm)	464	7.06 (6.15)	3.00	5.00	9.50
Pelvic height difference right (mm)	238	4.41(6.64)	0.00	0.10	7.30
Pelvic height difference left (mm)	226	4.08 (6.44)	0.00	0.00	6.10
Shoulder height difference right (mm)	155	6.48 (5.24)	2.70	5.40	8.60
Shoulder height difference left (mm)	309	11.59 (8.47)	4.80	10.10	16.40
Lateral inclination right (degree)	198	0.54 (0.88)	0.00	0.00	0.85
Lateral inclination left (degree)	266	0.78 (1.07)	0.00	0.20	1.40
The content of adipose tissue in % (Fat %)	464	21.61 (6.10)	16.80	20.40	25.40
The content of adipose tissue in kg (Fat kg)	464	10.25 (6.09)	5.65	8.65	13.25
Lean tissue in kg (FFM kg)	464	35.11(12.01)	24.35	34.80	43.50
Total water content in % (TBW %)	464	25.70 (8.79)	17.85	25.50	31.85
Muscle tissue in kg (PMM kg)	464	33.29 (11.44)	22.95	33.00	41.30
		Girls			
Age (years)	230	11.55 (2.96)	9.00	12	14.00
Height (cm)	230	150.46 (15.69)	136.00	155	163.00
Weight (kg)	230	43.63 (15.38)	29.70	44.35	54.60
BMI (kg/m^2)	230	18.64 (3.87)	15.90	18.00	20.90
Total length of spine (mm)	230	426.77 (46.97)	391.00	433.50	461.00
Thoracic Length (mm)	230	304.83 (33.52)	280.00	306.00	328.00
Lumbar Length (mm)	230	89.57 (18.79)	77.00	90.00	104.00
Pelvic torsion (degree)	230	3.72 (3.36)	1.40	2.90	4.80

Table 1. *Cont.*

Variables	*n*	Mean (SD)	Q_1	Median	Q_3
Pelvic obliquity (degree)	230	2.04 (1.83)	0.70	1.60	2.80
Pelvic/shoulder obliquity (degree)	230	2.38 (1.85)	1.00	1.90	3.40
Scapula distance right (mm)	230	42.94 (12.85)	34.00	44.00	52.00
Scapula distance left (mm)	230	45.56 (44.56)	34.00	43.00	51.00
Scapula distance difference (mm)	230	6.94 (5.83)	3.00	5.00	10.00
Pelvic height difference right(mm)	112	3.85 (6.24)	0.00	0.00	5.80
Pelvic height difference left(mm)	118	4.17 (6.53)	0.00	0.20	6.10
Shoulder height difference right (mm)	83	6.24 (5.50)	2.50	4.80	8.40
Shoulder height difference left (mm)	147	11.83 (8.04)	5.30	10.25	17.30
Lateral inclination right (degree)	98	0.51 (0.85)	0.00	0.00	0.80
Lateral inclination left (degree)	132	0.73 (1.00)	0.00	0.20	1.40
The content of adipose tissue in % (Fat %)	230	23.96 (5.52)	19.50	23.40	27.20
The content of adipose tissue in kg (Fat kg)	230	11.07 (6.24)	6.20	9.55	14.40
Lean tissue in kg (FFM kg)	230	32.56 (9.76)	22.90	34.00	39.80
Total water content in % (TBW %)	230	23.83 (7.14)	16.80	24.90	29.10
Muscle tissue in kg (PMM kg)	230	30.88 (9.28)	21.70	32.25	37.80
		Boys			
Age (years)	234	11.49 (3.02)	9.00	12	14.00
Height (cm)	234	154.46 (19.43)	138.00	156.50	172.00
Weight (kg)	234	47.13 (17.97)	31.80	45	60.00
BMI (kg/m^2)	234	18.94 (3.63)	16.40	18.56	21.05
Total length of spine (mm)	234	433.93 (57.52)	385.00	433.00	483.00
Thoracic Length (mm)	234	310.89 (37.87)	281.00	307.00	337.00
Lumbar Length (mm)	234	90.63 (21.23)	74.00	91.50	106.00
Pelvic torsion (degree)	234	3.59 (3.50)	1.30	2.85	4.80
Pelvic obliquity (degree)	234	2.18 (1.71)	0.80	1.95	3.00
Pelvic/shoulder obliquity (degree)	234	3.55 (10.86)	1.20	2.50	4.10
Scapula distance right (mm)	234	49.42 (14.99)	39.00	49.00	60.00
Scapula distance left (mm)	234	48.54 (13.57)	40.00	48.00	59.00
Scapula distance difference (mm)	234	7.18 (6.46)	3.00	6.00	9.00
Pelvic height difference right(mm)	126	4.95 (6.98)	0.00	0.40	9.00
Pelvic height difference left (mm)	108	3.98 (6.37)	0.00	0.00	6.10
Shoulder height difference right (mm)	72	6.79 (4.93)	3.10	6.60	9.40
Shoulder height difference left (mm)	162	11.38 (8.84)	4.50	9.40	16.20
Lateral inclination right (degree)	100	0.57 (0.90)	0.00	0.00	0.90
Lateral inclination left (degree)	134	0.83 (1.13)	0.00	0.20	1.40
The content of adipose tissue in % (Fat %)	234	19.30 (5.77)	15.40	17.60	21.70
The content of adipose tissue in kg (Fat kg)	234	9.44 (5.85)	5.30	8.20	11.20
Lean tissue in kg (FFM kg)	234	37.62 (13.43)	25.90	36.35	49.30
Total water content in % (TBW %)	234	27.54 (9.83)	19.00	26.60	36.10
Muscle tissue in kg (PMM kg)	234	35.65 (12.82)	24.50	34.45	46.80

M: mean; Me: median; n: number of subjects; Q_1: first quartile; Q_3: third quartile SD: standard deviation.

First, a check was made as to whether the age of the children correlated with the parameters of posture and body weight. Using Pearson's correlation, several statistically significant correlations were obtained. Older children were characterized by a greater total length of the spine, longer individual sections of the spine (a natural feature), but greater asymmetry in the right scapula, right shoulder and right side of the pelvis.

In the case of body mass composition. total fat decreased with age. Taking into account individual body segments, we can see that limb fatness decreases with age, while it increases in the trunk.

Taking into account the sex of the respondents, it turned out that older girls had more body fat, older boys had more muscle, and both older girls and boys had more water.

In the case of body posture parameters, most of the correlations by gender are consistent with the correlations in the whole group (Table 2).

Table 2. Correlation between age and body posture and body composition parameters.

Age &	All		Girls		Boys	
Body Posture	**R**	*p*	**R**	*p*	**R**	*p*
Total length (mm)	0.839	<0.001	0.823	<0.001	0.864	<0.001
Thoracic Length (mm)	0.696	<0.001	0.665	<0.001	0.730	<0.001
Lumbar Length (mm)	0.718	<0.001	0.664	<0.001	0.767	<0.001
Pelvic torsion (degree)	−0.072	0.122	−0.181	0.006	0.028	0.669
Pelvic obliquity (degree)	−0.151	0.001	−0.164	0.013	−0.137	0.036
Pelvic/shoulder obliquity (degree)	0.046	0.319	0.005	0.941	0.066	0.314
Scapula distance RIGHT (mm)	0.343	<0.001	0.182	0.006	0.500	<0.001
Scapula distance LEFT (mm)	0.055	0.235	−0.059	0.376	0.451	<0.001
Scapula distance difference (mm)	−0.072	0.120	−0.055	0.402	−0.087	0.187
Pelvic height difference RIGHT (mm)	0.092	0.047	0.173	0.009	0.025	0.700
Pelvic height difference LEFT (mm)	−0.085	0.068	−0.171	0.009	0.000	0.998
Shoulder height difference RIGHT (mm)	0.230	0.005	0.270	0.013	0.176	0.154
Shoulder height difference LEFT (mm)	−0.024	0.667	−0.039	0.644	−0.014	0.856
Lateral inclination LEFT (degree)	−0.063	0.173	−0.083	0.213	−0.047	0.476
Lateral inclination RIGHT (degree)	0.005	0.912	0.043	0.517	−0.028	0.665
Body Composition						
The content of adipose tissue in % (Fat %)	−0.121	0.009	0.383	<0.001	−0.109	0.096
Lean tissue in % (FFM %)	0.126	0.007	−0.381	<0.001	0.092	0.160
Total water content in % (TBW %)	0.127	0.006	0.856	<0.001	0.881	<0.001
Muscle tissue in % (PMM %)	0.102	0.027	−0.372	<0.001	0.133	0.043
Right leg FAT %	−0.088	0.057	0.263	<0.001	−0.390	<0.001
Left leg FAT %	−0.107	0.021	0.233	<0.001	−0.388	<0.001
Right arm FAT %	−0.163	<0.001	0.064	0.333	−0.404	<0.001
Left arm FAT %	−0.027	0.560	0.291	<0.001	−0.342	<0.001
Total trunk FAT %	0.203	<0.001	0.359	<0.001	0.052	0.429

R—Pearson's correlation ratio.

Taking into account that the demonstrated correlations are not too strong, it was decided to check the influence of other factors, including anthropometric factors. For this purpose, step regressions were performed to assess the influence of factors on the parameters of posture and body weight. In the first case, individual body posture parameters were selected as the dependent variable, and the independent variables were: age, sex, height, weight, BMI, FAT%, FFM%, TBW%, PMM%. Thus, the influence of both anthropometric factors and body mass composition on body posture parameters was checked. These data are presented in Table 3 and Table S1.

Table 3. Regression of factors influencing body posture parameters in children.

Model		Non-Standard Coefficient		Standardized Coefficient	t	p	95% CI (B)		R^2-Corrected
		B	**SE**	**Beta**					
Pelvic torsion (degree)				no variables in the model					
Pelvic obliquity (degree)	Age	−0.09	0.03	−0.15	−3.28	<0.001	−0.14	−0.04	0.021
Pelvic/shoulder obliquity (degree)				no variables in the model					
Scapula distance RIGHT (mm)	Body height	0.34	0.03	0.42	10.33	0.001	0.28	0.41	0.245
	FAT %	−0.56	0.10	−0.24	−5.36	<0.001	−0.76	−0.35	
	Sex	2.52	1.27	0.09	1.98	0.048	0.02	5.01	
Scapula distance LEFT (mm)	PMM %	16.66	7.48	2.94	2.23	0.026	1.96	31.35	0.017
	FFM %	−15.12	7.03	−2.84	−2.15	0.032	−28.94	−1.30	
Scapula distance difference (mm)	PMM %	−0.12	0.05	−0.12	−2.51	0.013	−0.22	−0.03	0.011
Pelvic height difference RIGHT (mm)	Body height	0.16	0.04	0.42	3.99	<0.001	0.08	0.23	0.037
	Age	−0.63	0.23	−0.28	−2.71	0.007	−1.09	−0.17	
Pelvic height difference LEFT (mm)	Body height	−0.04	0.02	−0.10	−2.22	0.027	−0.07	0.00	0.008
Shoulder height difference RIGHT (mm)	Age	0.40	0.14	0.23	2.87	0.005	0.12	0.67	0.046
Shoulder height difference LEFT (mm)	FAT %	0.18	0.07	0.14	2.44	0.015	0.04	0.33	0.016
Lateral inclination LEFT (degree)				no variables in the model					
Lateral inclination RIGHT (degree)				no variables in the model					

95% confidence interval for the B index (Non-standard coefficient); B—non-standard coefficient; Beta—standard coefficient; R^2-—coefficient of determination; SE—standard error.

The pelvic obliquity in relation to the ground was lower in older children (beta = −0.15). We also see that age played a significant role in the difference in the height of the right pelvis (beta = −0.28), and the difference in the height of the right shoulder (beta = 0.23). The notation "no variables in the model" means that none of the independent variables significantly influenced the given parameter of body posture and the model could not be built.

In the second regression analysis, body composition parameters were selected as dependent variables, and age, sex, height, weight and BMI were selected as independent variables. These data are presented in Table 4.

Table 4. Regression of factors influencing the parameters of body mass composition in children.

Model		Non-Standard Coefficient		Standardized Coefficient	t	p	95% CI (B)		R^2-Corrected
		B	SE	Beta					
FAT %	BMI	2.28	0.15	1.40	15.65	<0.001	1.99	2.57	0.810
	Sex	−4.73	0.26	−0.39	−18.33	<0.001	−5.24	−4.23	
	Body weight	−0.29	0.06	−0.80	−5.13	<0.001	−0.40	−0.18	
	Age	−0.46	0.10	−0.23	−4.65	<0.001	−0.66	−0.27	
	Body height	0.09	0.04	0.27	2.48	0.013	0.02	0.16	
FFM %	BMI	−2.22	0.15	−1.35	−14.48	<0.001	−2.52	−1.92	0.795
	Sex	4.70	0.27	0.38	17.34	<0.001	4.17	5.24	
	Body weight	0.26	0.06	0.71	4.40	<0.001	0.14	0.38	
	Age	0.48	0.10	0.23	4.54	<0.001	0.27	0.68	
	Body height	−0.08	0.04	−0.23	−2.07	0.039	−0.16	0.00	
TBW %	BMI	−1.62	0.11	−1.35	−14.48	<0.001	−1.84	−1.40	0.795
	Sex	3.44	0.20	0.38	17.34	<0.001	3.05	3.83	
	Body weight	0.19	0.04	0.71	4.40	<0.001	0.11	0.28	
	Age	0.35	0.08	0.23	4.52	<0.001	0.20	0.50	
	Body height	−0.06	0.03	−0.23	−2.06	0.040	−0.11	0.00	
PMM %	BMI	−1.87	0.08	−1.21	−22.76	<0.001	−2.03	−1.71	0.786
	Sex	0.15	0.03	0.45	5.90	<0.001	3.76	4.78	
	Body weight	4.27	0.26	0.37	16.41	<0.001	0.10	0.21	
	Age	0.35	0.09	0.18	4.16	<0.001	0.19	0.52	

95% confidence interval for the B index (Non-standard coefficient); B—non-standard coefficient; Beta—standard coefficient; R^2-—coefficient of determination; SE—standard error.

Regression analysis showed that FAT% increased with BMI and decreased with increasing weight, age, and height. Moreover, the FAT% was lower in boys than in girls (beta negative equal to −0.39). Taking into account the total body water, it increased with the age of the respondents and was higher in boys. Importantly, water content significantly decreased with the BMI of the respondents (beta = −1.35). Constant values were immediately removed from the table below.

As a complement to the analyzes, it was checked whether the puberty period differentiates the parameters of posture and body mass composition. It turned out that older children (puberty), in addition to greater spine length parameters (which is natural), also had greater asymmetry in the right shoulder blade ($p < 0.001$) and right shoulder ($p = 0.003$). On the other hand, younger children (who were still before puberty) had greater anomalies in the left trunk inclination ($p = 0.048$) as well as in the pelvic obliquity ($p = 0.008$). Additionally, it has been noticed that during adolescence, the development of adipose tissue is dominant in girls, and that in boys there is higher total body water and lean tissue content.

It was also noticed that younger children had lower body fat ($p = 0.019$) and an automatically higher content of total body water ($p = 0.015$). Older children had greater body fat ($p < 0.001$).

Girls in puberty were characterized by greater asymmetry on the right side, including the shoulders ($p = 0.001$), the scapula ($p = 0.001$) and the pelvis ($p < 0.001$). In boys, the problem related only to the asymmetry of the shoulder blades ($p < 0.001$). Interestingly, when comparing girls and boys before puberty, boys show greater asymmetries in body posture. These data are presented in Table 5. Detailed data presenting descriptive statistics for the following results are presented in additional materials (Tables S2–S6).

Table 5. The age of puberty and the parameters of posture and body weight in children.

Variables	Girls before vs. after Puberty	Boys before vs. after Puberty	Girls vs. Boys before Puberty	Girls vs. Boys after Puberty	All Children before vs. after Puberty
	p	*p*	*p*	*p*	*p*
Body posture					
Total length [mm]	<0.001	<0.001	0.008	<0.001	<0.001
Thoracic Length [mm]	<0.001	<0.001	0.008	<0.001	<0.001
Lumbar Length [mm]	<0.001	<0.001	0.241	<0.001	<0.001
Pelvic torsion	0.015	0.592	0.053	0.377	0.264
Pelvic obliquity	0.188	0.022	0.605	0.942	0.008
Pelvic/shoulder obliquity	0.690	0.403	0.139	0.139	0.617
Scapula distance RIGHT	0.001	<0.001	0.079	<0.001	<0.001
Scapula distance LEFT	0.492	<0.001	0.348	<0.001	0.470
Scapula distance difference	0.301	0.364	0.966	0.838	0.156
Pelvic height difference RIGHT	<0.001	0.950	<0.001	0.780	0.065
Pelvic height difference LEFT	0.039	0.901	0.136	0.588	0.136
Shoulder height difference RIGHT	0.001	0.251	0.033	0.995	0.003
Shoulder height difference LEFT	0.070	0.571	0.101	0.567	0.579
Lateral inclination LEFT	0.025	0.638	0.528	0.154	0.048
Lateral inclination RIGHT	0.008	0.644	0.016	0.594	0.288
Body composition					
The content of adipose tissue in % (Fat %)	0.092	0.063	0.016	<0.001	0.019
Lean tissue in % (FFM %)	0.234	0.103	0.015	<0.001	0.015
Total water content in % (TBW %)	0.726	0.105	0.015	<0.001	0.015
Muscle tissue in % (PMM %)	0.016	0.030	0.033	<0.001	0.054
Right leg FAT %	0.001	0.657	0.132	<0.001	0.246
Left leg FAT %	<0.001	<0.001	<0.001	<0.001	0.157
Right arm FAT %	<0.001	<0.001	<0.001	<0.001	0.005
Left arm FAT %	<0.001	<0.001	<0.001	<0.001	0.870
Total trunk FAT %	<0.001	<0.001	<0.001	<0.001	<0.001

p—*t*-test for independent groups (taking into account the correction resulting from the homogeneity of variance (homogeneity was calculated using Levene's test).

4. Discussion

The studies performed form a part of a major project which shows dependencies between the parameters describing body posture and age, puberty and anthropometric parameters in children from rural areas [25,26]. The project is important for practical purposes, as well as for the prevention of postural defects and excessive weight in children and adolescents. There are many studies in the literature assessing body posture, but there are relatively few that take into account factors such as body composition, age and sex. More importantly, in literature there are no scientific publications assessing the components of body weight, position of shoulder blades and the pelvis in puberty in children from rural areas. Accordingly, our research project is one of the first studies to evaluate these parameters.

The authors' own research has shown that asymmetry on the right side increased with age. This concerns the shoulders, pelvis and the deviation of the scapula from the frontal plane. A similar tendency was shown by Yang et al. in their research. They showed that the asymmetry in the area of the shoulder blades and shoulders more often affects older children (>15 years of age) [32]. Pelvic asymmetry in children was confirmed by Drnach et al. [33]. The study of Kapo et al. showed that there is a negative trend of increasing body mass index within the first and youngest age group. The trend of increasing deformity of the shoulder belt has been noted, often inclining towards the formation of milder forms of kyphotic posture. Other forms of deformity brought to light in the survey results are the negative trend of increasing pelvic rotation as well as pelvis rotation which inclines towards the formation of lordotic posture for all three age groups [34].

Considering the components of body weight, our own results showed a reduction in body fat and an increase in water content. Research by Wilczyński et al. and Leskinen et al. show a tendency for the percentage of adipose tissue to decrease with age and for increased water content in both girls and boys [35,36].

Wyszyńska et al. indicate no differences in body posture between girls and boys. In our study, gender only played a role in the right scapula position, where boys had higher scores [37,38]. Rusek et al. showed that sex differentiates the position of the pelvis and the occurrence of asymmetry within it. In addition, adipose tissue content affects the asymmetry of the scapulae and pelvis in the frontal plane, which is also affected by the muscle tissue. [25]. Children with the lowest content of muscle tissue showed the highest difference in the height of the inferior angles of the scapulas in the coronal plane [37]. This is consistent with the results of our own research, where lower muscle tissue content was associated with greater asymmetry of the shoulder blades.

There were significant differences concerning body composition indices containing soft lean mass (SLM) and lean body mass (LBM) in female adolescents with postural deformities in comparison with normal posture. These indices (LBM and SLM) are protective from postural deformities in female normal weight adolescents [39]. In our own study, only one negative relationship was found between lean body mass and the position of the left scapula.

The regression analysis of the body mass composition of children in relation to their sex, age, BMI, height and body weight showed the influence of all these factors on the parameters of body mass composition. However, the most interesting seem to be taking into account the results of puberty. Loncar-Dusek and Pećina report similar findings in their research, which shows more frequent occurrence of scoliosis in adolescence in relation to the period before puberty [40]. Similarly, the research of Demirbüken el al. proves, that body posture is related to age and weight in early adolescence. Adolescence is an important period for identification of postural disorders and one in which it is possible to take precautions for the later ages [41]. Body composition during puberty is a marker of metabolic changes that occur during this period of growth and maturation, and, thus, holds key information regarding current and future health. During puberty, the main components of body composition (total body fat, lean body mass, bone mineral content) all increase; however, considerable sexual dimorphism exists [42]. Both sexes experience rapid increases in total body fat, although the proportion of body fat increases more slowly in boys as a result of a simultaneous rapid increase in fat-free mass [43]. In our study, we found differences in body weight composition between boys and girls across all parameters. The girls were characterized by a higher content of adipose tissue in both general and segmental terms. The boys, on the other hand, had a higher content of lean tissue, muscle tissue, and total body water content.

The research focused on changes in children's body composition and posture during puberty. Changes in body posture in adolescence are a topic often described by various authors, but studies in this field should still be carried out because, as our research shows, there are still many factors influencing body posture that have not been studied.

Strengths and Limitations of the Study

The research was performed only among children from rural areas, which can also be seen as a positive factor, because such specialized research is rarely performed in this study group.

In addition, we cannot exclude the influence of confounding factors on changes in the composition of body weight or body posture (e.g., type of diet, genetic background, etc.).

The influence of physical activity on posture and body weight composition has not been investigated either, therefore it should be taken into account when planning further research.

Among the strengths of the study is undoubtedly the homogeneity and size of the study group as well as the use of objective measuring tools to assess posture and body weight composition.

5. Conclusions

The research conducted has shown that the period of puberty is a time of significant changes in the body of a child. These changes vary depending on the sex of the respondents. Significant differences appeared in the body posture and body composition of the examined children. Greater asymmetry within scapula and shoulders were seen in children during puberty. Girls were characterized by a greater increase in adipose tissue and boys by muscle tissue. Older children (both girls and boys) were characterized by a higher water content in the body. Therefore, a growing child should be closely monitored to protect them from the adverse consequences of poor posture or excessive accumulation of adipose tissue in the body.

Supplementary Materials: The following are available online at https://www.mdpi.com/article/10.3390/children8040288/s1, Table S1: Regression of factors influencing body posture parameters in children. Table S2: Differences between girls and boys in body posture and body composition parameters before puberty, Table S3: Differences between girls and boys in body posture and body composition parameters after puberty. Table S4: Differences in girl's body posture and body composition parameters before and after puberty. Table S5: Differences in boy's body posture and body composition parameters before and after puberty. Table S6: Differences in body posture and body composition parameters before and after puberty in all children.

Author Contributions: Conceptualization, W.R. and T.P.; methodology, W.R., T.P., M.A., J.B.; formal analysis, R.B., J.L., G.I., E.C.-L.; investigation, R.B., M.A., J.B., J.L., W.R., A.W.; resources, G.I., A.W., E.C.-L.; writing—original draft preparation, J.B., J.L.; writing—review and editing, W.R., J.B., J.L., R.B., A.W., M.A., G.I., E.C.-L., T.P.; supervision T.P. All authors have read and agreed to the published version of the manuscript.

Funding: This research received no external funding.

Institutional Review Board Statement: The study was conducted according to the guidelines of the Declaration of Helsinki, and approved of the Bioethics Committee of the University of Rzeszów No. 2016/06/28 from 28 June 2016 was obtained for conducting the research.

Informed Consent Statement: Informed consent was obtained from all subjects involved in the study.

Data Availability Statement: The data presented in this study is available upon request of the respective author. Due to the protection of personal data, the data is not publicly available.

Conflicts of Interest: The authors declare no conflict of interest.

References

1. Wawrzyniak, A.; Tomaszewski, M.; Mews, J.; Jung, A.; Kalicki, B. Postural defects in children and teenagers as one of the major issues in psychosomatic development. *Pediatr. Med. Rodz.* **2017**, *13*, 72–78. [CrossRef]
2. Walicka-Cupryś, K.; Drzał-Grabiec, J.; Rachwał, M.; Piwoński, P.; Perenc, L.; Przygoda, Ł.; Zajkiewicz, K. Body Posture Asymmetry in Prematurely Born Children at Six Years of Age. *BioMed Res. Int.* **2017**, *2017*, 1–8. [CrossRef]
3. Motylewski, S.; Zientala, A.; Pawlicka-Lisowska, A.; Poziomska-Piątkowska, E. Assessment of body posture in 12- and 13-year-olds attending primary schools in Pabianice. *Polski Merkur. Lek. Organ. Polskiego Towar. Lek.* **2015**, *39*, 368–371.

4. Yang, L.; Lu, X.; Yan, B.; Huang, Y. Prevalence of Incorrect Posture among Children and Adolescents: Finding from a Large Population-Based Study in China. *iScience* **2020**, *23*, 101043. [CrossRef]
5. Kasperczyk, T. *Wady Postawy Ciała—Diagnostyka i Leczenie*; Kasper: Kraków, Poland, 2004; pp. 9–10.
6. Maciałczyk-Paprocka, K.; Stawińska-Witoszyńska, B.; Kotwicki, T.; Sowińska, A.; Krzyżaniak, A.; Walkowiak, J.; Krzywińska-Wiewiorowska, M. Prevalence of incorrect body posture in children and adolescents with overweight and obesity. *Eur. J. Nucl. Med. Mol. Imaging* **2017**, *176*, 563–572. [CrossRef] [PubMed]
7. Schmidt, C.P.; Zwingenberger, S.; Walther, A.; Reuter, U.; Kasten, P.; Seifert, J.; Günther, K.P.; Stiehler, M. Prevalence of low back pain in adolescent athletes—an epidemiological investigation. *Int. J. Sports Med.* **2014**, *35*, 684–689. [CrossRef] [PubMed]
8. Kounter, T.D. The Prevalence and Consequences of Poor Posture in Children and Adolescents. 2019, pp. 1–41. Available online: https://digitalcommons.liberty.edu/cgi/viewcontent.cgi?article=1975&context=honors (accessed on 16 January 2021).
9. Simmonds, M.; Llewellyn, A.; Owen, C.G.; Woolacott, N. Predicting adult obesity from childhood obesity: A systematic review and meta-analysis. *Obes. Rev.* **2015**, *17*, 95–107. [CrossRef] [PubMed]
10. Baran, J.; Weres, A.; Czenczek-Lewandowska, E.; Leszczak, J.; Kalandyk-Osinko, K.; Łuszczki, E.; Sobek, G.; Mazur, A. Excessive Gestational Weight Gain: Long-Term Consequences for the Child. *J. Clin. Med.* **2020**, *9*, 3795. [CrossRef]
11. Guo, L.; Liu, J.; Ye, R.; Liu, J.; Zhuang, Z.; Ren, A. Gestational Weight Gain and Overweight in Children Aged 3–6 Years. *J. Epidemiology* **2015**, *25*, 536–543. [CrossRef]
12. Starling, A.P.; Brinton, J.T.; Glueck, D.H.; Shapiro, A.L.; Harrod, C.S.; Lynch, A.M.; Siega-Riz, A.M.; Dabelea, D. Associations of maternal BMI and gestational weight gain with neonatal adiposity in the Healthy Start study. *Am. J. Clin. Nutr.* **2015**, *101*, 302–309. [CrossRef]
13. Herman, A.N.; Malhotra, K.; Wright, G.; Fisher, J.O.; Whitaker, R.C. A qualitative study of the aspirations and challenges of low-income mothers in feeding their preschool-aged children. *Int. J. Behav. Nutr. Phys. Act.* **2012**, *9*, 132. [CrossRef]
14. Suzuki, K.; Sato, M.; Ando, D.; Kondo, N.; Yamagata, Z. Differences in the effect of maternal smoking during pregnancy for childhood overweight before and after 5 years of age. *J. Obstet. Gynaecol. Res.* **2013**, *39*, 914–921. [CrossRef]
15. Bernhardsen, G.P.; Stensrud, T.; Nystad, W.; Dalene, K.E.; Kolle, E.; Ekelund, U. Early life risk factors for childhood obesity—Does physical activity modify the associations? The MoBa cohort study. *Scand. J. Med. Sci. Sports* **2019**, *29*, 1636–1646. [CrossRef]
16. Karastergiou, K.; Smith, S.R.; Greenberg, A.S.; Fried, S.K. Sex differences in human adipose tissues—the biology of pear shape. *Biol. Sex. Differ.* **2012**, *3*, 13. [CrossRef] [PubMed]
17. Pierson, R.N.; Wang, J.; Thornton, J.C.; Heymsfield, S.B. The quality of the body cell mass—1996. Are we ready to measure it? *Appl. Radiat. Isot.* **1998**, *49*, 429–436. [CrossRef]
18. Tanner, J.M. Growth and Maturation during Adolescence. *Nutr. Rev.* **2009**, *39*, 43–55. [CrossRef]
19. Malmström, E.-M.; Karlberg, M.; Melander, A.; Magnusson, M. Zebris Versus Myrin: A Comparative Study Between a Three-Dimensional Ultrasound Movement Analysis and an Inclinometer/Compass Method, Intradevice Reliability, Concurrent Validity, Intertester Comparison, Intratester Reliability, and Intraindividual Variability. *Spine* **2003**, *28*, E433–E440. [CrossRef] [PubMed]
20. Cagnie, B.; Cools, A.; De Loose, V.; Cambier, D.; Danneels, L. Reliability and Normative Database of the Zebris Cervical Range-of-Motion System in Healthy Controls with Preliminary Validation in a Group of Patients with Neck Pain. *J. Manip. Physiol. Ther.* **2007**, *30*, 450–455. [CrossRef] [PubMed]
21. Dvir, Z.; Prushansky, T. Reproducibility and instrument validity of a new ultrasonography-based system for measuring cervical spine kinematics. *Clin. Biomech.* **2000**, *15*, 658–664. [CrossRef]
22. Rusek, W.; Pop, T.; Glista, J.; Skrzypiec, J. Assessment of student's body posture with the use of ZEBRIS system. *Med. J. Rzesz. Univ. Natl. Inst. Med. Wars.* **2010**, *3*, 277–288.
23. Walicka-Cupryś, K.; Kużdżał, A. Objectified criteria for the assessment of body posture using an ultrasonic system Zebris. *Man. Med.* **2013**, *17*, 19–23.
24. Walicka-Cupryś, K.; Skalska-Izdebska, R.; Rachwał, M.; Truszczyńska, A. Influence of the Weight of a School Backpack on Spinal Curvature in the Sagittal Plane of Seven-Year-Old Children. *BioMed Res. Int.* **2015**, *2015*, 1–6. [CrossRef] [PubMed]
25. Rusek, W.; Baran, J.; Leszczak, J.; Adamczyk, M.; Weres, A.; Baran, R.; Inglot, G.; Pop, T. The Influence of Body Mass Composition on the Postural Characterization of School-Age Children and Adolescents. *BioMed Res. Int.* **2018**, *2018*, 1–7. [CrossRef] [PubMed]
26. Rusek, W.; Leszczak, J.; Baran, J.; Adamczyk, M.; Weres, A.; Baran, R.; Inglot, G.; Czenczek-Lewandowska, E.; Porada, S.; Pop, T. Role of body mass category in the development of faulty postures in school-age children from a rural area in south-eastern Poland: A cross-sectional study. *BMJ Open* **2019**, *9*, e030610. [CrossRef]
27. Verney, J.; Schwartz, C.; Amiche, S.; Pereira, B.; Thivel, D. Comparisons of a Multi-Frequency Bioelectrical Impedance Analysis to the Dual-Energy X-Ray Absorptiometry Scan in Healthy Young Adults Depending on their Physical Activity Level. *J. Hum. Kinet.* **2015**, *47*, 73–80. [CrossRef]
28. Lee, L.-W.; Liao, Y.-S.; Lu, H.-K.; Hsiao, P.-L.; Chen, Y.-Y.; Chi, C.-C.; Hsieh, K.-C. Validation of two portable bioelectrical impedance analyses for the assessment of body composition in school age children. *PLoS ONE* **2017**, *12*, e0171568. [CrossRef] [PubMed]
29. Kelly., J.S.; Metcalfe., J. Validity and Reliability of Body Composition Analysis Using the Tanita BC418-MA. *J. Exer. Physiol.* **2012**, *15*, 74–83.

30. Thomson, R.; Brinkworth, G.D.; Buckley, J.D.; Noakes, M.; Clifton, P.M. Good agreement between bioelectrical impedance and dual-energy X-ray absorptiometry for estimating changes in body composition during weight loss in overweight young women. *Clin. Nutr.* **2007**, *26*, 771–777. [CrossRef] [PubMed]
31. Beeson, W.; Batech, M.; Schultz, E.; Salto, L.; Firek, A.; DeLeon, M.; Balcazar, H.; Cordero-MacIntyre, Z. Comparison of body composition by bioelectrical impedance analysis and dual-energy X-ray absorptiometry in Hispanic diabetics. *Int. J. Body Compos. Res.* **2010**, *8*, 45–50.
32. Huang, Y.; Yang, L.; Lu, X.; Yan, B. Prevalence of Abnormal Body Posture Among Chinese Children and Adolescents: A Large Population-Based Cross-Sectional Study. *Biomed. J. Sci. Tech. Res.* **2020**, *25*, 18779–18786. [CrossRef]
33. Drnach, M.; Janovich, M.; Capdeville, K.; Gehring, S.; Wheeling, W.V. The Prevalence of Pelvic Asymmetry and Sacroiliac Joint Dysfunction in Children. Pediatric Physical Therapy. *Pediatr. Phys. Ther.* **2004**, *16*, 64. [CrossRef]
34. Kapo, S.; Rađo, I.; Smajlović, N.; Kovač, S.; Talović, M.; Doder, I.; Čović, N. Increasing postural deformity trends and body mass index analysis in school-age children. *Slov. J. Public Health* **2018**, *57*, 25–32. [CrossRef]
35. Wilczyński, J.; Lipińska-Stańczak, M.; Wilczyński, I. Body Posture Defects and Body Composition in School-Age Children. *Children* **2020**, *7*, 204. [CrossRef]
36. Leskinen., T.; Eloranta., A.-M.; Tompuri., T.; Saari., A.; Ollila., H.; Mäkelä., J.; Niinikoski., H.; Lagström., H. Changes in body composition by age and obesity status in preschool-aged children: The STEPS study. *Eur. J. Clin. Nutr.* **2020**, 1–9. [CrossRef] [PubMed]
37. Wyszyńska, J.; Podgórska-Bednarz, J.; Drzał-Grabiec, J.; Rachwał, M.; Baran, J.; Czenczek-Lewandowska, E.; Leszczak, J.; Mazur, A. Analysis of Relationship between the Body Mass Composition and Physical Activity with Body Posture in Children. *BioMed Res. Int.* **2016**, *2016*, 1–10. [CrossRef]
38. Coelho, J.J.; Graciosa, M.D.; De Medeiros, D.L.; Pacheco, S.C.D.S.; Da Costa, L.M.R.; Ries, L.G.K. Influência da flexibilidade e sexo na postura de escolares. *Rev. Paul. de Pediatr.* **2014**, *32*, 223–228. [CrossRef] [PubMed]
39. Golalizadeh, D.; Toopchizadeh, V.; Fasaie, N.; Dolatkhah, N. Body composition indices in a sample of female adolescents with postural deformity: A case control study. *BMC Res. Notes* **2019**, *12*, 1–5. [CrossRef]
40. Loncar-Dusek, M.; Pećina, M. Pubertetski zamah rasta i skolioza kraljesnice. *Lijec Vjesn.* **1990**, *112*, 85–89. [PubMed]
41. Demirbüken, I.; Özgül, B.; Timurtaş, E.; Şahin, E.; Çekin, M.D.; Yurdalan, S.U.; Polat, M.G. Demographic characteristicsrelated to body posture in early adolescence. *J. Exerc. Ther. Rehabil.* **2016**, *3*, 84–89.
42. Siervogel, R.M.; Demerath, E.W.; Schubert, C.; Remsberg, K.E.; Chumlea, W.C.; Sun, S.; Czerwinski, S.A.; Towne, B. Puberty and Body Composition. *Horm. Res. Paediatr.* **2003**, *60*, 36–45. [CrossRef]
43. Siervogel, R.M.; Maynard, L.M.; Wisemandle, W.A.; Roche, A.F.; Guo, S.S.; Chumlea, W.C.; Towne, B. Annual changes in total body fat (TBF) and fat free mass (FFM) in children from 8 to 18 years in relation to changes in body mass index (BMI). The Fels Longitudinal Study. *Ann. N. Y. Acad. Sci.* **2000**, *904*, 420–423. [CrossRef] [PubMed]

children

MDPI

Article

Can Anthropometry and Body Composition Explain Physical Fitness Levels in School-Aged Children?

Chih-Yu Hsu [1], Liang-Sien Chen [1], I-Jen Chang [1], Wei-Ching Fang [1], Sun-Weng Huang [2,3], Rong-Ho Lin [2], Steve Wen-Neng Ueng [4,5] and Hai-Hua Chuang [1,2,4,6,*]

1 Department of Family Medicine, Linkou and Taipei Branches, Chang Gung Memorial Hospital, Taoyuan 33305, Taiwan; b85079@gmail.com (C.-Y.H.); m7090@cgmh.org.tw (L.-S.C.); Drchangijen@gmail.com (I.-J.C.); winds75526@gmail.com (W.-C.F.)
2 Department of Industrial Engineering and Management, National Taipei University of Technology, Taipei 10608, Taiwan; t107379004@ntut.org.tw (S.-W.H.); rhlin@mail.ntut.edu.tw (R.-H.L.)
3 Graduate Institute of Urban Planning, College of Public Affairs, National Taipei University, New Taipei City 23741, Taiwan
4 College of Medicine, Chang Gung University, Taoyuan 33302, Taiwan; wenneng@adm.cgmh.org.tw
5 Department of Orthopedic Surgery, Linkou Main Branch, Chang Gung Memorial Hospital, Taoyuan 33305, Taiwan
6 Obesity Institute, Geisinger, Danville, PA 17837, USA
* Correspondence: chhaihua@gmail.com; Tel.: +886-975-366-509

Citation: Hsu, C.-Y.; Chen, L.-S.; Chang, I-J.; Fang, W.-C.; Huang, S.-W.; Lin, R.-H.; Ueng, S.W.-N.; Chuang, H.-H. Can Anthropometry and Body Composition Explain Physical Fitness Levels in School-Aged Children? *Children* **2021**, *8*, 460. https://doi.org/10.3390/children8060460

Academic Editors: Odysseas Androutsos and Antonis Zampelas

Received: 30 April 2021
Accepted: 26 May 2021
Published: 31 May 2021

Abstract: Physical fitness (PF) is closely related to various health outcomes and quality of life among children. However, the associations between anthropometry, body composition (BC), and PF are not fully elucidated. This cross-sectional study aimed to investigate the associations between demographic metrics (age, sex), anthropometric measures (body mass index z-score (BMI z-score) waist/height ratio (WHtR)), BC parameters (body-fat percentage (BF%), muscle weight), and PF levels (800-m run, sit-and-reach, 1-min sit-ups, standing long jump) in school-aged children. Continuous variables were dichotomized by median splits. The results of 180 girls and 180 boys (mean age: 10.0 ± 0.7 years; mean BMI z-score: 0.366 ± 1.216) were analyzed. Multivariable linear regressions revealed that BF% (regression coefficient (B) = 3.4, 95% confidence interval (CI) = 2.5–4.3) was independently correlated with the 800-m run. Sex (B = 4.6, 95% CI = 3.0–6.3), age (B = 3.1, 95% CI = 1.9–4.3), and BMI z-score (B = −0.7, 95% CI = −1.4–−0.1) were independently related to sit-and-reach. Age (B = 3.3, 95% CI = 2.0–4.7), BF% (B = −0.3, 95% CI = −0.4–−0.2), and muscle weight (B = 0.7, 95% CI = 0.2–1.2) were independently associated with 1-min sit-ups. In addition to demography, anthropometry and BC provided additional information concerning some PF levels in school-aged children. Weight management and PF promotion should be addressed simultaneously in terms of preventive medicine and health promotion for children.

Keywords: anthropometry; body composition; children; epidemiography; physical fitness

1. Introduction

Physical fitness (PF) is defined as one's capability to work effectively, enjoy leisure time, be healthy, resist hypokinetic diseases, and meet emergencies [1,2]. PF also displays the body's ability to function efficiently. A substantial volume of research has been conducted in adults on the relationships between PF and various health outcomes [3]. The literature suggests that a lack of physical activity (PA) or poor PF is associated with a wide range of chronic diseases [4]. PF is an important predictor for adiposity, bone health, premature mortality, diabetes, cardiovascular diseases, metabolic syndrome, and mental health [3,5–8].

Among all the PF metrics, cardiorespiratory and muscular fitness seem to be the two most influential parameters for disease and health [9]. Low cardiorespiratory fitness is associated with metabolic syndrome, diabetes, and an increased prevalence of cardiovascular disease [7,10]. A recent review elucidates that good cardiorespiratory fitness mitigates stroke hazard in both sexes, and the effect is comparable to well-established stroke risk factors such as blood pressure, obesity, and smoking [4,11]. Muscular strength is a commonly accepted parameter for cardiometabolic risk, sarcopenia, functional disabilities, and frailty connected to all-cause mortality [12–14].

PF is also related to mental health. Previous studies have observed an association between the PF component and cognitive development [8,9,15]. Evidence supports that reduced cardiorespiratory fitness and muscular strength are linked to lower levels of mental well-being, higher levels of depressive symptoms and stress-related exhaustion, and increased health-risk behavior such as suicidal attempts [9]. Silverman and Deuster have proposed two possible mechanisms to explain how PF can buffer against stress-related diseases. The first one is the "blunting effects" of PF on stress responses from the hypothalamic–pituitary–adrenal axis or the sympathetic nervous system. The other one is that improving PF may minimize excessive systemic inflammation. Moreover, PF enhances the neural plasticity and the expression of growth factor, and therefore improves an individual's mood and cognition [16].

Similar observations have been reported among children. Children with suboptimal PF are at higher risk of pediatric obesity, type 2 diabetes, hypercholesterolemia, hypertension, metabolic syndrome, and cardiovascular diseases [17–19]. Poor childhood PF is linked to increased all-cause mortality in adulthood [5,20]. A low level of PF is associated with excessive inflammatory cytokine production [21], with promoting adipose tissue aggregate [22], disrupting platelet function [23], and increasing insulin resistance [24,25]. Moreover, one cross-sectional study suggests that PF plays a crucial role and may ameliorate the negative influence of obesity on children's academic performance [15,26].

Notably, the effects of PF on health outcomes are independent of weight status. Individuals with poor PF, even normal-weighted, have higher mortality than individuals with optimal PF status [27]. Obesity (defined with anthropometric measures), suboptimal body composition (BC), and poor PF have various comorbidities in common, such as metabolic alterations, cardiovascular risks, and mental–cognitive impairment. However, most studies have focused on the medical impacts of each of the three, and the relationships between anthropometry, BC, and PF have not been studied as extensively. Mendoza-Muñoz et al. investigated the influence of BC on PF in adolescents and reported that overweight and obese adolescents had lower levels of PF performance [28]. Ortega, F.B., et al. and Hussey, J., et al. demonstrated that cardiorespiratory fitness was negatively correlated with body mass index (BMI), waist circumference and abdominal adiposity in children [5,29]. However, connections between anthropometric measures, BC, and PF in school-aged children had not yet been studied and elucidated thoroughly.

We hypothesized that not only anthropometry but also BC was associated with PF performance among children. The aim of this study was to investigate the associations between body mass index (BMI) z-score, waist/height ratio (WHtR), body-fat percentage (BF%), muscle weight (MW), 800-m run, sit-and-reach, 1-min sit-ups, and standing long jump in a sample of elementary students in Taiwan.

2. Materials and Methods

2.1. Study Design and Subjects

The study was quantitative and cross-sectional. Data were collected between 2013 and 2016 from a school-based health promotion project conducted by the Chang Gung Memorial Hospital, Linkou Main Branch, Taoyuan, Taiwan for several elementary schools in Northern Taiwan. The details of the project have been reported elsewhere [30]. Anonymous data was retrieved and analyzed in 2020. All sensitive pieces of information that could be linked to a specific person were deidentified. Demographic data (age, sex), anthropometric measures (BMI z-score, WHtR), BC measures (BF, MW), and PF outcomes (cardiorespiratory fitness (800-m run), flexibility (sit-and-reach), speed/agility (1-min sit-ups), lower body power (long-jump)) were analyzed. This study was approved by the Institutional Review Board of our hospital (101-4158A3). Written informed consent was obtained from all participants and their parents.

2.2. Anthropometric Measures

Anthropometric measures were obtained collaboratively by school teachers, nurses, and investigators in elementary schools. Each student was asked to take off his or her shoes before body height and weight were measured according to standard protocols [31]. Body height was assessed to the nearest 0.1 cm by using a wall-mounted stadiometer with head held in horizontal plane [32]. Body weight was measured to the nearest 0.1 kg by using a metric balance scale [33]. BMI was calculated by dividing the weight (kg) by the height squared (m^2) [31]. To make weight status comparable across children at different ages, BMI z-scores were calculated based on sex and age in months according to the United States Centers for Disease Control and Prevention 2000 growth charts [34]. Waist circumference (cm) was measured as the circumference in the horizontal plane midway between the lowest ribs and the iliac crest, and it was further divided by the height (cm) to calculate WHtR [35,36].

2.3. Body Composition Parameters

The detailed BC measurement protocol has been described elsewhere [30]. Briefly, BF% and MW were automatically obtained using bioelectrical impedance analysis (X-scan model, Jawon Medical, Co., Ltd., Seoul, Korea). This segmental multi-frequency bioelectrical impedance analysis module can estimate not only fat-free mass/fat mass but also muscle weight, intracellular water weight and extracellular water weight across limbs and trunks [37]. We chose MW rather than fat-free mass as our parameter of interest since fat-free mass is more complex in its components, including not only muscles but also bone mineral, extracellular water, intracellular water, and visceral protein [37]. Jensen, B., et al. indicated that skeletal muscle mass accounted for 45% to 49% of fat-free mass across different sexes [38].

2.4. Physical Fitness Levels

PF levels were measured by four exercise tests, including 800-m run, sit-and-reach, 1-min sit-ups, and standing long jump. Cardiorespiratory fitness (endurance run) was assessed by 800-m run (sec), defined as the time required to sprint 800-m run [39,40]. Flexibility was evaluated using sit-and-reach (cm), defined by the maximum distance between fingertips and feet reached as the participants slide their hands forward as far as possible towards their feet without bending the hamstring, and maintain this maximum position for at least two seconds [41]. Speed/agility was measured by 1-min sit-ups (times), defined as the maximum number of correct sit-ups achieved in one minute [42]. Lower body power was evaluated using standing long jump, defined as the distance between the starting line and the heel of the closest foot [43]. All PF tests were performed following the methods reported by the Health Promotion Administration, Ministry of Health and Welfare in Taiwan. Standard and corrected stopwatches were used to measure time. Professional physical educators were enrolled to conduct the tests and document the results [44].

2.5. Statistical Analysis

The sample size was calculated using a priori calculation (correlation test, coefficient of determination $\rho 2 = 0.11$, two-tailed $\alpha = 0.05$, power = 0.85; sample size for each age–sex subgroup = 60), as reported in a previous study of BF% and standing long jump in 9-year-old boys [43]. For boys and girls aged 9–11 years old (six age–sex subgroups), the sample size was estimated to be 360.

Most of the distributions of the variables were non-normal, assessed by using the Kolmogorov–Smirnov test. Therefore, non-normally distributed continuous variables were transformed using a two-step approach: the fractional rank and inverse–normal transformation [45]. Thereafter, continuous variables were expressed as means and standard deviations, and categorical variables were expressed as numbers with percentages. Student *t*-tests were used to compare continuous variables, and chi-square tests were used to compare categorical variables in different subgroups, as appropriate. Pearson's correlation test was used to investigate the relationships between continuous variables, whereas Spearman's correlation test was used to analyze associations between continuous variables and categorical variables. The variables with a significance level of $p < 0.05$ in univariate analyses were further assessed using multivariable linear regression models with the forward selection for examining independent variables.

SPSS software (version 25; International Business Machines Corp., Armonk, NY, USA) and G*Power software (version 3.1.9.2; University of Kiel, Kiel, Germany) were used for the statistical analysis. A two-tailed *p*-value lower than 0.05 was considered statistically significant.

3. Results

3.1. Demographic and Clinical Characteristics of the Overall Cohort and Two Subgroups Stratified by Sex

Table 1 displayed demographic and clinical characteristics of the overall cohort and two subgroups stratifies by sex. A total of 360 Taiwanese children from Han ancestry (180 (50.0%) girls and 180 (50.0%) boys); mean age: 10.0 ± 0.7 years; mean BMI z-score: 0.366 ± 1.216) were analyzed.

Table 1. Participant characteristics of the overall cohort as well as two subgroups stratified by sex.

Variables	Overall	Boys	Girls	*p*-Value [a]
Participants	$n = 360$	$n = 180$	$n = 180$	
		Demographic measures		
Age (years)	10.0 ± 0.6	10.0 ± 0.7	10.0 ± 0.6	0.72
		Anthropometric measures		
Body mass index z-score	0.366 ± 1.216	0.543 ± 1.269	0.190 ± 1.137	0.01
Waist/height ratio	0.46 ± 0.06	0.47 ± 0.06	0.45 ± 0.06	<0.001
		Body composition measures		
Body-fat percentage	16.9 ± 7.6	14.1 ± 8.0	19.5 ± 6.1	<0.001
Muscle weight (kg)	29.0 ± 6.1	30.7 ± 5.6	27.5 ± 6.2	<0.001
		Physical fitness levels		
800-m run (sec)	306.9 ± 69.4	297.1 ± 74.2	316.7 ± 63.0	0.01
Sit-and-reach (cm)	25.6 ± 8.4	23.1 ± 8.1	28.1 ± 7.9	<0.001
1-min sit-ups (time)	27.8 ± 8.8	28.7 ± 9.6	26.9 ± 7.9	0.049
Standing long jump (cm)	137.1 ± 25.6	141.1 ± 27.0	133.1 ± 23.5	0.003

Data are summarized as means ± standard deviations. [a] Data were compared using the Student *t*-test.

In the overall cohort, the boys had a significantly higher BMI z-score, higher WHtR, lower BF%, higher MW, lower time in the 800-m run, lower distance of sit-and-reach, higher number of 1-min sit-ups, and higher distance of standing long jump compared to those of the girls. The boys and girls were comparable in age.

3.2. Associations of Physical Fitness Levels and Variables of Interest in the Overall Cohort

Table 2 displayed associations of PF and variables of interest in the overall cohort. The 800-m run was positively associated with female sex, BMI z-score, WHtR, BF%, and MW. Sit-and-reach was positively correlated with female sex and age, and inversely related to BMI z-score and WHtR. The 1-min sit-ups score was positively related to age and inversely associated with BMI z-score, WHtR, and BF%. Standing long jump was positively associated with age and MW, and inversely correlated with female sex, WHtR, and BF%.

3.3. Variables Independently Associated with Physical Fitness Levels in the Overall Cohort Using Logistic Regression Models

Table 3 displayed variables independently associated with PF in the overall cohort. Using univariate linear regression models, we determined that girls, BMI z-score, WHtR, BF%, and MW were significant variables of the 800-m run. However, only the BF% (regression coefficient (B) = 3.4, 95% confidence interval (CI) = 2.5–4.3, $p < 0.001$) was independently correlated with the 800-m run using multivariable linear regression models.

Female sex, age, BMI z-score, and WHtR were significant variables of sit-and-reach. Using multivariable analysis, girls (B = 4.6, 95% CI = 3.0–6.3, $p < 0.001$), age (B = 3.1, 95% CI = 1.9–4.3, $p < 0.001$), and BMI z-score (B = −0.7, 95% CI = −1.4–−0.1, $p = 0.03$) were still significantly and independently related to the sit-and-reach.

Female sex, age, BMI z-score, WHtR, and BF% were significant factors of 1-min sit-ups. Furthermore, age (B = 3.3, 95% CI = 2.0–4.7, $p < 0.001$) and BF% (B = −0.3, 95% CI = −0.4–−0.2, $p < 0.001$) were still significant variables of 1-min sit-and-stand-ups using multivariable analysis.

Female sex, age, WHtR, BF%, and MW were significant factors of the standing long jump. Their associations with the standing long jump remained significant using multivariable analysis: age (B = 14.1, 95% CI = 9.9–18.3, $p < 0.001$), BF% (B = −1.1, 95% CI = −1.4–1.2, $p < 0.001$), and MW (B = 0.7, 95% CI = 0.2–1.2, $p = 0.01$).

Table 2. Associations [a] of dichotomized variables of interest.

Predictors	Female Sex [a]	Age [b]	Body Mass Index z-Score [b]	Waist/Height Ratio [b]	Body-Fat Percentage [b]	Muscle Weight [b]	800-m Run [b]	Sit-and-Reach [b]	1-min Sit-Ups [b]	Standing Long Jump [b]
Female sex	–									
Age	0.02 (0.72)	–								
Body mass index z-score	−0.14 (0.01 *)	−0.03 (0.63)	–							
Waist/height ratio	−0.19 (<0.001 *)	−0.05 (0.37)	0.80 (<0.001 *)	–						
Body-fat percentage	0.36 (<0.001 *)	0.03 (0.57)	0.69 (<0.001 *)	0.62 (<0.001 *)	–					
Muscle weight	−0.24 (<0.001 *)	0.44 (<0.001 *)	0.68 (<0.001 *)	0.43 (<0.001 *)	0.43 (<0.001 *)	–				
800-m run	0.15 (0.004 *)	−0.04 (0.42)	0.28 (<0.001 *)	0.27 (<0.001 *)	0.37 (<0.001 *)	0.11 (0.04 *)	–			
Sit-and-reach	0.32 (<0.001 *)	0.26 (<0.001 *)	−0.15 (0.004 *)	−0.15 (0.01 *)	0.02 (0.69)	−0.07 (0.22)	−0.04 (0.51)	–		
1-min sit-ups	−0.10 (0.06)	0.22 (<0.001 *)	−0.14 (0.01 *)	−0.18 (<0.001 *)	−0.28 (<0.001 *)	0.01 (0.92)	−0.48 (<0.001 *)	0.08 (0.11)	–	
Standing long jump	−0.15 (0.004 *)	0.42 (<0.001 *)	−0.10 (0.06)	−0.11 (0.04 *)	−0.24 (<0.001 *)	0.18 (0.001 *)	−0.39 (<0.001 *)	0.27 (<0.001 *)	0.44 (<0.001 *)	–

Data are summarized as Spearman correlation coefficients (p-values). [a] Data were analyzed using the Spearman correlation test. [b] Data were analyzed using the Pearson correlation test. * p-value < 0.05.

Table 3. Univariate and multivariable regression models of physical fitness levels in the overall cohort.

Predictors	B (95%CI)	*p*-Value [a]	B (95%CI)	*p*-Value [a]
	Univariate Model		Multivariate Model	
		800-m run		
Female sex	19.6 (5.3–33.9)	0.01		NS
Age	−4.5 (−15.5–6.5)	0.42		NI
Body mass index z-score	15.9 (10.2–21.6)	<0.001		NS
Waist/height ratio	290.5 (180.7–400.2)	<0.001		NS
Body-fat percentage	3.3 (2.4–4.2)	<0.001	3.4 (2.5–4.3)	<0.001
Muscle weight	1.3 (0.1–2.5)	0.04		NS
		Sit-and-reach		
Female sex	5.0 (3.4–6.7)	<0.001	4.6 (3.0–6.3)	<0.001
Age	3.3 (2.0–4.6)	<0.001	3.1 (1.9–4.3)	<0.001
Body mass index z-score	−1.1 (−1.8–−0.3)	0.004	−0.7 (−1.4–−0.1)	0.03
Waist/height ratio	−19.0 (−32.5–−5.5)	0.01		NS
Body-fat percentage	0.02 (−0.10–0.14)	0.69		NI
Muscle weight	−0.1 (−0.2–0.1)	0.22		NI
		1-min sit-ups		
Female sex	−1.8 (−3.7–−0.1)	0.049		NS
Age	2.9 (1.6–4.3)	<0.001	3.3 (2.0–4.7)	<0.001
Body mass index z-score	−1.0 (−1.8–−0.3)	0.01		NS
Waist/height ratio	−24.8 (−38.9–−10.7)	0.001		NS
Body-fat percentage	−0.3 (−0.4–−0.2)	<0.001	−0.3 (−0.4–−0.2)	<0.001
Muscle weight	0.1 (−0.1–0.2)	0.92		NI
		Standing long jump		
Female sex	−8.0 (−13.2–−2.7)	0.003		NS
Age	16.3 (12.6–20.0)	<0.001	14.1 (9.9–18.3)	<0.001
Body mass index z-score	−2.1 (−4.3–0.1)	0.06		NS
Waist/height ratio	−42.5 (−83.8–−1.1)	0.04		NS
Body-fat percentage	−0.8 (−1.2–−0.5)	<0.001	−1.1 (−1.4–−0.7)	<0.001
Muscle weight	0.8 (0.3–1.2)	0.001	0.7 (0.2–1.2)	0.01

Abbreviations: B: regression coefficient; CI: confidence interval; NI: not included; NS: not significant. [a] Data were compared using linear regression models.

4. Discussion

The prevalence of pediatric obesity has been increasing alarmingly worldwide in the past 20 years [46–48]. It is generally believed that a decrease in PA and excessive calorie consumption are the two most important factors responsible for this obesity pandemic [49,50]. The detrimental effects of obesity can result in a wide range of negative impacts on health and quality of life [51]. The literature has elucidated various comorbidities of pediatric obesity, including early sexual maturation, polycystic ovary syndrome, nonalcoholic fatty liver disease, obstructive sleep apnea, asthma, left ventricular hypertrophy, hypertension, lower extremity malalignment and joint pain, acanthosis nigricans, higher risk of idiopathic intracranial hypertension, poor self-esteem, anxiety, and depression [51].

On the other hand, a decrease in PF performance has been observed in many developing and developed countries [50]. Young populations are becoming more obese and less fit at the same time, and some research has suggested a linkage between the changes in PF and weight status [50,52,53]. Overweight and obese adolescents had inferior PF compared to their normal-weight peers [28]. Children with higher cardiorespiratory fitness had less total and abdominal adiposity [5,28]. Tomkinson et al. demonstrated that the variability in fatness was accounted for about 20% of the variability in running performance [54]. A cross-sectional study also showed that fatness attributed to the decline of PF by 29–61%, more than other factors such as PA did [52].

In our results, and consistent with the literature, the PF performance of children was associated with their demographic variables [55]. Older children had better flexibility, speed/agility, and lower body power. Children grow up with their height, weight, lean mass, fat tissue, and organs increasing in size [56,57]. These changes are the results of cellular hyperplasia, hypertrophy, and intercellular accretion, enhancing the functional utility of skeletal muscles [58]. Moreover, children develop better motor performance with a better integration of their central nervous system and skeletal–muscular system [59]. Previous investigations report that older age is associated with higher running speed from increased stride length, frequency, and neuromuscular coordination [60,61].

Interesting sex differences were observed. Boys were higher in z-BMI and WHtR, consistent with the well-documented higher prevalence of overweight and obesity in boys in Taiwan [62,63]. Girls were higher in BF% and lower in MW, also consistent with the literature [64,65]. After adjusting for other confounders, girls were better with flexibility. Previous studies show that girls have more total and subcutaneous fat, more connective tissue, and less paraspinous musculature compared to boys [64,66]. This difference could be explained by the difference in sex hormone levels between prepubertal children [67]. Boys have a greater capacity of skeletal muscle because of a higher level of testosterone [57,58]. In comparison, girls have a higher BF% and lower percentage of muscle mass because of a higher amount of estrogen, which contributes to lower tissue density and better flexibility [66]. The divergence of sexual hormone level and BC causes distinct flexibility performance between different sexes [68,69].

As expected and consistent with some previous research [29,70,71], our data demonstrated connections between weight status, BC, and PF. In general, higher BMI-z score and BF% were linked with lower levels of PF.

First, higher z-BMI was associated with worse flexibility. Larger trunk mass increased mechanical work and moment of inertia. As a result, there was higher propulsion and extra load during dynamic activities [70]. Casonatto et al. indicated that abdominal obesity might affect the lower back and hamstring flexibility and hamper the trunk to the extreme reach position [72].

Second, excessive BF% was related to poorer muscle endurance, cardiorespiratory fitness, and lower-body power. Similar findings have been reported by other investigators [28,73–75]. One systematic review from the European Childhood Obesity Group suggests that insulin resistance causes mitochondrial dysfunction in the skeletal muscle, which leads to muscle fatigability and delayed post-effort muscle recovery [76]. Another systematic review shows the negative effects of childhood obesity on spirometric variables, namely reductions in forced expiratory volume in one second, forced vital capacity, and forced expiratory volume in one second/forced vital capacity ratio [77,78]. An increase in adipose tissue directly negatively impacts one's pulmonary function by reducing functional residual capacity, expiratory reserve volume, and residual volume [79,80]. Previous studies have reported that children with obesity perform less well in weight-bearing activities such as jumping and running [53,75,81]. Children with higher BMI usually have higher muscle mass, which increases absolute muscle strength; however, relative muscle strength is the key component of muscle function for daily activities. Excessive BF% increases inert load, which impacts negatively on relative muscle strength and results in poorer lower-body power [75].

Third, children with higher muscle weight had better lower-body power. There are two types of skeletal muscle, type I (red fibers) and type II (white fibers). Type II fibers have higher force, power, and speed. Most one-burst moves, such as the long jump or badminton smash, are largely contributed by type II fibers due to their higher contractility and fast reactions. On the other hand, type I fibers are slow movers but have better ability in utilizing oxygen and higher endurance [82]. Gaining muscle weight is largely related to type II fiber rather than type I fibers [83]. Our results were consistent with the physiology, showing that muscle weight was significant in relation to lower-body explosive power, but not muscle endurance.

The main contribution of this study was to provide scientific evidence on how anthropometry and BC may affect the PF performance, in addition to age and sex. The study had some limitations. First, all the participants were Han and confined to age of 9–11 years, which might limit the generalizability of this study to other populations. Second, the information on pubertal status was not collected, which might have been an important confounder on weight status, BC, and PF. Third, BC was measured by BIA; BIA modules are less costly and more available, but less accurate than dual energy X-ray absorptiometry modules. Also, the study was cross-sectional and therefore unable to conduct causation in predicting the future performance. Further studies with a prospective design will be of interest.

5. Conclusions

Overall, younger age, higher BMI z-score, and higher BF% were associated with poorer PF performance among elementary school students; girls had better flexibility. Our study concluded that other than demographic variables, anthropometric measures and BC parameters provided additional information in concerning PF levels in school-aged children. PF testing should be considered in addition to clinical measures to understand children's health more comprehensively. Health promotion for children should focus not only on anthropometry but also BC and PF.

Author Contributions: Conceptualization, S.W.-N.U. and H.-H.C.; methodology, S.-W.H. and R.-H.L.; software, S.-W.H. and R.-H.L.; validation, L.-S.C., I.-J.C. and W.-C.F.; formal analysis, H.-H.C.; investigation, C.-Y.H. and H.-H.C.; resources, L.-S.C., I.-J.C., W.-C.F. and H.-H.C.; data curation, S.-W.H. and R.-H.L.; writing—original draft preparation, C.-Y.H., L.-S.C. and H.-H.C.; writing—review and editing, L.-S.C., I.-J.C. and H.-H.C.; visualization, S.-W.H. and R.-H.L.; supervision, S.W.-N.U.; project administration, S.W.-N.U. and H.-H.C.; funding acquisition, S.W.-N.U. and H.-H.C. All authors have read and agreed to the published version of the manuscript.

Funding: The study was funded by Chang Gung Medical Foundation, Grant number: CMRPG1H0061, CMRPG1H0062, CMRPG1H0063, CMRPG1J041, CMRPG1J042 (H.-H.C.).

Institutional Review Board Statement: The study was conducted according to the guidelines of the Declaration of Helsinki and approved by the Institutional Review Board of Chang Gung Memorial Hospital, Linkou (101-4158A3).

Informed Consent Statement: Informed consent was obtained from all patients involved in the study.

Data Availability Statement: Data available on request due to privacy/ethical restrictions.

Acknowledgments: The authors would like to thank Pei-Chi Yang, Yu-Chan Liao, and Yi-Ting Lin, Chang Gung Memorial Hospital, Linkou and Taipei Branches, Taoyuan, Taiwan, for their help in the data collection, analysis, and manuscript editing. The authors would also like to express our respects and appreciations for Kuang-Hung Hsu from Chang Gung University, Taoyuan, Taiwan for his great help and support for the project.

Conflicts of Interest: The authors declare no conflict of interest.

References

1. Corbin, C.B.; Lindsey, R. *Concepts of Physical Fitness*; McGraw-Hill Education: New York, NY, USA, 1997.
2. Pate, R.R. The Evolving Definition of Physical Fitness. *Quest* **1988**, *40*, 174–179. [CrossRef]
3. Lin, Y.-T.; Lee, P.-F.; Lee, T.-S.; Ho, C.-C. Poor Physical Fitness Performance as a Predictor of General Adiposity in Taiwanese Adults. *Int. J. Environ. Res. Public Health* **2020**, *17*, 2686. [CrossRef]
4. Booth, F.W.; Roberts, C.K.; Laye, M. Lack of Exercise Is a Major Cause of Chronic Diseases. *Compr. Physiol.* **2012**, *2*, 1143–1211. [CrossRef]
5. Ortega, F.B.; Ruiz, J.; Castillo, M.J.; Sjöström, M. Physical fitness in childhood and adolescence: A powerful marker of health. *Int. J. Obes.* **2007**, *32*, 1–11. [CrossRef]
6. Henriques-Neto, D.; Magalhães, J.P.; Hetherington-Rauth, M.; Santos, D.A.; Baptista, F.; Sardinha, L. Physical Fitness and Bone Health in Young Athletes and Nonathletes. *Sports Health* **2020**, *12*, 441–448. [CrossRef] [PubMed]

7. Mäestu, E.; Harro, J.; Veidebaum, T.; Kurrikoff, T.; Jürimäe, J.; Mäestu, J. Changes in cardiorespiratory fitness through adolescence predict metabolic syndrome in young adults. *Nutr. Metab. Cardiovasc. Dis.* **2020**, *30*, 701–708. [CrossRef]
8. Åvitsland, A.; Leibinger, E.; Haugen, T.; Lerum, Ø.; Solberg, R.B.; Kolle, E.; Dyrstad, S.M. The association between physical fitness and mental health in Norwegian adolescents. *BMC Public Health* **2020**, *20*, 776. [CrossRef] [PubMed]
9. Appelqvist-Schmidlechner, K.; Vaara, J.P.; Vasankari, T.; Häkkinen, A.; Mäntysaari, M.; Kyröläinen, H. Muscular and cardiorespiratory fitness are associated with health-related quality of life among young adult men. *BMC Public Health* **2020**, *20*, 842. [CrossRef] [PubMed]
10. Kim, B.; Ku, M.; Kiyoji, T.; Isobe, T.; Sakae, T.; Oh, S. Cardiorespiratory fitness is strongly linked to metabolic syndrome among physical fitness components: A retrospective cross-sectional study. *J. Physiol. Anthr.* **2020**, *39*, 1–9. [CrossRef] [PubMed]
11. Wang, Y.; Li, F.; Cheng, Y.; Gu, L.; Xie, Z. Cardiorespiratory fitness as a quantitative predictor of the risk of stroke: A dose–response meta-analysis. *J. Neurol.* **2020**, *267*, 491–501. [CrossRef]
12. García-Hermoso, A.; Cavero-Redondo, I.; Ramírez-Vélez, R.; Ruiz, J.R.; Ortega, F.B.; Lee, D.-C.; Martínez-Vizcaíno, V. Muscular Strength as a Predictor of All-Cause Mortality in an Apparently Healthy Population: A Systematic Review and Meta-Analysis of Data from Approximately 2 Million Men and Women. *Arch. Phys. Med. Rehabil.* **2018**, *99*, 2100–2113. [CrossRef]
13. Lee, S.H.; Gong, H.S. Measurement and Interpretation of Handgrip Strength for Research on Sarcopenia and Osteoporosis. *J. Bone Metab.* **2020**, *27*, 85–96. [CrossRef] [PubMed]
14. Zaccagni, L.; Toselli, S.; Bramanti, B.; Gualdi-Russo, E.; Mongillo, J.; Rinaldo, N. Handgrip Strength in Young Adults: Association with Anthropometric Variables and Laterality. *Int. J. Environ. Res. Public Health* **2020**, *17*, 4273. [CrossRef]
15. Moradi, A.; Damirchi, E.S.; Narimani, M.; Esmaeilzadeh, S.; Dziembowska, I.; Azevedo, L.B.; Prado, W.L.D. Association between Physical and Motor Fitness with Cognition in Children. *Medicina* **2019**, *55*, 7. [CrossRef]
16. Silverman, M.N.; Deuster, P.A. Biological mechanisms underlying the role of physical fitness in health and resilience. *Interface Focus* **2014**, *4*, 20140040. [CrossRef] [PubMed]
17. Dwyer, T.; Magnussen, C.G.; Schmidt, M.D.; Ukoumunne, O.C.; Ponsonby, A.-L.; Raitakari, O.T.; Zimmet, P.Z.; Blair, S.N.; Thomson, R.; Cleland, V.; et al. Decline in Physical Fitness from Childhood to Adulthood Associated with Increased Obesity and Insulin Resistance in Adults. *Diabetes Care* **2008**, *32*, 683–687. [CrossRef]
18. García-Hermoso, A.; Alonso-Martinez, A.M.; Ramírez-Vélez, R.; Izquierdo, M. Effects of Exercise Intervention on Health-Related Physical Fitness and Blood Pressure in Preschool Children: A Systematic Review and Meta-Analysis of Randomized Controlled Trials. *Sports Med.* **2020**, *50*, 187–203. [CrossRef] [PubMed]
19. Pazzianotto-Forti, E.M.; Moreno, M.A.; Plater, E.; Baruki, S.B.S.; Junior, I.R.; Reid, W.D. Impact of Physical Training Programs on Physical Fitness in People with Class II and III Obesity: A Systematic Review and Meta-Analysis. *Phys. Ther.* **2020**, *100*, 963–978. [CrossRef] [PubMed]
20. Sato, M.; Kodama, S.; Sugawara, A.; Saito, K.; Sone, H. Physical Fitness During Adolescence and Adult Mortality. *Epidemiology* **2009**, *20*, 463–464. [CrossRef]
21. Sacheck, J. Pediatric Obesity: An Inflammatory Condition? *J. Parenter. Enter. Nutr.* **2008**, *32*, 633–637. [CrossRef] [PubMed]
22. Dumuid, D.; Olds, T.; Lewis, L.K.; Martin-Fernández, J.A.; Barreira, T.; Broyles, S.; Chaput, J.-P.; Fogelholm, M.; Hu, G.; Kuriyan, R.; et al. The adiposity of children is associated with their lifestyle behaviours: A cluster analysis of school-aged children from 12 nations. *Pediatr. Obes.* **2018**, *13*, 111–119. [CrossRef] [PubMed]
23. Buchan, D.S.; Thomas, N.E.; Baker, J.S. Novel risk factors of cardiovascular disease and their associations between obesity, physical activity and physical fitness. *J. Public Health Res.* **2012**, *1*, 11–66. [CrossRef] [PubMed]
24. Schmitz, K.; Jacobs, D.R.; Hong, C.-P.; Steinberger, J.; Moran, A.; Sinaiko, A. Association of physical activity with insulin sensitivity in children. *Int. J. Obes.* **2002**, *26*, 1310–1316. [CrossRef]
25. Chae, H.-W.; Kwon, Y.-N.; Rhie, Y.-J.; Kim, H.-S.; Kim, Y.-S.; Paik, I.-Y.; Suh, S.-H.; Kim, D.-H. Effects of a Structured Exercise Program on Insulin Resistance, Inflammatory Markers and Physical Fitness in Obese Korean Children. *J. Pediatr. Endocrinol. Metab.* **2010**, *23*, 1065–1072. [CrossRef]
26. Muntaner-Mas, A.; Palou, P.; Vidal-Conti, J.; Esteban-Cornejo, I. A Mediation Analysis on the Relationship of Physical Fitness Components, Obesity, and Academic Performance in Children. *J. Pediatr.* **2018**, *198*, 90–97. [CrossRef]
27. Barry, V.W.; Baruth, M.; Beets, M.W.; Durstine, J.L.; Liu, J.; Blair, S.N. Fitness vs. Fatness on All-Cause Mortality: A Meta-Analysis. *Prog. Cardiovasc. Dis.* **2014**, *56*, 382–390. [CrossRef] [PubMed]
28. Mendoza-Muñoz, M.; Adsuar, J.; Pérez-Gómez, J.; Muñoz-Bermejo, L.; Garcia-Gordillo, M.; Carlos-Vivas, J. Influence of Body Composition on Physical Fitness in Adolescents. *Medicina* **2020**, *56*, 328. [CrossRef] [PubMed]
29. Hussey, J.; Bell, C.; Bennett, K.; O'Dwyer, J.; Gormley, J. Relationship between the intensity of physical activity, inactivity, cardiorespiratory fitness and body composition in 7-10-year-old Dublin children. *Br. J. Sports Med.* **2007**, *41*, 311–316. [CrossRef]
30. Chuang, H.-H.; Lin, R.-H.; Chen, J.-Y.; Yeh, W.-C.; Lin, H.-F.; Ueng, S.W.-N.; Hsu, K.-H. Effectiveness of a multi-faceted intervention among elementary school children. *Medicina* **2019**, *98*, e15079. [CrossRef] [PubMed]
31. Sekgala, M.D.; Monyeki, K.D.; Mogale, M.A.; Ramoshaba, N.E. Performance of blood pressure to height ratio as a screening tool for elevated blood pressure in rural children: Ellisras Longitudinal Study. *J. Hum. Hypertens.* **2017**, *31*, 591–595. [CrossRef]
32. Nafiu, O.O.; Burke, C.; Lee, J.; Voepel-Lewis, T.; Malviya, S.; Tremper, K.K. Neck Circumference as a Screening Measure for Identifying Children with High Body Mass Index. *Pediatrics* **2010**, *126*, e306–e310. [CrossRef]

33. Chung, K.; Chiou, H.; Chang, J.; Chen, Y. Associations of nitric oxide with obesity and psychological traits among children and adolescents in Taiwan. *Pediatr. Obes.* **2019**, *15*, e12593. [CrossRef]
34. Flegal, K.M.; Cole, T.J. Construction of LMS parameters for the Centers for Disease Control and Prevention 2000 growth charts. *Natl. Health Stat. Rep.* **2013**, *2013*, 1–3.
35. Ma, W.-Y.; Yang, C.-Y.; Shih, S.-R.; Hsieh, H.-J.; Hung, C.S.; Chiu, F.-C.; Lin, M.-S.; Liu, P.-H.; Hua, C.-H.; Hsein, Y.-C.; et al. Measurement of Waist Circumference: Midabdominal or iliac crest? *Diabetes Care* **2013**, *36*, 1660–1666. [CrossRef] [PubMed]
36. Lohman, T.G.; Roche, A.F.; Martorell, R. *Anthropometric Standardization Reference Manual*; Human Kinetics Books: Champaign, IL, USA, 1998.
37. Grossi, M.; Riccò, B. Electrical impedance spectroscopy (EIS) for biological analysis and food characterization: A review. *J. Sens. Sens. Syst.* **2017**, *6*, 303–325. [CrossRef]
38. Jensen, B.; Braun, W.; Geisler, C.; Both, M.; Klückmann, K.; Müller, M.J.; Bosy-Westphal, A. Limitations of Fat-Free Mass for the Assessment of Muscle Mass in Obesity. *Obes. Facts* **2019**, *12*, 307–315. [CrossRef]
39. Hu, J.; Li, Z.; Li, S.; Li, H.; Wang, S.; Wang, S.; Han, L. Skipping breakfast and physical fitness among school-aged adolescents. *Clinics* **2020**, *75*, e1599. [CrossRef]
40. Hsieh, P.-L.; Chen, M.-L.; Huang, C.-M.; Chen, W.-C.; Li, C.-H.; Chang, L.-C. Physical Activity, Body Mass Index, and Cardiorespiratory Fitness among School Children in Taiwan: A Cross-Sectional Study. *Int. J. Environ. Res. Public Health* **2014**, *11*, 7275–7285. [CrossRef]
41. Hartman, J.G.; Looney, M. Norm-Referenced and Criterion-Referenced Reliability and Validity of the Back-Saver Sit-and-Reach. *Meas. Phys. Educ. Exerc. Sci.* **2003**, *7*, 71–87. [CrossRef]
42. Lovecchio, N.; Novak, D.; Sedlacek, J.; Hamar, P.; Milanovic, I.; Radisavljevic-Janic, S.; Emeljanovas, A.; Eid, L.; Zago, M. Physical fitness for sedentary students: A common trend from six European countries. *J. Sports Med. Phys. Fit.* **2019**, *59*, 1389–1396. Available online: https://www.minervamedica.it/index2.php?show=R40Y2019N08A1389 (accessed on 4 May 2020). [CrossRef] [PubMed]
43. Reid, R.E.; Fillon, A.; Thivel, D.; Henderson, M.; Barnett, T.A.; Bigras, J.-L.; Mathieu, M.-E. Can anthropometry and physical fitness testing explain physical activity levels in children and adolescents with obesity? *J. Sci. Med. Sport* **2020**, *23*, 580–585. [CrossRef] [PubMed]
44. Ministry of Education. Measures for the Implementation of Physical Fitness Testing in Taiwan. 2018. Available online: https://edu.law.moe.gov.tw/LawContent.aspx?id=GL000769 (accessed on 28 March 2021).
45. Templeton, G.F. A Two-Step Approach for Transforming Continuous Variables to Normal: Implications and Recommendations for IS Research. *Commun. Assoc. Inf. Syst.* **2011**, *28*, 41–58. [CrossRef]
46. Wang, Y.; Lim, H. The global childhood obesity epidemic and the association between socio-economic status and childhood obesity. *Int. Rev. Psychiatry* **2012**, *24*, 176–188. [CrossRef]
47. NCD Risk Factor Collaboration (NCD-RisC). Heterogeneous contributions of change in population distribution of body mass index to change in obesity and underweight. *Elife* **2021**, *10*, e60060. [CrossRef] [PubMed]
48. Rodriguez-Martinez, A.; Zhou, B.; Sophiea, M.K.; Bentham, J.; Paciorek, C.J.; Iurilli, M.L.; Carrillo-Larco, R.M.; Bennett, J.E.; Di Cesare, M.; Taddei, C.; et al. Height and body-mass index trajectories of school-aged children and adolescents from 1985 to 2019 in 200 countries and territories: A pooled analysis of 2181 population-based studies with 65 million participants. *Lancet* **2020**, *396*, 1511–1524. [CrossRef]
49. Ostojic, S.M.; Stojanovic, M.D.; Stojanovic, V.; Maric, J.; Njaradi, N. Correlation between Fitness and Fatness in 6-14-year Old Serbian School Children. *J. Health Popul. Nutr.* **2011**, *29*, 53–60. [CrossRef]
50. Tomkinson, G.R.; Léger, L.A.; Olds, T.S.; Cazorla, G. Secular Trends in the Performance of Children and Adolescents (1980–2000). *Sports Med.* **2003**, *33*, 285–300. [CrossRef]
51. Kumar, S.; Kelly, A.S. Review of Childhood Obesity: From Epidemiology, Etiology, and Comorbidities to Clinical Assessment and Treatment. *Mayo Clin. Proc.* **2017**, *92*, 251–265. [CrossRef]
52. Olds, T.S.; Ridley, K.; Tomkinson, G.R. Declines in aerobic fitness: Are they only due to increasing fatness? *Med. Sport Sci.* **2007**, *50*, 226–240. [PubMed]
53. Sacchetti, R.; Ceciliani, A.; Garulli, A.; Masotti, A.; Poletti, G.; Beltrami, P.; Leoni, E. Physical fitness of primary school children in relation to overweight prevalence and physical activity habits. *J. Sports Sci.* **2012**, *30*, 633–640. [CrossRef] [PubMed]
54. Tomkinson, G.R.; Olds, T.S. Secular Changes in Pediatric Aerobic Fitness Test Performance: The Global Picture. *Med. Sport Sci.* **2007**, *50*, 46–66.
55. Zaqout, M.; Vyncke, K.; Moreno, L.A.; De Miguel-Etayo, P.; Lauria, F.; Molnar, D.; Lissner, L.; Hunsberger, M.; Veidebaum, T.; Tornaritis, M.; et al. Determinant factors of physical fitness in European children. *Int. J. Public Health* **2016**, *61*, 573–582. [CrossRef]
56. Lämmle, L.; Worth, A.; Bös, K. Socio-demographic correlates of physical activity and physical fitness in German children and adolescents. *Eur. J. Public Health* **2012**, *22*, 880–884. [CrossRef]
57. Rosa-Guillamón, A.; Carrillo-López, P.J.; García-Cantó, E. Analysis of physical fitness according to sex, age, body mass index and level of physical activity in Spanish elementary school students. *Rev. Fac. Med.* **2020**, *68*, 92–99.
58. Malina, R.M.; Bouchard, C. Growth, Maturation, and Physical Activity. *Med. Sci. Sports Exerc.* **1992**, *24*, 841. [CrossRef]
59. Haywood, K.M.; Getchell, N. *Life Span Motor Development*; Human Kinetics: Champaign, IL, USA, 2019.

60. Milanese, C.; Sandri, M.; Cavedon, V.; Zancanaro, C. The role of age, sex, anthropometry, and body composition as determinants of physical fitness in nonobese children aged 6-12. *PeerJ* **2020**, *8*, e8657. [CrossRef]
61. Praagh, E.V. Anaerobic fitness tests: What are we measuring? *Med. Sport Sci.* **2007**, *50*, 26–45.
62. Chu, N.-F. Prevalence and trends of obesity among school children in Taiwan—The Taipei Children Heart Study. *Int. J. Obes.* **2001**, *25*, 170–176. [CrossRef]
63. Chu, N.-F.; Pan, W.-H. Prevalence of obesity and its comorbidities among schoolchildren in Taiwan. *Asia Pac. J. Clin. Nutr.* **2007**, *16*, 601.
64. Arfai, K.; Pitukcheewanont, P.D.; Goran, M.I.; Tavaré, C.J.; Heller, L.; Gilsanz, V. Bone, Muscle, and Fat: Sex-related Differences in Prepubertal Children. *Radiology* **2002**, *224*, 338–344. [CrossRef]
65. Leonard, M.B.; Elmi, A.; Mostoufi-Moab, S.; Shults, J.; Burnham, J.M.; Thayu, M.; Kibe, L.; Wetzsteon, R.J.; Zemel, B.S. Effects of Sex, Race, and Puberty on Cortical Bone and the Functional Muscle Bone Unit in Children, Adolescents, and Young Adults. *J. Clin. Endocrinol. Metab.* **2010**, *95*, 1681–1689. [CrossRef] [PubMed]
66. Graf, C.; Schierz, O.; Steinke, H.; Körner, A.; Kiess, W.; Kratzsch, J.; Hirsch, C. The LIFE Child study team; LIFE Child study team Sex hormones in association with general joint laxity and hypermobility in the temporomandibular joint in adolescents—results of the epidemiologic LIFE child study. *J. Oral Rehabil.* **2019**, *46*, 1023–1030. [CrossRef] [PubMed]
67. Garnett, S.P.; Högler, W.; Blades, B.; Baur, L.A.; Peat, J.; Lee, J.; Cowell, C.T. Relation between hormones and body composition, including bone, in prepubertal children. *Am. J. Clin. Nutr.* **2004**, *80*, 966–972. [CrossRef]
68. Valdivia, O.D.; Ortega, F.Z.; Rodríguez, J.J.A.; Sánchez, M. Changes in flexibility according to gender and educational stage. *Apunt. Med. l'esport* **2009**, *161*, 10–17. [CrossRef]
69. Golle, K.; Muehlbauer, T.; Wick, D.; Granacher, U. Physical Fitness Percentiles of German Children Aged 9–12 Years: Findings from a Longitudinal Study. *PLoS ONE* **2015**, *10*, e0142393. [CrossRef] [PubMed]
70. Artero, E.G.; España-Romero, V.; Ortega, F.B.; Jimenez-Pavon, D.; Ruiz, J.R.; Vicente-Rodriguez, G.; Bueno, M.; Marcos, A.; Gómez-Martinez, S.; Urzanqui, A.; et al. Health-related fitness in adolescents: Underweight, and not only overweight, as an influencing factor. The AVENA study. *Scand. J. Med. Sci. Sports* **2010**, *20*, 418–427. [CrossRef]
71. Xu, Y.; Mei, M.; Wang, H.; Yan, Q.; He, G. Association between Weight Status and Physical Fitness in Chinese Mainland Children and Adolescents: A Cross-Sectional Study. *Int. J. Environ. Res. Public Health* **2020**, *17*, 2468. [CrossRef]
72. Casonatto, J.; Fernandes, R.A.; Batista, M.B.; Cyrino, E.S.; Coelho-E-Silva, M.J.; De Arruda, M.; Ronque, E.R.V. Association between health-related physical fitness and body mass index status in children. *J. Child. Health Care* **2016**, *20*, 294–303. [CrossRef] [PubMed]
73. Silva, D.A.S.; Nunes, H.E.G. Prevalence and Factors Associated with Low Aerobic Performance Levels in Adolescents: A Systematic Review. *Curr. Pediatr. Rev.* **2015**, *11*, 56–70. [CrossRef]
74. Zanini, D.; Kuipers, A.; Somensi, I.V.; Pasqualotto, J.F.; Quevedo, J.D.G.; Teo, J.C.; Antes, D.L. Relationship between body composition and physical capacities in junior soccer players. *Rev. Bras. Cineantropometria Desempenho Humano* **2020**, *22*, 22. [CrossRef]
75. Ceschia, A.; Giacomini, S.; Santarossa, S.; Rugo, M.; Salvadego, D.; Da Ponte, A.; Driussi, C.; Mihaleje, M.; Poser, S.; Lazzer, S. Deleterious effects of obesity on physical fitness in pre-pubertal children. *Eur. J. Sport Sci.* **2016**, *16*, 271–278. [CrossRef] [PubMed]
76. Thivel, D.; Ring-Dimitriou, S.; Weghuber, D.; Frelut, M.-L.; O'Malley, G. Muscle Strength and Fitness in Pediatric Obesity: A Systematic Review from the European Childhood Obesity Group. *Obes. Facts* **2016**, *9*, 52–63. [CrossRef] [PubMed]
77. Winck, A.D.; Heinzmann-Filho, J.P.; Soares, R.B.; da Silva, J.S.; Woszezenki, C.T.; Zanatta, L.B. Effects of obesity on lung volume and capacity in children and adolescents: A systematic review. *Rev. Paul Pediatr.* **2016**, *34*, 510–517. [CrossRef] [PubMed]
78. Jones, R.L.; Nzekwu, M.-M.U. The Effects of Body Mass Index on Lung Volumes. *Chest* **2006**, *130*, 827–833. [CrossRef]
79. Davidson, W.J.; Mackenzie-Rife, K.A.; Witmans, M.B.; Montgomery, M.D.; Ball, G.D.; Egbogah, S.; Eves, N.D. Obesity negatively impacts lung function in children and adolescents. *Pediatr. Pulmonol.* **2014**, *49*, 1003–1010. [CrossRef]
80. Dubern, B.; Tounian, P.; Medjadhi, N.; Maingot, L.; Girardet, J.-P.; Boulé, M. Pulmonary function and sleep-related breathing disorders in severely obese children. *Clin. Nutr.* **2006**, *25*, 803–809. [CrossRef]
81. Brunet, M.; Chaput, J.-P.; Tremblay, A. The association between low physical fitness and high body mass index or waist circumference is increasing with age in children: The 'Québec en Forme' Project. *Int. J. Obes.* **2006**, *31*, 637–643. [CrossRef]
82. Hopkins, P.M. Skeletal muscle physiology. *Contin. Educ. Anaesth. Crit. Care Pain* **2006**, *6*, 1–6. [CrossRef]
83. Ogborn, D.; Schoenfeld, B.J. The role of fiber types in muscle hypertrophy: Implications for loading strategies. *Strength Cond. J.* **2014**, *36*, 20–25. [CrossRef]

children

MDPI

Article

Oral Diseases and Quality of Life between Obese and Normal Weight Adolescents: A Two-Year Observational Study

Tengku Nurfarhana Nadirah Tengku H [1,*], Wei Ying Peh [2], Lily Azura Shoaib [1], Nor Adinar Baharuddin [3], Rathna Devi Vaithilingam [3] and Roslan Saub [4]

1 Department of Paediatric Dentistry and Orthodontics, Faculty of Dentistry, University Malaya, Kuala Lumpur 50350, Malaysia; lilyazura@um.edu.my
2 Klinik Pergigian Cheras, Jalan Yaacob Latiff, Cheras, Kuala Lumpur 56000, Malaysia; wying18@hotmail.com
3 Department of Restorative Dentistry, Faculty of Dentistry, University Malaya, Kuala Lumpur 50350, Malaysia; noradinar@um.edu.my (N.A.B.); rathna@um.edu.my (R.D.V.)
4 Department of Community Oral Health & Clinical Prevention, Faculty of Dentistry, University Malaya, Kuala Lumpur 50350, Malaysia; roslans@um.edu.my
* Correspondence: tengkunurfarhana@um.edu.my; Tel.: +(603)-79674802

Abstract: This study aimed to investigate the association between oral disease burden and oral health related quality of life (OHRQoL) among overweight/obese (OW/OB) and normal weight (NW) Malaysian adolescents. A total of 397 adolescents were involved in the two-year prospective observational cohort study. OHRQOL was measured through a self-administered questionnaire containing the short version of the Malaysian Oral Health Impact Profile (OHIP[M]). Body mass index (BMI) was used for anthropometric measurement. Whilst, decayed, missing, and filled teeth (DMFT) index, Significant Caries Index (SiC), simplified basic periodontal examination (S-BPE), and gingival bleeding index (GBI) were used for clinical assessment tools. Higher dental caries prevalence was observed in the NW group while higher SiC was reported in the OW/OB group. Regardless of the obesity status, the prevalence of gingivitis (BPE code 1 and 2) was high in this study. A reduction of GBI prevalence was observed in the two-year follow-up results with an increased prevalence of OHRQoL impact in the OW/OB group compared to the NW group ($p > 0.05$). The findings from this study suggested that obesity status did not have influence over the burden of oral diseases and OHRQoL. It offers insights referring to the changes in adolescents' oral diseases burden and OHRQoL.

Keywords: obesity; oral disease; oral health related quality of life (OHRQoL)

Citation: Tengku H, T.N.N.; Peh, W.Y.; Shoaib, L.A.; Baharuddin, N.A.; Vaithilingam, R.D.; Saub, R. Oral Diseases and Quality of Life between Obese and Normal Weight Adolescents: A Two-Year Observational Study. *Children* **2021**, *8*, 435. https://doi.org/10.3390/children8060435

Academic Editors: Odysseas Androutsos and Antonis Zampelas

Received: 30 April 2021
Accepted: 17 May 2021
Published: 22 May 2021

1. Introduction

Overweight and obesity are defined as abnormal or excessive fat accumulation in the body, which can impact one's health [1]. The current global trend showed that obesity has become a serious public concern regardless of a country's development status [2]. Globally, over 340 million children and adolescents aged five to nineteen were either overweight or obese in 2016 [1]. The Malaysian National and Health Morbidity Survey (NHMS) reported that 29.8 percent of adolescents (aged 13–17 years old) were overweight/obese (OW/OB) in Malaysia [3]. This is a situation that we cannot afford to neglect as 40 percent of the current Malaysian demographic profile range from 0 to 24 years of age.

Research showed obesity is likely to progress from adolescence to adulthood [4]. This is rather a worrying phenomenon as it will potentially impair general and oral health in both adolescence and adulthood. To date, research on oral health has been primarily conducted on OW/OB adults, with less attention given to adolescents. Different studies are required in adolescence, as they are incomparable with existing studies that have been carried out in adults. Unlike with adults, the puberty hormonal interaction alters the body composition and regional distribution of body fat in adolescents.

The most common oral health problems are dental caries, while periodontal disease is known as the eleventh most prevalent condition globally [1]. In general, obesity, dental caries, and periodontal disease share common risk factors, namely: sugary diet, dental plaque, socioeconomic status, and behavioral and genetic issues [5–7]. Although, there is scarce evidence of a long-term relationship among OW/OB, poor oral health status, and oral health-related quality of life (OHRQoL), some studies suggest positive associations [5,8]. Compromised health-related quality of life (HRQoL) associated with increased age and gender in OW/OB adolescents was also reported [9].

It is essential to further scrutinize the relationship between obese adolescence and changes in oral health status through longitudinal studies as the establishment between obesity and oral health will lead to a huge global economic impact due to the rising costs attributed to obesity and oral diseases [10]. This, in a way, will prompt health policymakers in prioritizing action and evaluating preventive measures for the OW/OB adolescence oral health. Thus, this longitudinal study aimed to investigate the association of OW/OB in adolescence with oral disease burden (dental caries and periodontal disease) and OHRQoL over a two-year period.

2. Materials and Methods

2.1. Study Design and Study Population

This is a prospective observational cohort study on overweight/obese and normal weight schoolchildren residing in the districts of Klang Valley, Selangor. A stratified multiple-stage random sampling technique was employed to recruit secondary school adolescents aged fourteen for the baseline study in the year 2015. The follow-up study was conducted in the year 2017 within the same group of participants at the age of sixteen. Malaysian citizens who were fit and healthy, able to understand and communicate well in Malay language, and attending multi-racial co-educational daily public school were recruited. Participants undergoing orthodontics treatment, or had underweight BMI status, or had a change of BMI status during the two-year follow-up were excluded.

A total of 202 OW/OB participants were cross matched by age and gender with 195 NW participants who fulfilled the inclusion criteria. Anthropometric measurement and clinical examination including caries experience and periodontal status were performed following the completion of self-administered questionnaires consisting of sociodemographic data, oral health related behaviors, and the short version of the Malaysian Oral Health Impact Profile (OHIP[M]).

2.2. Ethics, Consents, and Permission

Ethical approval (DF CD1706/0031(U)) and permission to conduct the study at public school was obtained from the Faculty of Dentistry Medical Ethics Committee, University of Malaya, Ministry of Education Malaysia, Selangor State Education Department, and authorities of the selected secondary schools. Written informed consent from the parents and caregivers of the participants were obtained at both baseline and follow-up studies. Participations for both studies were on a voluntary basis.

2.3. Anthropometric Measurement

The 2007 WHO reference was used to determine each participant's body mass index (BMI) status [1].

The participants were instructed to only wear standard school uniforms and socks upon standing on the weight scale platform during body height and body weight measurement. The body weight was measured to the nearest 0.5 kg and height to the nearest 0.1 cm using a calibrated digital scale with body mass index (BMI) function and integrated stadiometer (Tanita, IL, USA), where routine calibrations were conducted based on manufacturer recommendations.

2.4. Clinical Examinations

2.4.1. Caries Experiences

The caries experienced was recorded following the WHO criteria [11], where the DMFT score was calculated by adding the number of decayed (D), missing (M), and filled (F) permanent teeth. The Significant Caries Index (SiC) was used to identify individuals with the highest caries values in the population. SiC was determined by selecting one-third of the samples with highest DMFT scores and the mean DMFT for this subgroup was calculated [12].

2.4.2. Periodontal Assessments

The Simplified Basic Periodontal Examination (S-BPE) by the British Society of Periodontology (BSP) and British Society of Pediatric Dentistry (BSPD) was used for the screening of periodontal health. The examination was performed on the following six index teeth, 16, 11, 26, 36, 31, and 46, using the WHO probe with the ball end tip. The BPE codes (Code 0 to Code 2) were used to determine each participant's periodontal status. The Ainamo and Bay Gingival Bleeding Index (GBI) [13] was used as an adjunct to the BPE to determine each participant's gingival health by gently probing on the orifice of the gingival crevice to the bottom of the pocket. Positive findings were recorded if bleeding occurred within 10 s.

2.4.3. Training and Calibration

All participating dentists involved in this study were trained and calibrated to minimize inter-examiner variability in the data collection. The calibration sessions were carried out prior to the data collection at baseline and the follow-up study for caries diagnoses, periodontal and gingival examination to a single gold standard examiner of each field (pediatric dentist and periodontist). All examiners needed to achieve the minimum required kappa value of 0.75 prior to the conduct of the study. The kappa values of the examiners compared to the gold standard ranged from 0.75 to 0.80. As the standard operating procedure for this study, the intra-examiner's reliability was conducted at each data collection session.

2.5. Questionnaires

A self-administered questionnaire was used to collect participants' information on demographics, oral health related behaviors, and oral health related quality of life (OHRQoL) using the short form of the Malaysian version of Oral Health Impact Profile OHIP[M]-14 that contained 14 items in seven domains. The two parameters obtained from OHIP[M]-14 were prevalence and severity. Prevalence of impact is the percentage of respondents reporting one or more impacts of "fairly often" or "often", whereas severity impact was the summation of response codes for the fourteen items. The content of the questionnaire was validated during the baseline study by two dental specialists from pediatric dentistry and dental public health. It was pre-tested for clarity, simplicity, sequencing, and ease of understanding instructions and questions. The internal consistency measured by Cronbach's alpha was recorded at 0.75, which indicated acceptable reliability.

2.6. Statistical Analysis

All data collected were treated confidentially and verified by the examiner for its completeness prior to data entry and data analysis. The data were coded and recorded using IBM SPSS Version 26.0 (SPSS Inc., Chicago, IL, USA). The p-value was set at $\alpha = 0.05$ with 95% confidence interval.

Differences between baselines and follow-up distribution of gender, race, mother level of education, and body mass index were compared between participants who successfully completed each phase of data collection through the Cohen effect size (r). The characteristics of participants were described using frequency distribution for categorical variables while continuous data were described using mean (M) and standard deviation (±SD).

The chi square statistic (X^2) was used for comparison of oral hygiene practice and self-rated perception of health and oral health between obese and normal weight groups and the association was reported as statistically significance if the p value < 0.05.Comparison for mean DMFT, mean SiC, and severity of OHRQoL impact between obese and normal weight groups were calculated using the independent t-test. The chi square test was used to compare the prevalence of dental caries, periodontal disease, gingival bleeding, and impact of OHRQoL between the two groups.

3. Results

Out of 397 participants (NW and OW/OB) recruited at the baseline study (age fourteen), only 238 (60%) participants who fulfilled the inclusion criteria managed to be traced and agreed to participate in the follow-up study (age sixteen). The sociodemographic and anthropometric characteristics of the follow-up participants were comparable to the dropout participants (Cohen's effect size ≤ 0.5) (Table 1). The reasons for participants' dropout included change of schools, not willing to participate, being absent during data collection day, and not fulfilling the criterion. The final analysis of this study involved the total number of those who participated at both baseline and follow-up study.

Table 1. Background information of participants at baseline.

Characteristics	Baseline *n* (%)		Follow-Up *n* (%)		*p* Value	Cohen's Effect Size
	OW/OB *n* = 195	NW *n* = 202	OW/OB *n* = 122	NW *n* = 116		
Gender						
Male	124 (31.2)	118 (29.7)	73 (30.7)	73 (30.7)		*0.03* [1]
Female	71 (17.9)	84 (21.2)	49 (20.6)	43 (18.1)	*0.81*	
Ethnicity						
Malays	117 (29.5)	105 (26.4)	72 (30.3)	68 (28.6)	*0.05* *	*0.05* [1]
Non-Malays	78 (19.7)	97 (24.5)	50 (21.0)	48 (20.2)		
Mother level of education						
High	57 (14.4)	67 (16.9)	32 (13.4)	35 (14.7)	*0.79*	*0.02* [1]
Low	138 (34.8)	135 (34.1)	90 (37.8)	81 (34.0)		
BMI (kg/m^2) (Mean) (SD)	28.5 (4.01)	19.2 (1.84)	29.6 (4.30)	19.6 (1.67)		*0.02* [1]

OW = overweight, OB = obese, NW = normal weight. SD = standard deviation, p values were calculated using chi square tests, $p < 0.05$ significance values. [1] Effects size based on Cohen's r: 0.2 for small; 0.5 for medium; 0.8 for large effect.

3.1. Sociodemographic

The sociodemographic characteristics of participants in the baseline and follow-up study were almost similar across BMI groups. Participants for the follow-up study consisted of 146 males and 92 females with 60% of the participants from Malay ethnicity. It was found that the majority of the participants' mothers came from lower educational backgrounds. The mean baseline BMI of the overweight/obese group was 28.5 kg/m^2 and the normal weight group was 19.2 kg/m^2. The mean BMI of OW/OB (29.6 kg/m^2 ± 4.30) and NW (19.6 kg/m^2 ± 1.67) increased at the follow-up study compared to the baseline study.

3.2. Oral Health Practices

Good tooth brushing habits were seen in more than 40% of the participants across all groups. The majority of the participants in the SiC group brushed their teeth twice daily by using fluoridated toothpaste and using mouth rinses (Table 2). However, a two-fold increment in the number of participants not using fluoridated toothpaste was seen in OW/OB and NW groups during the follow-up study.

Table 2. Oral hygiene practice and perception on health and oral health in obese, normal weight, and SiC groups at baseline and follow-up.

	Baseline *n* (%)			**Follow-Up *n* (%)**		
	OW/OB *n* = 122	**NW** *n* = 116	**SIC** *n* = 79	**OW/OB** *n* = 122	**NW** *n* = 116	**SIC** *n* = 79
Frequency of tooth brushing		$X^2 = 0.26, p = 0.87$			$X^2 = 0.49, p = 0.77$	
2x/day	89 (37.0)	95 (39.9)	64 (81.0)	91 (38.4)	91 (38.4)	65 (82.3)
≤1x/day	33 (14.2)	21 (8.8)	15 (19.0)	31 (12.6)	25 (10.5)	13 (16.5)
Used fluoride toothpaste		$X^2 = 2.28, p = 0.13$			$X^2 = 0.26, p = 0.87$	
Yes	120 (50.4)	110 (46.2)	76 (96.2)	78 (35.5)	69 (29.5)	55 (69.6)
No	2 (0.8)	6 (2.5)	3 (3.8)	14 (5.9)	15 (6.3)	24 (30.4)
Used mouth rinse		$X^2 = 0.00, p = 0.94$			$X^2 = 0.49, p = 0.77$	
Yes	120 (50.4)	110 (46.2)	21 (26.6)	25 (10.5)	22 (9.3)	15 (19.0)
No	2 (0.8)	6 (2.5)	58 (73.4)	97 (40.9)	93 (39.2)	64 (81.0)
Perception of health		$X^2 = 5.46, p = 0.17$			$X^2 = 2.98, p = 0.08$	
Poor	43 (18.1)	25 (10.5)	24 (30.4)	61 (26.1)	46 (19.7)	32 (40.5)
Good	79 (33.2)	91 (38.2)	55 (69.6)	58 (24.8)	69 (29.5)	46 (58.2)
Perception of oral health		$X^2 = 0.00, p = 0.97$			$X^2 = 3.13, p = 0.77$	
Poor	45 (18.9)	43 (18.1)	33 (41.8)	65 (27.9)	50 (21.5)	36 (45.6)
Good	77 (32.4)	73 (30.7)	46 (58.2)	53 (22.7)	65 (27.9)	41 (51.9)
Visited dentist for the past 1 year		$X^2 = 1.14, p = 0.56$			$X^2 = 0.56, p = 0.45$	
Yes	16 (6.7)	26 (10.9)	12 (15.2)	17 (7.2)	17 (7.2)	8 (10.1)
No	106 (44.5)	90 (37.8)	67 (84.8)	104 (43.9)	99 (41.8)	71 (89.9)

OW = overweight, OB = obese, NW = normal weight, SIC = significant index caries. X^2 = chi square statistic, *p* values were calculated using chi square tests with $p < 0.05$ is the significance value.

Irrespective of BMI status, an increase in the number of participants who rated poor on perception of health and oral health was observed in the follow-up study. The same pattern was also observed among participants in the SiC group. In contrast, the percentages of participants visiting dentists within one-year remained low in both baseline and follow-up studies.

3.3. Caries and Periodontal Health

The mean (M ± SD) caries experiences among OW/OB and NW groups in the baseline were 1.80 ± 3.16, 1.84 ± 2.62, with a slight increment to 2.37 ± 3.30 and 2.47 ± 3.05, respectively at the two-year follow-up study (Table 3). Likewise, the number of teeth affected by carious lesions showed an increasing trend. High mean of caries experiences increased from 5.22 to 6.49 per participants in the OW/OB of SIC group at the baseline and follow-up study. In regards to the SiC participants, nearly 90% did not visit the dentist for the past one year at the time when the study was conducted.

The distribution of BPE scores are shown in Table 3 for OW/OB and NW of the total population and SIC group. The highest score observed in the study population was "Code 2", which can be described as presentation of calculus or plaque retention factor with no periodontal pocket ≥3.5 mm detected. At the baseline study, 77% of participants were presented with "Code 2" (Table 3). During the follow-up study, it was observed that the BPE scores reduced more than 20% across all participants, including in the SIC group. Gingival bleeding can be observed in more than 90% of participants at baseline and follow-up study.

Table 3. Oral health status in obese and normal weight groups of total participants and SiC group at baseline and follow-up.

Total Participants, *n* = 238	**Baseline *** M ± SD		**Follow-Up **** M ± SD		***p* Value**
	OW/OB	**NW**	**OW/OB**	**NW**	
Caries experience	1.80 (3.16)	1.84 (2.62)	2.37 (3.30)	2.47 (3.05)	0.91 *, 0.79 **
Based on teeth					
Decayed (D)	1.66 (3.10)	1.53 (2.35)	2.11 (3.16)	2.07 (2.70)	0.71 *, 0.92 **
Missing (M)	0	0.03 (0.18)	0.01 (0.09)	0.06 (0.24)	0.03 *, 0.02 **
Filled (F)	0.13 (0.48)	0.27 (0.92)	0.25 (0.72)	0.34 (1.10)	0.57 *, 0.45 **
Sound dentition *n* (%)	73 (59.8)	60 (51.7)	56 (45.9)	42 (36.2)	
Periodontal status *n* (%)					
BPE screening					
Code 0	8 (6.6)	6 (5.2)	7 (6.0)	10 (8.2)	
Code 1	20 (16.4)	15 (12.9)	49 (42.2)	39 (32.0)	
Code 2	94 (77.0)	95 (81.9)	60 (51.7)	73 (59.8)	
Gingival bleeding *n* (%)					>0.05 +
Yes	118 (96.7)	112 (96.6)	111 (91.0)	106 (91.4)	
No	4 (3.3)	4 (3.4)	11 (9.0)	10 (8.6)	
Significant Caries Index (SiC) *n* =79	5.22 (3.90)	*4.48 (2.70)*	6.49 (3.08)	5.80 (2.73)	0.32 *, 0.28 **
Based on teeth					
Decayed (D)					
Missing (M)	4.89 (4.0)	*3.70 (2.70)*	5.80 (3.46)	4.83 (2.67)	0.11 *, 0.17 **
Filled (F)	0	*0.07 (0.26)*	0.03 (0.16)	0.12 (0.33)	0.10 *, 0.12 **
	0.32 (0.75)	*0.71 (1.42)*	0.68 (1.13)	0.83 (1.68)	0.12 *, 0.63 **
Periodontal status *n* (%)					
BPE screening					
Code 0	0				
Code 1	5 (13.5)	*1 (2.4)*	3 (8.1)	2 (4.8)	
Code 2	32 (86.5)	*5 (11.9)*	19 (51.4)	18 (42.9)	
		36 (85.7)	15 (40.5)	22 (52.4)	
Gingival bleeding *n* (%)					
Yes	37 (100)				
No	0	*41 (97.6)*	34 (91.9)	38 (90.5)	>0.05 +
		1 (2.4)	3 (8.1)	4 (9.5)	

OW = overweight, OB = obese, NW = normal weight. M = mean, ±SD = standard deviation. * = at baseline, ** = at follow up. *p* values were calculated using independent *t*-test, $p < 0.05$ significance values. + *p* values were calculated using chi square tests, $p < 0.05$ significance values.

3.4. Oral Health Related Quality of Life

The oral health related quality of life among OB/OW, NW, and SIC groups at baseline and follow-up are reported in Table 4. The Oral Health Impact Profile 14 (OHIP 14) scores ranged from 0 to 56, with a high score indicating poor oral health related quality of life. The OW/OB participants in the SIC group presented with the highest mean score of 9.83 ± 6.68 at baseline and 8.18 ± 7.47 during the follow-up study ($p > 0.05$). In general, the prevalence of impact was low in all domains at baseline and the follow-up study. There were four most affected domains, with at least 10% of OW/OB groups involved. The "self-consciousness" and "felt tense" items in the psychological discomfort domain were the two most affected items among the SIC group at baseline. The prevalence increase between 4% and 10% was observed at the follow-up study in the same group. The social disability domain (been irritable with others) was also reported as the affected domain at 11.4%, followed by functional limitation (worsening in the sense of taste).

Table 4. Oral health related quality of life in obese, normal weight, and SIC groups at baseline and follow-up.

	Baseline * M ± SD			Follow-Up ** M ± SD			*p* Value
	OW/O	NW	SiC +	OW/OB	NW	SiC +	
OHIP severity	9.14 (6.37)	8.63 (5.97)	9.83 (6.68)	6.80 (7.49)	7.52 (6.65)	8.18 (7.47)	0.53 *, 0.75 *+, 0.45 **, 0.57 **+
OHIP Impact based on items (*n*, %)							
Functional limitation (trouble pronouncing words)	0	1 (1.3)	1 (1.3)	1 (0.4)	2 (0.8)	3 (3.8)	
Functional limitation (worsened sense of taste)	1 (1.3)	0	1 (1.3)	4 (1.7)	1 (0.4)	8 (10.1)	
Physical pain (painful aching in the mouth)	1 (1.3)	1 (1.3)	2 (2.5)	2 (0.8)	2 (0.8)	6 (7.6)	
Physical pain (uncomfortable eating)	2 (2.5)	0	2 (2.5)	6 (2.5)	1 (0.4)	6 (7.6)	
Psychological discomfort (self-consciousness)	5 (6.3)	8 (10.1)	13 (16.5)	12 (5.0)	16 (6.7)	15 (19.0)	
Psychological discomfort (felt tense)	4 (5.1)	2 (2.5)	6 (7.6)	7 (2.9)	9 (3.8)	8 (10.1)	>0.05 ***
Physical disability (diet has been unsatisfactory)	2 (2.5)	2 (2.5)	4 (5.1)	5 (2.1)	5 (2.1)	7 (8.9)	
Physical disability (meals interrupted)	1 (1.3)	1 (1.3)	2 (2.5)	5 (2.1)	8 (3.4)	6 (7.6)	
Psychological disability (difficult to relax)	0	1 (1.3)	4 (5.1)	3 (1.3)	1 (0.4)	3 (3.8)	
Psychological disability (been embarrassed)	3 (3.8)	1 (1.3)	1 (1.3)	0	2 (0.8)	4 (5.1)	
Social disability (been irritable with others)	8 (10.1)	1 (1.3)	9 (11.4)	17 (7.1)	6 (2.5)	2 (2.5)	
Social disability (difficulty doing schoolwork)	0	1 (1.3)	1 (1.3)	0	2 (0.8)	2 (2.5)	
Handicap (felt life is less satisfying)	0	0	0	0	1 (0.4)	2 (2.5)	
Handicap (unable to perform usual function)	1 (1.3)	2 (2.5)	3 (3.8)	3 (1.3)	4 (1.7)	5 (6.3)	

OW = overweight, OB = obese, NW = normal weight, SIC = significant index caries. M = mean, ±SD = standard deviation. *, **, + *p* values were calculated using independent *t*-test, $p < 0.05$ significance values. *** *p* values were calculated using independent *t*-test, $p < 0.05$ significance values. * = at baseline, ** = at follow up.

4. Discussion

The aim of this study was to present the results of a longitudinal study on the severity of dental caries and periodontal health among obese and normal weight Malaysian adolescents and the impact on oral health related quality of life. Despite the good response rate in the present study, the refusal to participate in the follow-up study may represent a source of selection bias. However, the effect size of participants that turned up and lost to follow-up was low (Cohen effect size < 0.5), not affecting the overall result of this study.

Severe obese individuals were reported with risk of developing dental caries, periodontal disease, and health morbidity [7,8,14]. However, it was not reflected in the findings of this study. It can be explained by the recruitment of obese participants in this study—it was not a true or severe obese population by definition (BMI > 25.9 kg/m^2 for boys and BMI > 27.3 kg/m^2 for girls at the age of 14 years old). Even though, most participants were overweight, it was important to start planning for an early prevention for one-third of the participants who presented with a high burden of oral diseases. It is evident that BMI provides an excellent indicator for obesity, but studies proved that waist circumference is the best index for predictor of obesity related health risks [14].

This study adopted the WHO growth reference, which was also employed in the National Physical Fitness Standard for Malaysia School Students (SEGAK) program. SEGAK was first implemented by the Malaysian Ministry of Education in 2008 for Malaysian students aged 10 to 17 years old [15]. It was designed to evaluate each student's level of fitness and, at the same time, monitor their BMI. By adopting the same reference, it allowed comparison of results to be made if similar future studies were to be conducted in other states in Malaysia.

From the sociodemographic background related to maternal education, the majority were from lower education levels. The majority were reported to have at least attended secondary school. Therefore, it can be suggested that most of the participants came from low socioeconomic status backgrounds. Growing interest in studies related to low socioeconomic status and development of childhood obesity can be seen in few areas, particularly in mental health, impact on health-related quality of life, and financial hardship. These areas affect healthy lifestyle choices that are less accessible and indirectly promote calorie dense foods, lifestyles that reflect a lack of physical activity, poor oral hygiene practices, and less opportunities for self-growth and development, especially in education [16,17]. The number of participants who rated good on 'perception of health and oral health' also showed a reduction in the follow-up study compared to the baseline study across the BMI status. In contrast, the percentages of participants visiting dentists within one year remained low in both baseline and follow-up studies. This can be postulated from the statistically non-significant ($p > 0.05$) association between BMI status with the burden of oral diseases and impact on oral health related quality of life.

Good tooth brushing habits were also being observed with the majority of participants brushing their teeth twice daily and using fluoridated toothpaste. However, a two-fold increment in the number of participants not using fluoridated toothpaste was seen in both groups at the follow-up study. Providentially, the increment in the mean of caries was not significant. This might be due to the misconceptions by some participants and their parents, regarding the fluoride concentration levels in the toothpaste. Despite using non-fluoridated toothpaste, the participants still benefited (i.e., via a caries-preventive effect) through public water fluoridation, provided by the government for Malaysian households, which might be one of the contributing factors for insignificant increments in the mean caries.

Lacking oral health awareness is another possible risk factor for caries and obesity. These oral health habits include the frequency of tooth brushing and dental visits, which affect the prevalence of dental caries and gingivitis among children and adolescents. It is suggested in the literature that, apart from variation of weight among participants, positive oral health behavior plays an influential role in participants' caries experiences [18]. It can be observed in this study—regardless of body weight—that the majority of participants reported "good" in the perception of health and oral health, and practiced tooth brushing twice daily, and the changes in the burden of oral diseases were found not statistically significant ($p > 0.05$).

Study results highlighted the severity of dental caries and gingival bleeding in the Malaysian adolescent population. An increased trend in the prevalence of dental caries was observed with more than half of the participants, regardless of body mass index presented with at least two teeth affected, at the age of 16 years old. The severity of caries was high in the OW/OB group of the SiC subgroup, with at least five teeth affected, with no significant association between dental caries and body mass index. In line with the Malaysia National Oral Health Survey of School Children (NOHSS) 2007 report, about 60% of the Malaysian adolescents, aged 16 year old, were diagnosed with dental caries [19]. This showed that dental caries was a common oral health problem in the population, regardless of body mass index and age. Results from this study also suggested that the association between obesity and dental caries was more complex and a simple explanation based on unhealthy dietary habits and sugar consumption were inadequate. Less than 10% of the OW/OB group from the total study population and in the SiC subgroup presented with healthy gingiva. It is in accordance with the prevalence of healthy periodontium in Malaysian schoolchildren, reported in theNational Health and Morbidity survey (NHMS) 2017 findings [20].

The most prevalent periodontal disease in adolescents was found to be plaque induced gingivitis. Although gingivitis is an irreversible disease in nature, obese adolescents, even without concomitant metabolic syndrome, are more likely to be diagnosed with gingivitis and to develop periodontal disease earlier compared to healthy normal weight adolescents [13]. Therefore, based on the findings of this study, the OW/OB participants, especially in the SiC subgroup, was at a higher risk of being diagnosed with periodontitis in the future. In view of this, further studies should be carried out to further understand the impact of obesity on periodontal disease in healthy adolescents.

The DMFT index was employed to determine the prevalence of dental caries. Despite its limitation of overestimation of dental caries, it has been reliably used in research studies.

Recent systematic review and meta analysis also supported the used of DMFT index to standardize the definition of dental caries in the population for further studies[21]. Moreover, the assessment of caries severity at various stages might be suitable to be used in addition to the DMFT index in the study for OW/OB children and adolescents. An earlier study recommended the use of ICDAS criteria as a strong predictor of lesion progression, due to the slow progression of dentinal caries compared to initiation of enamel caries noted from radiographic examination [22]. Observation of differences and development of caries through intraoral examination in obese adolescents can be done through a longer duration of study.

A review also reported no standardization in the diagnostic criteria within the published studies of caries and obesity [23]. It was suggested that, without a gold standard in research methods, it might affect the results and potential findings of the relationship between caries and obesities.

Interestingly, this study reported low impact in the oral health related quality of life, which was consistent with good self-rated perception of oral health and health among participants regardless of their BMI and severity of oral diseases. In contrast, a study by Banu et al. reported high impact on the OHRQoL [24]. It might be related to the subjective measure by young adolescents regarding their body weight and oral diseases, which indirectly may have influence in their quality of life.

The psychological discomfort was the most affected domain among OW/OB participants with a high burden of dental caries. This is an important finding, which supported other evidence-based studies on mental health, showing that obese adolescents have higher incidence of depression, anxiety, low self-esteem, and poor quality of life [25]. The possible explanation would be victimization by peers, parents, or siblings who shape the belief of having an ideal body weight, defined as beautiful and good for one's health and oral health. Although OW/OB presented with good perception of health and oral health with low impact on quality of life, it affected the psychological discomfort domain the most. Thus, it is important for policymakers to strengthen health promotion among adolescents, which, in the long run, may help to prevent diseases and health problems, such as obesity, dental caries, and periodontal diseases.

A selection bias was identified as one of the limitations in our study. The cluster random sampling was used to determine samples. This allowed schools located near the Faculty of Dentistry University Malaya to be selected. University Malaya is located in a well-developed area in the Klang Valley. Hence, all participants basically originated from an urban area, which had higher access to healthcare services, with a better education system. Thus, it will be important to have a more diversified environment to determine population impact measures. Although this is a longitudinal study, the duration for the follow-up review was insufficient for the causal relationship between the burden of oral diseases, body mass index, and impact on quality of life to be established. Another follow-up study may be needed to explore further relationships after the peak of growth spurt is achieved. This is because the effect of hormonal changes during puberty varies from one person to another, which may have led to the negative findings from this study.

5. Conclusions

In conclusion, this two-year follow-up study suggested that obesity status did not influence the burden of oral disease and OHRQoL. This statement agrees with systematic reviews that report inconclusive and conflicting evidence on the association between obesity and dental caries [20]. However, this study offers insights into changes in oral disease burden and OHRQoL among adolescents. Study limitations highlighted by this study could be discussed further and improved for future studies.

Author Contributions: Conceptualization, methodology, data analysis, editing, and revision, T.N.N.T.H.; first draft and revision, L.A.S.; data curation, W.Y.P.; conceptualization, revision, and approval, R.S., N.A.B., and R.D.V. All authors have read and agreed to the published version of the manuscript.

Funding: This research was funded by the University of Malaya research grant (UMRG RP031B-14HTM).

Institutional Review Board Statement: The study was conducted according to the guidelines of the Declaration of Helsinki, and approved by the Medical Ethics Committee, Faculty of Dentistry, University of Malaya, "(DF CD1706/0031(U) and 20/04/2017)".

Informed Consent Statement: Informed consent was obtained from all subjects involved in the study and written informed consent has also been obtained from the patient(s) and legal guardian(s) to publish this paper.

Acknowledgments: Special thanks to resources from Faculty of Dentistry, University of Malaya, for their assistance, Ong Jie Yen for helping in data collections at the follow-up study, and all the students who participated in this study.

Conflicts of Interest: The authors declare no conflict of interest. The funders had no role in the design of the study; in the collection, analyses, or interpretation of data; in the writing of the manuscript, or in the decision to publish the results.

References

1. World Health Organisation. World Health Organisation. Obesity-and-Overweight. Who.int. 2020. Available online: https://www.who.int/news-room/fact-sheets/detail/obesity-and-overweight (accessed on 3 March 2021).
2. Ng, M.; Fleming, T.; Robinson, M.; Thomson, B.; Graetz, N.; Margono, C.; Mullany, E.C.; Biryukov, S.; Abbafati, C.; Abera, S.F.; et al. Global, regional, and national prevalence of overweight and obesity in children and adults during 1980–2013: A systematic analysis for the Global Burden of Disease Study 2013. *Lancet* **2014**, *384*, 766–781. [CrossRef]
3. Institute of Public Health. *National Health and Morbidity Survey 2019: Technical Report Volume I*; Ministry of Health: Putrajaya, Malaysia, 2020.
4. Simmonds, M.; Burch, J.; Llewellyn, A.; Griffiths, C.; Yang, H.; Owen, C.; Duffy, S.; Woolacott, N. The use of measures of obesity in childhood for predicting obesity and the development of obesity-related diseases in adulthood: A systematic review and meta-analysis. *Health Technol. Assess.* **2015**, *19*, 1–336. [CrossRef] [PubMed]
5. Kaul, R.; Angrish, P.; Saha, S.; Halder, S.; Bhattacharya, B.; Mitra, M. Association between Obesity and Oral Health Status in Schoolchildren: A Survey in Five Districts of West Bengal, India. *Int. J. Clin. Pediatr. Dent.* **2018**, *11*, 233–237. [CrossRef] [PubMed]
6. Killedar, A.; Lung, T.; Petrou, S.; Teixeira-Pinto, A.; Tan, E.J.; Hayes, A. Weight status and health-related quality of life during childhood and adolescence: Effects of age and socioeconomic position. *Int. J. Obes.* **2020**, *44*, 637–645. [CrossRef] [PubMed]
7. Drummond, B.K.; Brosnan, M.G.; Leichter, J.W. Management of periodontal health in children: Pediatric dentistry and periodontology interface. *Periodontology 2000* **2017**, *74*, 158–167. [CrossRef] [PubMed]
8. Franchini, R.; Petri, A.; Migliario, M.; Rimondini, L. Poor oral hygiene and gingivitis are associated with obesity and overweight status in paediatric subjects. *J. Clin. Periodontol.* **2011**, *38*, 1021–1028. [CrossRef] [PubMed]
9. D'Avila, H.F.; Poll, F.A.; Reuter, C.P.; Burgos, M.S.; Mello, E.D. Health-related quality of life in adolescents with excess weight. *J. Pediatr. (Versão em Português)* **2019**, *95*, 495–501. [CrossRef]
10. Listl, S.; Galloway, J.; Mossey, P.; Marcenes, W. Global Economic Impact of Dental Diseases. *J. Dent. Res.* **2015**, *94*, 1355–1361. [CrossRef] [PubMed]
11. Petersen, P.E.; Baez, R.J. *Oral Health Surveys: Basic Methods*, 5th ed.; World Health Organization: Geneva, Switzerland, 2013. Available online: https://apps.who.int/iris/handle/10665/97035 (accessed on 3 March 2021).
12. Nishi, M.; Bratthall, D.; Stjernswärd, J. *How to Calculate the Significant Caries Index (SiC Index)*; WHO Collaborating Centre/Faculty of Odontology, University of Malmö: Malmö, Sweden, 2001.
13. Ainamo, J.; Bay, I. Problems and proposals for recording gingivitis and plaque. *Int. Dent. J.* **1975**, *25*, 229–235. [CrossRef] [PubMed]
14. Eckel, R.H.; Matsuzawa, Y.; Després, J.P. Waist circumference as a vital sign in clinical practice: A Consensus Statement from the IAS and ICCR Working Group on Visceral Obesity. *Nat. Rev. Endocrinol.* **2020**, *16*, 177–189. [CrossRef]
15. *Bahagian Pembangunan Kurikulum Kementerian Pendidikan Malaysia*; Panduan Standard Kecergasan Fizikal Kebangsaan untuk Murid Sekolah Malaysia (SEGAK); Ministry of Education: Putrajaya, Malaysia, 2016.
16. Li, L.-W.; Wong, H.M.; Gandhi, A.; McGrath, C.P. Caries-related risk factors of obesity among 18-year-old adolescents in Hong Kong: A cross-sectional study nested in a cohort study. *BMC Oral Health* **2018**, *18*, 188. [CrossRef] [PubMed]
17. Hemmingsson, E. Early Childhood Obesity Risk Factors: Socioeconomic Adversity, Family Dysfunction, Offspring Distress, and Junk Food Self-Medication. *Curr. Obes. Rep.* **2018**, *7*, 204–209. [CrossRef] [PubMed]
18. Nihtila, A.; West, N.; Lussi, A.; Bouchard, P.; Ottolenghi, L.; Senekola, E.; Llodra, J.C.; Viennot, S.; Bourgeois, D. Oral Health Behavior and Lifestyle Factors among Overweight and Non-Overweight Young Adults in Europe: A Cross-Sectional Questionnaire Study. *Health* **2016**, *4*, 21. [CrossRef] [PubMed]
19. Oral Health Division, Ministry of Health Malaysia. National Oral Health Survey of Schoolchildren 2007 (NOHSS 2007): 16-Year-olds. November 2010. Available online: http://ohd.moh.gov.my/images/pdf/nohp20112020.pdf (accessed on 3 March 2021).

20. Oral Health Division, Ministry of Health Malaysia. *National Health Morbidity Survey 2017: Oral Health Status of Schoolchildren in Malaysia: National Oral Health Survey of Schoolchildren (NOHSS) 2017 Survey Findings*; Ministry of Health: Putrajaya, Malaysia, 2019.
21. Chen, D.; Zhi, Q.; Zhou, Y.; Tao, Y.; Wu, L.; Lin, H. Association between dental caries and BMI in Children: A systematic review and meta-analysis. *Caries Res.* **2018**, *52*. [CrossRef] [PubMed]
22. Zandoná, A.F.; Santiago, E.; Eckert, G.; Katz, B.; De Oliveira, S.P.; Capin, O.; Mau, M.; Zero, D. The Natural History of Dental Caries Lesions. *J. Dent. Res.* **2012**, *91*, 841–846. [CrossRef] [PubMed]
23. Alshihri, A.A.; Rogers, H.J.; Alqahtani, M.A.; Aldossary, M.S. Association between Dental Caries and Obesity in Children and Young People: A Narrative Review. *Int. J. Dent.* **2019**, *2019*, 1–8. [CrossRef] [PubMed]
24. Banu, A.; Şerban, C.; Pricop, M.; Urechescu, H.; Vlaicu, B. Dental health between self-perception, clinical evaluation and body image dissatisfaction—A cross-sectional study in mixed dentition pre-pubertal children. *BMC Oral Health* **2018**, *18*, 1–9. [CrossRef] [PubMed]
25. Van Vuuren, C.L.; Wachter, G.G.; Veenstra, R.; Rijnhart, J.J.M.; Van Der Wal, M.F.; Chinapaw, M.J.M.; Busch, V. Associations between overweight and mental health problems among adolescents, and the mediating role of victimization. *BMC Public Health* **2019**, *19*, 1–10. [CrossRef] [PubMed]

children MDPI

Article

Positive Associations between Body Mass Index and Hematological Parameters, Including RBCs, WBCs, and Platelet Counts, in Korean Children and Adolescents

Hwal Rim Jeong [1], Hae Sang Lee [2], Young Suk Shim [2,*] and Jin Soon Hwang [2]

[1] Department of Pediatrics, School of Medicine, Soonchunhyang University, Cheonan Hospital, Cheonan 31151, Korea; hrjeong@schmc.ac.kr

[2] Department of Pediatrics, School of Medicine, Ajou University Hospital, Suwon 16499, Korea; seaon98@naver.com (H.S.L.); pedhwang@ajou.ac.kr (J.S.H.)

* Correspondence: royjays@aumc.ac.kr; Tel.: +82-31-219-4446; Fax: +82-31-219-5169

Abstract: We conducted this study to investigate the associations between hematological parameters and obesity in children and adolescents. The levels of hematological parameters (including white blood cells [WBCs], red blood cells [RBCs], hemoglobin [Hb], hematocrit [Hct], and platelets) of 7997 participants (4259 boys and 3738 girls) aged 10–18 years were recorded. The parameters were compared among participants with normal weight, overweight, and obesity. Significantly higher mean levels of WBCs (7.16 vs. 6.16 × 103/mm^3, $p < 0.001$), RBCs (4.90 vs. 4.82 × 106/mm^3, $p < 0.001$), Hb (14.07 vs. 13.99 g/dL, $p < 0.05$), Hct (42.31 vs. 41.91%, $p < 0.001$), and platelets (311.87 vs. 282.66 × 103/mm^3, $p < 0.001$) were found in the obese than normal weight group, respectively, after adjusting for body mass index (BMI) and sex. BMI SDS had significant positive associations with the levels of WBCs ($\beta = 0.275$, $p < 0.001$), RBCs ($\beta = 0.028$, $p < 0.001$), Hb ($\beta = 0.034$, $p < 0.001$), Hct ($\beta = 0.152$, $p < 0.001$), and platelets ($\beta = 8.372$, $p < 0.001$) after adjusting for age, sex, and socioeconomic factors in a multiple linear regression analysis. A higher BMI was associated with elevated WBC, RBC, Hb, Hct, and platelet counts in children and adolescents. Because higher levels of hematological parameters are potential risk factors for obesity-related diseases, hematological parameters should be evaluated in obese children and adolescents.

Keywords: obesity; children; WBC; RBC; platelets

Citation: Jeong, H.R.; Lee, H.S.; Shim, Y.S.; Hwang, J.S. Positive Associations between Body Mass Index and Hematological Parameters, Including RBCs, WBCs, and Platelet Counts, in Korean Children and Adolescents. *Children* **2022**, *9*, 109. https://doi.org/10.3390/children9010109

Academic Editors: Odysseas Androutsos and Antonis Zampelas

Received: 9 November 2021
Accepted: 10 January 2022
Published: 14 January 2022

1. Introduction

Obesity is characterized by excessive body fat mass or weight. The prevalence of obesity has increased significantly worldwide, including among children [1]. Obesity is associated with chronic inflammation, which contributes to atherosclerosis and metabolic syndrome (MS) [2–4]. MS refers to a constellation of abnormalities, including central obesity, glucose intolerance, hypertriglyceridemia, decreased high-density lipoprotein (HDL) cholesterol levels, and high blood pressure [5]. MS is a precursor of cardiovascular disease and type 2 diabetes mellitus (T2DM) [6]. Childhood obesity leads to adult obesity and MS [7,8] and is associated with increased risks of cardiovascular disease and mortality [9]. Therefore, prevention and early treatment of childhood obesity and related comorbidities are important challenges for public health care systems.

Obesity-induced inflammation leads to insulin resistance and T2DM [3,10], which contribute to MS [11]. The levels of inflammatory markers, including C-reactive protein, ferritin, and cytokines, are elevated in obesity [11,12]. The total white blood cell (WBC) count, an inflammatory marker, is elevated in obesity and MS [13–15]. In addition, increased platelet count and activation occur as part of chronic inflammation in obesity [16,17]. Furthermore, obesity is a risk factor for high blood pressure [18], which is associated with elevated red blood cell (RBC) parameters, such as hemoglobin (Hb) and hematocrit

(Hct) levels [19,20]. These findings suggest that changes in hematological parameters are accompanied by increased body mass and chronic inflammation in obesity.

Previous studies of the hematological changes in obesity have focused mainly on adults, and population studies of children are limited. The purpose of this study was to investigate the associations between body mass index (BMI) and hematologic parameters in children and adolescents.

2. Materials and Methods

We analyzed the data from the 2007–2018 Korea National Health and Nutrition Examination Survey (KNHANES), a cross-sectional and nationally representative survey with a multistage, stratified probability sampling design. The KNHANES survey has been conducted by the Division of Chronic Disease Surveillance, Korean Centers for Disease Control and Prevention, in 3-year cycles since 1998 to assess the health and nutritional statuses of the noninstitutionalized civilian population of Korea [21]. To enhance the statistical power of the analyses, we combined the data from the fourth (2007–2009), fifth (2010–2012), and sixth (2013–2015) survey, as well as the first and second years of the seventh survey (2017–2018). The details of the study design have been reported previously [22].

KNHANES enrolled 97,622 individuals between 2007 and 2017. Of these, our preliminary analyses included 10,734 (5670 boys, 5064 girls) aged 10–18 years (Figure 1). We attempted to include a large number of representative Korean children and adolescents. After obtaining informed consent, the participants and their parents were interviewed at their homes, and the participants underwent several examinations, including blood sampling, an oral examination, and pulmonary function tests. Different blood tests were conducted in each survey. The levels of glucose, HbA1c, total cholesterol, triglyceride, HDL cholesterol, LDL cholesterol (measured directly), HBs Ag, AST, ALT, HCV titer, BUN, and creatinine were measured annually. In the seventh survey from 2017 to 2018, the levels of hsCRP, uric acid, folate, vitamin A, and vitamin E were measured. In addition, the levels of urinary pH, nitrite, specific gravity, protein, glucose, ketone, cotinine, sodium, potassium, and heavy metals (including cadmium and mercury) were evaluated every year. The hematological parameters recorded were the WBC, RBC, Hb, Hct, and platelet count. Participants with incomplete physical examination records, including incomplete anthropometric measurements and laboratory test results (such as lipid profiles) or triglyceride levels ≥400 mg/dL, were excluded ($n = 17$). The KNHANES database is publicly available (http://knhabes.cdc.go.kr, accessed on 4 August 2021). The 2007–2018 KNHANES study protocols were approved by the Institutional Review Board of the Korean Centers for Disease Control and Prevention. Informed consent was provided by the KNHANES participants.

Anthropometric assessments, including height, weight, waist circumference (WC), and systolic and diastolic BP (SBP and DBP, respectively), were performed by an expert and recorded annually. Height was measured to the nearest 0.1 cm using the standard method on a flat floor without shoes or bulky clothing. The participants stood with their back toward the measuring rod, face forward, feet placed together, knees straight, heels touching the heel plate or wall, and shoulders, buttocks, and head inside the stadiometer (seca, Hamburg, Germany). Body weight was measured to the nearest 0.1 kg while wearing light clothing using an electronic scale (G-tech, Seoul, Korea). WC was measured to the nearest 0.1 cm at the level between the lower rib margin and iliac crest using a calibrated measuring tape (seca). SBP and DBP were measured three times at an interval of 2 min to the nearest 1 mmHg from the right upper arm using a calibrated sphygmomanometer. The mean values of the final two SBP and DBP measurements were used for analysis. The standard deviation scores (SDSs) for height, weight, WC, and BMI were calculated using age- and sex-specific least mean square parameters based on the 2017 growth reference values for Korean children and adolescents, developed by the Korean Pediatric Society and the Korea Centers for Disease Control and Prevention [23]. Based on the BMI, participants were categorized into normal weight (NW; BMI < 85th percentile), overweight (OW; BMI 85–95th percentile), and obese (OB; BMI ≥ 95th percentile) groups.

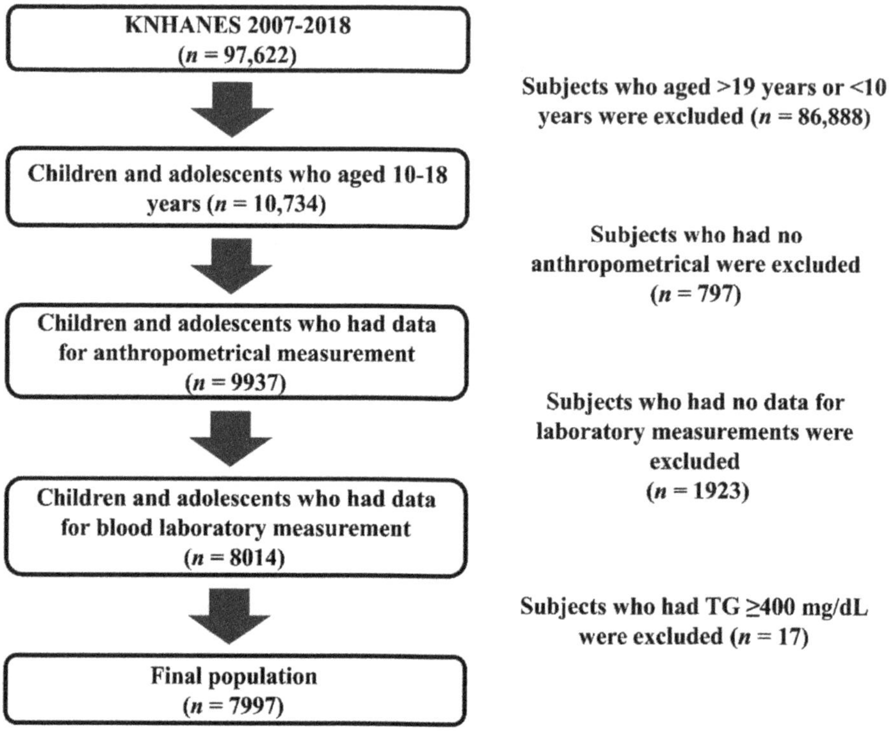

Figure 1. Flow chart of study population.

Lifestyle-related behaviors, such as alcohol consumption, smoking, household income, and residence area, were assessed using a questionnaire. Information about alcohol consumption (drinkers vs. nondrinkers) and smoking status (smokers vs. nonsmokers) was collected from participants aged ≥12 years using a self-administered questionnaire. Participants were also categorized based on physical activity (yes or no); those who performed intense physical activity for ≥20 min/day and ≥3 days/week or moderate physical activity for ≥30 min/day and ≥5 days/week, or who walked for ≥30 min/day and ≥5 days/week were included in the "yes" group.

Questionnaires related to household income, residence area (urban vs. rural), and previous diagnoses (T2DM, hypertension, and dyslipidemia) were administered by trained interviewers.

3. Statistical Analyses

R software (ver. 3.5.1; R Foundation for Statistical Computing, Vienna, Austria) was used for the statistical analysis. Continuous variables are expressed as means and standard deviations (SDs). Categorical variables are presented as numbers and percentages (%). Differences were analyzed using analysis of variance (ANOVA) for continuous variables and chi-square test for categorical variables. The adjusted means and standard errors (SEs) of the hematological indices (i.e., WBC, RBC, platelet, Hb, and Hct levels) were compared among the BMI groups using analysis of covariance (ANCOVA) after adjustment for possible confounders. In ANCOVA model 1, the adjusted means and SEs of hematological indices were estimated after controlling for age and sex. In ANCOVA models 2 and 3, the adjusted means and SEs of hematological indices were estimated for boys and girls after controlling for age. The pairwise differences among the BMI groups were tested for significance using *post-hoc* tests with Bonferroni correction in each ANCOVA model.

To evaluate the correlations between hematological parameters and BMI SDS, Pearson's correlation coefficient analysis with age and sex adjustments was performed. To evaluate the relationships between BMI SDS and hematological indices, multiple linear regression analyses were performed after adjusting for age, sex, alcohol consumption, smoking, physical activity, rural residence, household income, diagnosis of T2DM, hypertension, and dyslipidemia. The corresponding standardized regression coefficient (β) and SE were estimated. *p*-values < 0.05 were considered to indicate statistical significance.

4. Results

The final analyses included 7997 individuals: 4259 (53.26%) boys and 3738 girls. Participants were divided into the NW (6421, 80.29%), OW (782, 9.78%), and OB (794, 9.93%) groups. Among the 4259 boys, 3350 (78.66%), 443 (10.40%), and 466 (10.94%) were included in the NW, OW, and OB groups. Among the 3738 girls, 3071 (82.16%), 339 (9.07%), and 328 (8.77%) were included in the NW, OW, and OB groups.

4.1. Clinical Characteristics According to BMI

Table 1 shows the clinical characteristics of the study participants according to BMI. Mean SDSs for height, weight, BMI, and WC and mean SBP and DBP were significantly different among the subgroups ($p < 0.001$ for all). OB individuals had higher mean levels of hematological parameters and serum concentrations of glucose, total cholesterol, triglycerides, low-density lipoprotein cholesterol (LDL-C), and a lower serum concentration of HDL-C compared with non-OB individuals. The rate of alcohol use was higher in the OB group than in the other groups. The clinical characteristics of the study participants according to weight and sex are presented in Supplementary Table S1.

Table 1. Clinical characteristics of the study participants by body weight (n = 7997).

	Normal Weight	Overweight	Obesity	*p*
Total, *n* (%)	6421 (80.29%)	782 (9.78%)	794 (9.93%)	
Boys, *n* (%)	3350 (52.17%)	443(56.65%)	466 (58.69%)	<0.001
Age (year)	14.33 ± 2.51	14.24 ± 2.51	14.78 ± 2.54	<0.001
Height SDS	0.17 ± 1.03	0.47 ± 1.04	0.53 ± 1.11	<0.001
Weight SDS	−0.31 ± 0.89	1.29 ± 0.49	2.20 ± 0.74	<0.001
BMI SDS (kg/m^2)	−0.59 ± 0.88	0.86 ± 0.45	1.64 ± 0.75	<0.001
WC SDS	−0.49 ± 0.85	1.32 ± 0.18	2.40 ± 0.67	<0.001
SBP (mmHg)	105.65 ± 9.77	109.94 ± 10.28	113.68 ± 10.95	<0.001
DBP (mmHg)	65.59 ± 8.91	66.88 ± 8.88	69.07 ± 9.19	<0.001
WBC ($\times 10^3$/mm^3)	6.16 ± 1.49	6.65 ± 1.58	7.16 ± 1.65	<0.001
RBC ($\times 10^6$/mm^3)	4.81 ± 0.39	4.89 ± 0.40	4.93 ± 0.41	<0.001
Hemoglobin (g/dL)	13.97 ± 1.23	14.07 ± 1.22	14.19 ± 1.30	<0.001
Hematocrit (%)	41.86 ± 3.40	42.28 ± 3.42	42.65 ± 3.53	<0.001
Platelets ($\times 10^3$/mm^3)	282.96 ± 58.58	296.81 ± 59.92	309.03 ± 63.53	<0.001
Glucose (mg/dL)	90.01 ± 7.39	91.83 ± 11.05	92.45 ± 12.16	<0.001
T-C (mg/dL)	158.25 ± 26.27	163.00 ± 28.46	169.35 ± 29.50	<0.001
HDL-C (mg/dL)	52.16 ± 9.94	47.68 ± 8.63	44.99 ± 8.37	<0.001
TG (mg/dL)	79.29 ± 41.79	99.34 ± 54.29	112.75 ± 58.76	<0.001
LDL-C (mg/dL)	90.24 ± 22.42	95.46 ± 25.07	101.81 ± 25.63	<0.001
Alcohol consumption	1564 (24.36%)	188 (24.04%)	237 (29.85%)	0.003
Smoker	718 (11.18%)	88 (11.25%)	106 (13.35%)	0.191
Household income ≤ 1st quartile	687 (10.70%)	82 (10.49%)	95 (11.96%)	0.531
Rural residence	985 (15.34%)	113 (14.45%)	125 (15.74%)	0.757
Physical activity	2368 (36.88%)	279 (35.68%)	316 (39.80%)	0.194
Hypertension	1 (0.02%)	1 (0.13%)	1 (0.13%)	0.124
T2DM	0 (0%)	0 (0%)	0 (0%)	>0.999
Dyslipidemia	0 (0%)	0 (0%)	0 (0%)	>0.999

SDS, standard deviation score; BMI, body mass index; WC, waist circumference; SBP, systolic blood pressure; DBP, diastolic blood pressure; WBC, white blood cell; RBC, red blood cell; T-C, total cholesterol; HDL-C, high-density lipoprotein cholesterol; TG, triglycerides; LDL-C, low-density lipoprotein cholesterol; T2DM, type 2 diabetes mellitus.

The mean hematological parameters according to sex and BMI are shown in Figure 2. The mean WBC count was significantly elevated in boys and girls in all groups (Table 2). In both sexes, the OB group showed significantly higher mean RBC and Hct levels compared with the NW group. The mean Hb level was significantly higher in the OB than NW group in boys but not girls. Mean platelet counts in the OW and OB groups were significantly higher than that in the NW group for both sexes.

Table 2. Mean white blood cell (WBC), red blood cell (RBC), hemoglobin, hematocrit, and platelet levels according to sex and obesity.

	Boys			Girls		
	NW n = 3350	**OW** n = 443	**OB** n = 466	**NW** n = 3071	**OW** n = 339	**OB** n = 328
WBC ($\times 10^3/mm^3$)	6.13 ± 1.45	6.68 ± 1.53 [a]	7.13 ± 1.60 [b,c]	6.18 ± 1.53	6.61 ± 1.64 [a]	7.22 ± 1.71 [b,c]
RBC ($\times 10^6/mm^3$)	5.02 ± 0.34	5.10 ± 0.34 [a]	5.13 ± 0.33 [c]	4.59 ± 0.31	4.62 ± 0.30	4.63 ± 0.31 [c]
Hemoglobin (g/dL)	14.60 ± 1.11	14.64 ± 1.15	14.79 ± 1.20 [c]	13.28 ± 0.96	13.33 ± 0.85	13.34 ± 0.92
Hematocrit (%)	43.43 ± 3.30	43.73 ± 3.38	44.18 ± 3.34 [c]	40.15 ± 2.60	40.38 ± 2.42	40.49 ± 2.50 [c]
Platelet ($\times 10^3/mm^3$)	278.59 ± 58.48	295.33 ± 61.62 [a]	302.91 ± 61.66 [c]	287.73 ± 58.32	298.74 ± 57.65 [a]	317.73 ± 65.21 [b,c]

Data are presented as the mean ± standard deviation (SD). NW, underweight and normal weight; OW, overweight; OB, obesity; WBC, white blood cell; RBC, red blood cell. Bonferroni's *post-hoc* test after adjustment for age among girls.[a]: $p < 0.05$, NW group versus OW group after Bonferroni's post-hoc test. [b]: $p < 0.05$, OW group versus OB group after Bonferroni's post-hoc test. [c]: $p < 0.05$, NW group versus OB group after Bonferroni's post-hoc test.

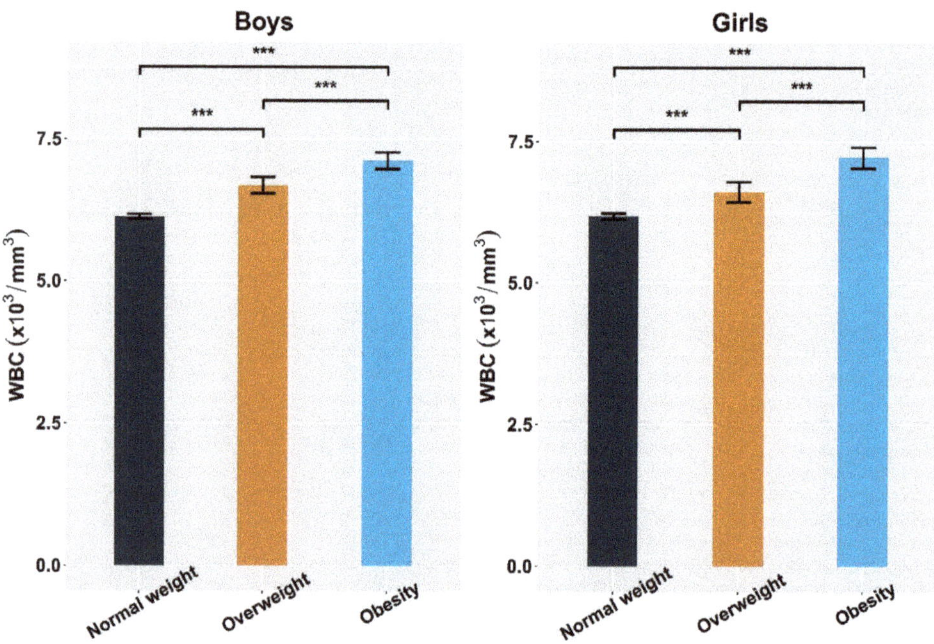

(a) Mean white blood cell (WBC) according to sex and obesity

Figure 2. *Cont.*

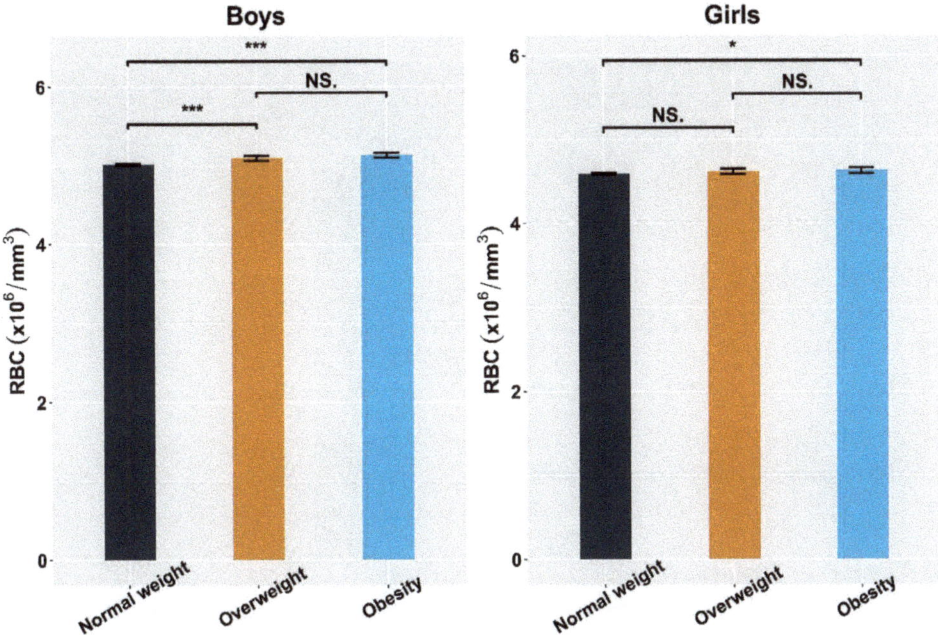

(b) Mean red blood cell (RBC) according to sex and obesity

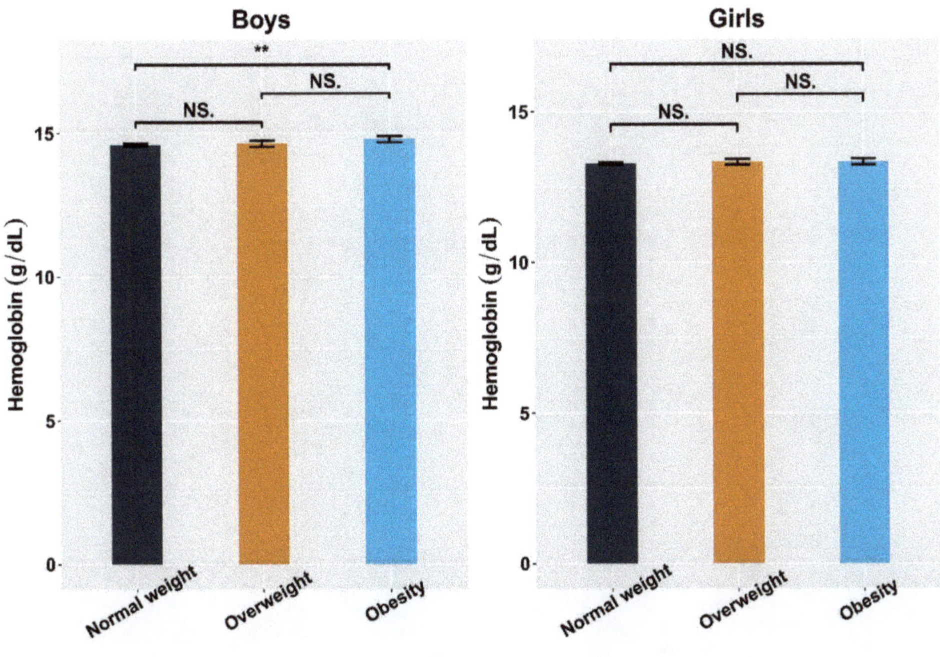

(c) Mean hemoglobin levels according to sex and obesity

Figure 2. *Cont.*

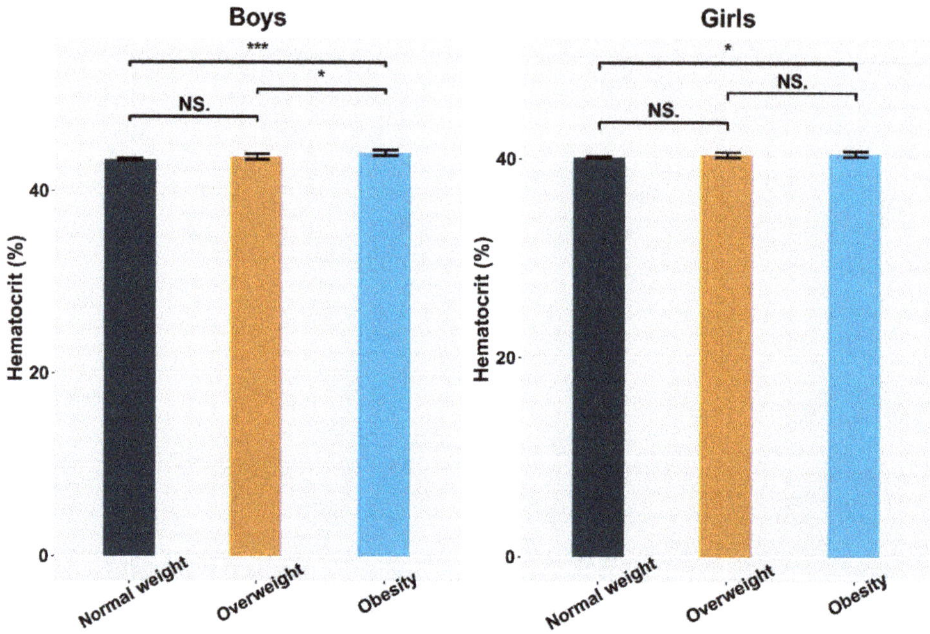

(d) Mean hematocrit levels according to sex and obesity

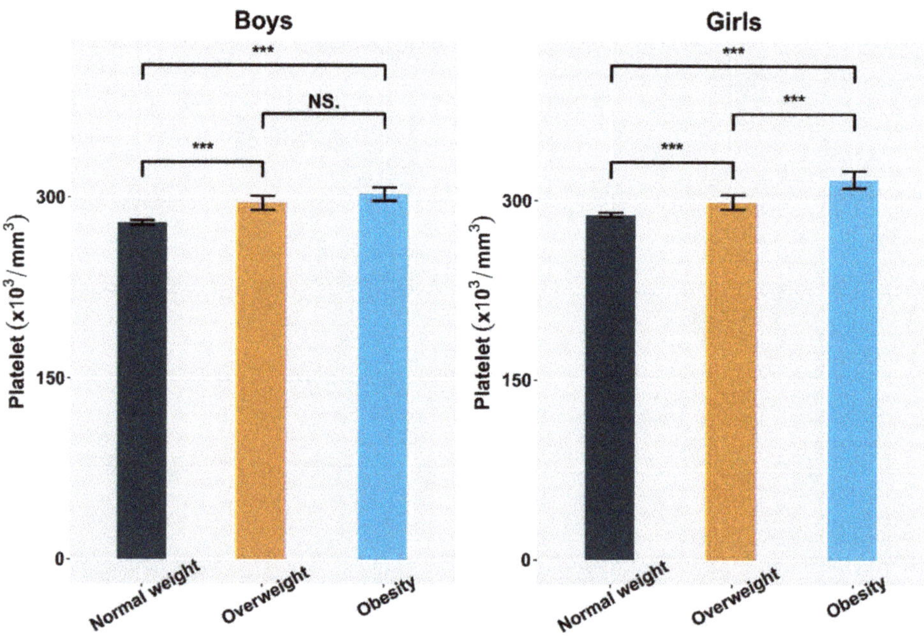

(e) Mean platelet levels according to sex and obesity

Figure 2. Mean white blood cell (WBC), red blood cell (RBC), hemoglobin, hematocrit, and platelet levels according to sex and obesity. (**a**), Mean white blood cell (WBC) according to sex and obesity;

(**b**), Mean red blood cell (RBC) according to sex and obesity; (**c**), Mean hemoglobin levels according to sex and obesity; (**d**), Mean hematocrit levels according to sex and obesity; (**e**), Mean platelet levels according to sex and obesity; *, $p < 0.05$; **, $p < 0.005$; ***, $p < 0.001$; NS., $p > 0.05$.

4.2. Hematological Parameters Adjusted by Sex and Obesity

The adjusted mean levels of hematological parameters were significantly different among the BMI groups (Table 3). OB participants had higher adjusted mean levels of WBCs (7.16 vs. 6.16 $\times$ $10^3/mm^3$), RBCs (4.90 vs. 4.82 $\times$ $10^6/mm^3$), Hct (42.31% vs. 41.91%), and platelets (311.87 vs. 282.66 $\times$ $10^3/mm^3$) ($p < 0.001$ for all) compared with NW participants. Similarly, OB girls had higher adjusted mean levels of WBCs, RBCs, Hct, and platelets ($p < 0.001$ for all) compared with NW girls. Although the Hb level was significantly different among the three groups, the post-hoc tests did not reveal a significant difference in all participants and girls. OB boys had higher adjusted mean levels of WBCs, RBCs, Hb, Hct, and platelets than those of NW boys ($p < 0.001$ for all).

Table 3. Adjusted mean levels of white blood cells, red blood cells, hemoglobin, hematocrit, and platelets by sex and BMI.

All Participants [1]	**Normal Weight**	**Overweight**	**Obesity**	***p* for Trend**
(*n* = 7997)	***n* = 6421**	***n* = 782**	***n* = 794**	
WBC ($\times 10^3/mm^3$)	6.16 ± 0.02	6.66 ± 0.05 [a]	7.16 ± 0.05 [b,c]	<0.001
RBC ($\times 10^6/mm^3$)	4.82 ± 0.01	4.88 ± 0.01 [a]	4.90 ± 0.01 [c]	<0.001
Hemoglobin (g/dL)	13.99 ± 0.01	14.04 ± 0.04	14.07 ± 0.04	0.016
Hematocrit (%)	41.91 ± 0.04	42.21 ± 0.10 [a]	42.31 ± 0.10 [c]	<0.001
Platelets ($\times 10^3/mm^3$)	282.66 ± 0.72	296.38 ± 2.05 [a]	311.87 ± 2.04 [b,c]	<0.001
		Boys [2]		
(*n* = 4259)	*n* = 3350	*n* = 443	*n* = 466	
WBC ($\times 10^3/mm^3$)	6.13 ± 0.03	6.69 ± 0.07 [a]	7.12 ± 0.07 [b,c]	<0.001
RBC ($\times 10^6/mm^3$)	5.02 ± 0.01	5.11 ± 0.01 [a]	5.12 ± 0.01 [c]	<0.001
Hemoglobin (g/dL)	14.60 ± 0.01	14.72 ± 0.04 [a]	14.72 ± 0.04 [c]	<0.001
Hematocrit (%)	43.43 ± 0.04	43.96 ± 0.12 [a]	44.00 ± 0.12 [c]	<0.001
Platelets ($\times 10^3/mm^3$)	278.64 ± 0.97	293.21 ± 2.66 [a]	304.58 ± 2.59 [b,c]	<0.001
		Girls [3]		
(*n* = 3738)	*n* = 3071	*n* = 339	*n* = 328	
WBC ($\times 10^3/mm^3$)	6.18 ± 0.03	6.61 ± 0.08 [a]	7.20 ± 0.09 [b,c]	<0.001
RBC ($\times 10^6/mm^3$)	4.59 ± 0.01	4.62 ± 0.02	4.66 ± 0.02 [c]	<0.001
Hemoglobin (g/dL)	13.27 ± 0.02	13.34 ± 0.05	13.39 ± 0.05	0.015
Hematocrit (%)	40.14 ± 0.05	40.39 ± 0.14	40.59 ± 0.14 [c]	<0.001
Platelets ($\times 10^3/mm^3$)	287.47 ± 1.05	298.90 ± 3.17 [a]	319.95 ± 3.23 [b,c]	<0.001

Data are presented as the mean ± standard error (SE). WBC, white blood cell; RBC, red blood cell. Model 1: Comparisons with BMI using analysis of covariance with Bonferroni's *post-hoc* test after adjustment for age and sex among all participants. Model 2: Comparisons with BMI using analysis of covariance with Bonferroni's *post-hoc* test after adjustment for age among boys. Model 3: Comparisons with BMI using analysis of covariance with Bonferroni's *post-hoc* test after adjustment for age among girls. [a]: $p < 0.05$ between normal weight vs. overweight group after Bonferroni's *post-hoc* test. [b]: $p < 0.05$ between overweight vs. obesity group after Bonferroni's *post-hoc* test. [c]: $p < 0.05$ between normal weight vs. obesity group after Bonferroni's *post-hoc* test.

4.3. Correlations between BMI SDS and Hematological Indices

The unadjusted and adjusted correlations between BMI SDS and hematological parameters are presented in Table 4. BMI SDS was positively correlated with the levels of WBCs ($r = 0.222$), RBCs ($r = 0.109$), Hb ($r = 0.042$), Hct ($r = 0.067$), and platelets ($r = 0.180$) after adjusting for age and sex ($p < 0.001$ for all). BMI SDS was not correlated with the Hb level in girls in the unadjusted model but was significantly correlated after adjusting for age. After adjusting for age, BMI SDS was positively correlated with hematological indices in both sexes. Pearson's coefficients for the correlations of BMI SDS with RBC, Hb, and Hct levels in girls were less than half of those in boys.

Table 4. Unadjusted and adjusted correlations between body mass index (BMI) standard deviation score (SDS) and white blood cell (WBC), red blood cell (RBC), hemoglobin, hematocrit, and platelet levels.

All Participants (n = 7997)	r [1]	p	r [2]	p
WBC ($\times 10^3/mm^3$)	0.223	<0.001	0.222	<0.001
RBC ($\times 10^6/mm^3$)	0.096	<0.001	0.109	<0.001
Hemoglobin (g/dL)	0.044	<0.001	0.042	<0.001
Hematocrit (%)	0.067	<0.001	0.067	<0.001
Platelets ($\times 10^3/mm^3$)	0.169	<0.001	0.180	<0.001
Boys (n = 4259)	r [3]	p	r [4]	p
WBC ($\times 10^3/mm^3$)	0.239	<0.001	0.240	<0.001
RBC ($\times 10^6/mm^3$)	0.157	<0.001	0.180	<0.001
Hemoglobin (g/dL)	0.070	<0.001	0.111	<0.001
Hematocrit (%)	0.090	<0.001	0.137	<0.001
Platelets ($\times 10^3/mm^3$)	0.175	<0.001	0.177	<0.001
Girls (n = 3738)	r [5]	p	r [6]	p
WBC ($\times 10^3/mm^3$)	0.206	<0.001	0.202	<0.001
RBC ($\times 10^6/mm^3$)	0.047	0.004	0.078	<0.001
Hemoglobin (g/dL)	0.014	0.383	0.033	0.043
Hematocrit (%)	0.042	0.009	0.054	<0.001
Platelets ($\times 10^3/mm^3$)	0.164	<0.001	0.176	<0.001

WBC, white blood cell; RBC, red blood cell. [1]: Pearson's correlation analyses were conducted between BMI SDS and WBC, RBC, hemoglobin, hematocrit, and platelet levels with no adjustments. [2]: Pearson's correlation analyses were conducted between BMI SDS and WBC, RBC, hemoglobin, hematocrit, and platelet levels after adjustments for age and sex for all participants. [3]: Pearson's correlation analyses were conducted between BMI SDS and WBC, RBC, hemoglobin, hematocrit, and platelet levels with no adjustments in boys. [4]: Pearson's correlation analyses were conducted between BMI SDS and WBC, RBC, hemoglobin, hematocrit, and platelet levels after adjustment for age in boys. [5]: Pearson's correlation analyses were conducted between BMI SDS and WBC, RBC, hemoglobin, hematocrit, and platelet levels with no adjustments. [6]: Pearson's correlation analyses were conducted between BMI SDS and WBC, RBC, hemoglobin, hematocrit, and platelet levels after adjustment for age in girls.

4.4. Multiple Linear Regression Analyses of the BMI Groups with Hematological Parameters

Table 5 shows that BMI SDS had significant positive associations with hematological parameters among all participants after adjusting for age, sex, alcohol consumption, smoking, physical activity, rural residence, household income, T2DM, hypertension, and dyslipidemia. BMI SDS was significantly associated with WBC ($\beta = 0.275$), RBC ($\beta = 0.028$), Hb ($\beta = 0.034$), Hct ($\beta = 0.152$), and platelet ($\beta = 8.372$) levels ($p < 0.001$ for all). In boys, BMI SDS was significantly associated with WBC ($\beta = 0.279$), RBC ($\beta = 0.043$), Hb ($\beta = 0.073$), Hct ($\beta = 0.267$), and platelet ($\beta = 7.658$) levels ($p < 0.001$ for all). In girls, BMI SDS was associated with WBC ($\beta = 0.270$, $p < 0.001$), RBC ($\beta = 0.019$, $p < 0.001$), Hb ($\beta = 0.027$, $p = 0.038$), Hct ($\beta = 0.118$, $p < 0.001$), and platelet ($\beta = 8.715$, $p < 0.001$) levels. The regression coefficients between BMI SDS and RBC, Hb, and Hct levels were less than half of those in boys.

Table 5. Multiple regression analysis of the associations of body mass index (BMI) standard deviation score (SDS) with white blood cell (WBC), red blood cell (RBC), hemoglobin, hematocrit, and platelet levels.

All Participants [1] (n = 7997)	B	SE	p
WBC ($\times 10^3/mm^3$)	0.275	0.013	<0.001
RBC ($\times 10^6/mm^3$)	0.028	0.003	<0.001
Hemoglobin (g/dL)	0.034	0.009	<0.001
Hematocrit (%)	0.152	0.025	<0.001
Platelet ($\times 10^3/mm^3$)	8.372	0.510	<0.001
Boys [2] (n = 4259)			
WBC ($\times 10^3/mm^3$)	0.279	0.017	<0.001
RBC ($\times 10^6/mm^3$)	0.043	0.004	<0.001
Hemoglobin (g/dL)	0.073	0.010	<0.001
Hematocrit (%)	0.267	0.030	<0.001
Platelet ($\times 10^3/mm^3$)	7.658	0.657	<0.001
Girls [3] (n = 3738)			
WBC ($\times 10^3/mm^3$)	0.270	0.021	<0.001
RBC ($\times 10^6/mm^3$)	0.019	0.004	<0.001
Hemoglobin (g/dL)	0.027	0.013	0.038
Hematocrit (%)	0.118	0.035	<0.001
Platelet ($\times 10^3/mm^3$)	8.715	0.798	<0.001

WBC, white blood cell; RBC, red blood cell. [1]: Multiple linear regression analysis was conducted between BMI SDS (independent variable) and WBC, RBC, hemoglobin, hematocrit, and platelet levels (dependent variables) after adjustments for age, sex, alcohol consumption, smoking, physical activity, rural residence, household income, type 2 diabetes mellitus (T2DM), hypertension, and dyslipidemia for all participants. [2]: Multiple linear regression analysis was conducted between BMI SDS (independent variable) and WBC, RBC, hemoglobin, hematocrit, and platelet levels (dependent variables) after adjustments for age, sex, alcohol consumption, smoking, physical activity, rural residence, household income, T2DM, hypertension, and dyslipidemia among boys. [3]: Multiple linear regression analysis was conducted between BMI SDS (independent variable) and WBC, RBC, hemoglobin, hematocrit, and platelet levels (dependent variables) after adjustments for age, sex, alcohol consumption, smoking, physical activity, rural residence, household income, diagnosis of T2DM, hypertension, and dyslipidemia among girls.

5. Discussion

The current study investigated the relationships between BMI and hematological parameters in Korean children and adolescents. OB children had higher blood pressure, glucose, LDL-C, WBC, RBC, and platelet levels. In particular, BMI SDS had an independent positive association with hematological parameters after adjusting for cofounding variables. To the best of our knowledge, this is the largest population-based study of the relationships between BMI and hematological parameters in Korean children and adolescents.

Chronic inflammation around adipocytes plays an important role in obesity-related diseases [3]. Because the WBC count is increased in inflammation, it is likely that the WBC count is elevated in OB patients. A plausible explanation for the increased WBC count in OB individuals is that adipose tissue produces IL-6, a proinflammatory cytokine involved in bone marrow granulopoiesis and WBC proliferation and differentiation [24,25]. In our study, the WBC count in children increased by $0.275 \times 10^3/mm^3$ for every 1-point increase in BMI SDS. The elevated WBC count observed in the present study is consistent with previous studies [26,27]. The elevated WBC count was associated with carotid atherosclerosis and impaired glucose tolerance in previous studies [28]. Tong et al. reported that patients with high WBC counts had adverse metabolic profiles, even when the WBC levels were high but within the normal range. The elevated WBC count is associated with macrovascular and microvascular complications of T2DM [29]. Furthermore, Veronelli et al. observed a significant decrease in the WBC count after bariatric surgery, suggesting the usefulness of weight loss in reducing the WBC count in morbidly OB individuals [27].

Our findings are consistent with those of previous studies that found higher mean WBC counts and MS prevalence with increasing BMI in boys and girls [13,30]. In Colombian children, the WBC count was associated with truncal adiposity [12]. An increased WBC count was associated with early derangement of glucose metabolism and preclinical signs of liver, vascular, and cardiac damage in Italian children [31]. Park et al. reported that a higher WBC count was positively associated with an increased risk of insulin resistance in Korean children and adolescents [32]. Lee et al. suggested that an elevated WBC count is a surrogate marker of MS in Korean children and adolescents [13]. Because it is inexpensive, the WBC count has been suggested as an effective tool for identifying OB children at risk of complications [31,32].

In this study, red cell indices, including RBC count Hb and Hct levels, were positively associated with BMI SDS, which was consistent with previous studies. Mărginean et al. reported a significantly higher RBC count but no significant difference in the Hb level in OB children than controls [33]. In addition, several studies suggested that the RBC count has a significant correlation with MS [34,35], and the Hb level is significantly associated with high blood pressure [36]. The mechanism underlying the increased RBC cell indices in obesity is not known. However, iron deficiency anemia occurs more frequently in OB and OW individuals compared with NW individuals [37], which is likely related to an obesity-induced chronic inflammatory state and effects of hepcidin. Bekri et al. reported that hepcidin, a proinflammatory adipokine, reduces iron bioavailability by controlling the ferroportin-1 exporter, resulting in severe iron deficiency anemia in OB individuals [38]. Therefore, obesity-induced chronic inflammation may influence the serum iron level. Ausk et al. reported that the serum Hb concentration was not significantly different between OB and NW individuals [39]; however, a higher serum ferritin level was associated with higher BMI and lower serum iron and transferrin saturation levels. The study researchers concluded that OW and OB individuals were no more likely to be anemic than were NW individuals. In our study, BMI SDS was independently associated with RBC indices after adjusting for multiple variables. However, the serum iron level, nutritional habits, and anemia prevalence were not compared among the BMI subgroups. Further studies conducted in different ethnic groups are needed to validate our results.

The changes in RBC indices differ between males and females. Kim et al. reported that the RBC count was significantly increased in men, but not women, with MS [40]. In the present study, all RBC indices increased with increasing BMI SDS, but the degree of increase differed between boys and girls. The RBC count increased by $0.043 \times 10^6/\text{mm}^3$ in boys and $0.019 \times 10^6/\text{mm}^3$ in girls with every 1-point increase in BMI SDS. The RBC count was over 2-fold higher in boys than girls. Similar patterns were observed for Hb and Hct levels. Obesity is a chronic hypoxia state that causes adipose tissue dysfunction, inflammation, and insulin resistance [41]. Chronic hypoxia may lead to increased production of RBCs and WBCs. In addition, the level of ferritin, an inflammatory marker, is increased in obesity [11], which may be involved in the changes in RBC indices. Although adolescent girls typically have more body fat and less muscle than boys, they periodically lose blood and iron due to menstruation. Consequently, the increase in RBC count with increasing BMI SDS may be relatively less for girls than boys.

Thrombocytosis suggests inflammation, and platelet activation leads to accelerated atherothrombosis [17]. Lim et al. reported that an elevated platelet count was associated with an increased prevalence and risk of MS in children and adolescents [42]. In our study, for every 1-point increase in BMI SDS, the platelet count increased by $8.372 \times 10^3/\text{mm}^3$ for all participants, $8.715 \times 10^3/\text{mm}^3$ for girls, and $7.658 \times 10^3/\text{mm}^3$ for boys. The increase in the platelet count also differed by sex. Dorit et al. reported that OB females had a significantly higher platelet count compared with NW females [16]; however, the platelet count was not elevated in males. In addition, Charles et al. reported positive associations of obesity with WBC and platelet counts among female officers, but not male officers [43]. In our study, BMI SDS was positively associated with hematological parameters; however, the changes in hematological parameters (especially RBC and platelet counts) showed differences between the sexes (i.e., sexual dimorphism). The adjusted mean RBC indices and changes thereof with BMI SDS were higher in boys than girls for all subgroups. In contrast, the mean platelet count and increase in platelet count with increasing BMI SDS were higher in girls than boys in all subgroups. The mechanisms underlying the sex differences in blood cell composition and their effects on the risk of obesity-related complications are not well understood. It is possible that these findings result from differences in body fat composition between males and females. Nuttall reported that BMI may not be an accurate marker of obesity because men tend to accumulate fat in the abdominal area, whereas women tend to accumulate it in the peripelvic area and thighs [44]. Before puberty, boys and girls have similar patterns of body fat deposition; however, during and after puberty, girls tend to accumulate a large quantity of fat, whereas boys accumulate a large quantity of lean mass (bone and muscle) but not fat mass. These changes lead to an increased BMI in both sexes. The hematological parameters of boys and girls are different between those tested before and after puberty. In the present study, information related to body fat mass and pubertal status was not available. Therefore, further studies are needed to determine the sex differences in hematological parameters before and after puberty.

This study had several limitations. First, because this was a cross-sectional study, we could not identify causal relationships between obesity and hematological parameters. Second, our study included children and adolescents aged 10–18 years; therefore, no data are available for children aged < 10 years. Third, because data regarding body composition and pubertal status were not available, the associations of body fat mass with hematological parameters were not analyzed. In addition, the effects of puberty on hematological changes and body composition were not evaluated. Importantly, BMI cannot differentiate between body lean and fat mass and poorly represents the body fat percentage and location [44]. Finally, we could not determine the mechanisms underlying the relationships between BMI SDS and hematological indices. Our results may not be applicable to specific populations, such as those with severe obesity and malnutrition. Several studies showed that OW and OB children have an increased risk of iron deficiency anemia [45,46]. Despite the limitations, this study showed that BMI SDS was independently associated with hematological parameters in a relatively large number of children and adolescents. Our findings provide insight into the hematologic changes in OB children and adolescents.

In conclusion, this nationally representative population-based study showed that higher BMI was independently associated with high levels of hematological parameters, including WBC, RBC, Hb, Hct, and platelet levels, in children and adolescents. Our results suggest that obesity is related to hematological changes, which may predispose to obesity-related diseases. When interpreting the complete blood count in children and adolescents, it is important to consider the effects of BMI and sex.

Supplementary Materials: The following supporting information can be downloaded at: https://www.mdpi.com/article/10.3390/children9010109/s1, Table S1. Clinical characteristics of the study participants by weight and sex (n = 7997).

Author Contributions: Conceptualization, Y.S.S. and J.S.H.; methodology, Y.S.S.; software, Y.S.S.; validation, Y.S.S., H.S.L. and J.S.H.; formal analysis, Y.S.S.; investigation, Y.S.S.; resources, Y.S.S.; data curation, Y.S.S.; writing—original draft preparation, H.R.J.; writing—review and editing, H.R.J. and H.S.L.; visualization, H.R.J. and H.S.L.; supervision, Y.S.S. and J.S.H.; project administration, Y.S.S.; funding acquisition, H.R.J. All authors have read and agreed to the published version of the manuscript.

Funding: This work was supported by the Soonchunhyang University Research Fund.

Institutional Review Board Statement: The 2007–2018 KNHANES study protocols were approved by the Institutional Review Board of the Korean Centers for Disease Control and Prevention. Informed consent was provided by the KNHANES participants.

Conflicts of Interest: The authors declare no conflict of interest.

References

1. Chia, D.J.; Boston, B.A. Childhood obesity and the metabolic syndrome. *Adv. Pediatr.* **2006**, *53*, 23–53. [CrossRef]
2. Hotamisligil, G.S. Inflammation and metabolic disorders. *Nature* **2006**, *444*, 860–867. [CrossRef] [PubMed]
3. Kim, J.; Lee, J. Role of obesity-induced inflammation in the development of insulin resistance and type 2 diabetes: History of the research and remaining questions. *Ann. Pediatr. Endocrinol. Metab.* **2021**, *26*, 1–13. [CrossRef] [PubMed]
4. Stępień, M.; Stępień, A.; Wlazeł, R.N.; Paradowski, M.; Banach, M.; Rysz, J. Obesity indices and inflammatory markers in obese non-diabetic normo- and hypertensive patients: A comparative pilot study. *Lipids Health Dis.* **2014**, *13*, 29. [CrossRef] [PubMed]
5. Monteiro, R.; Azevedo, I. Chronic inflammation in obesity and the metabolic syndrome. *Mediat. Inflamm.* **2010**, *2010*, 289645. [CrossRef]
6. Wilson, P.W.; D'Agostino, R.B.; Parise, H.; Sullivan, L.; Meigs, J.B. Metabolic syndrome as a precursor of cardiovascular disease and type 2 diabetes mellitus. *Circulation* **2005**, *112*, 3066–3072. [CrossRef] [PubMed]
7. Srinivasan, S.R.; Bao, W.; Wattigney, W.A.; Berenson, G.S. Adolescent overweight is associated with adult overweight and related multiple cardiovascular risk factors: The Bogalusa heart study. *Metabolism* **1996**, *45*, 235–240. [CrossRef]
8. Ogden, C.L.; Carroll, M.D.; Kit, B.K.; Flegal, K.M. Prevalence of childhood and adult obesity in the United States, 2011–2012. *JAMA J. Am. Med Assoc.* **2014**, *311*, 806–814. [CrossRef]
9. Twig, G.; Yaniv, G.; Levine, H.; Leiba, A.; Goldberger, N.; Derazne, E.; Ben-Ami Shor, D.; Tzur, D.; Afek, A.; Shamiss, A.; et al. Body-Mass Index in 2.3 Million Adolescents and Cardiovascular Death in Adulthood. *N. Eng. J. Med.* **2016**, *374*, 2430–2440. [CrossRef]
10. Castorani, V.; Polidori, N.; Giannini, C.; Blasetti, A.; Chiarelli, F. Insulin resistance and type 2 diabetes in children. *Ann. Pediatr. Endocrinol. Metab.* **2020**, *25*, 217–226. [CrossRef]
11. Ellulu, M.S.; Patimah, I.; Khaza'ai, H.; Rahmat, A.; Abed, Y. Obesity and inflammation: The linking mechanism and the complications. *Arch. Med. Sci.* **2017**, *13*, 851–863. [CrossRef] [PubMed]
12. Gilbert-Diamond, D.; Baylin, A.; Mora-Plazas, M.; Villamor, E. Chronic inflammation is associated with overweight in Colombian school children. *Nutr. Metab. Cardiovasc. Dis.* **2012**, *22*, 244–251. [CrossRef]
13. Lee, Y.-J.; Shin, Y.-H.; Kim, J.-K.; Shim, J.Y.; Kang, D.-R.; Lee, H.-R. Metabolic syndrome and its association with white blood cell count in children and adolescents in Korea: The 2005 korean national health and nutrition examination survey. *Nutr. Metab. Cardiovasc. Dis.* **2010**, *20*, 165–172. [CrossRef] [PubMed]
14. Nieto, F.J.; Szklo, M.; Folsom, A.R.; Rock, R.; Mercuri, M. Leukocyte count correlates in middle-aged adults: The atherosclerosis risk in communities (ARIC) study. *Am. J. Epidemiol.* **1992**, *136*, 525–537. [CrossRef] [PubMed]
15. Schwartz, J.; Weiss, S.T. Host and environmental factors influencing the peripheral blood leukocyte count. *Am. J. Epidemiol.* **1991**, *134*, 1402–1409. [CrossRef]
16. Samocha-Bonet, D.; Justo, D.; Rogowski, O.; Saar, N.; Abu-Abeid, S.; Shenkerman, G.; Shapira, I.; Berliner, S.; Tomer, A. Platelet counts and platelet activation markers in obese subjects. *Mediat. Inflamm.* **2008**, *2008*, 834153. [CrossRef]
17. Santilli, F.; Vazzana, N.; Liani, R.; Guagnano, M.T.; Davì, G. Platelet activation in obesity and metabolic syndrome. *Obes. Rev.* **2012**, *13*, 27–42. [CrossRef]
18. Hall, J.E.; Crook, E.D.; Jones, D.W.; Wofford, M.R.; Dubbert, P.M. Mechanisms of obesity-associated cardiovascular and renal disease. *Am. J. Med. Sci.* **2002**, *324*, 127–137. [CrossRef]
19. Göbel, B.O.; Schulte-Göbel, A.; Weisser, B.; Glänzer, K.; Vetter, H.; Düsing, R. Arterial blood pressure. Correlation with erythrocyte count, hematocrit, and hemoglobin concentration. *Am. J. Hypertens.* **1991**, *4*, 14–19. [CrossRef]
20. Atsma, F.; Veldhuizen, I.; de Kort, W.; van Kraaij, M.; Pasker-de Jong, P.; Deinum, J. Hemoglobin level is positively associated with blood pressure in a large cohort of healthy individuals. *Hypertension* **2012**, *60*, 936–941. [CrossRef]
21. Kweon, S.; Kim, Y.; Jang, M.-J.; Kim, Y.; Kim, K.; Choi, S.; Chun, C.; Khang, Y.-H.; Oh, K. Data resource profile: The Korea national health and nutrition examination survey (KNHANES). *Int. J. Epidemiol.* **2014**, *43*, 69–77. [CrossRef]
22. Park, H.K.; Shim, Y.S. Distribution of Tri-Ponderal Mass Index and its Relation to Body Mass Index in Children and Adolescents Aged 10 to 20 Years. *J. Clin. Endocrinol. Metab.* **2020**, *105*, e826–e834. [CrossRef]
23. Kim, J.H.; Yun, S.; Hwang, S.-S.; Shim, J.O.; Chae, H.W.; Lee, Y.J.; Lee, J.-H.; Kim, S.C.; Lim, D.; Yang, S.W.; et al. The 2017 Korean national growth charts for children and adolescents: Development, improvement, and prospects. *Korean J. Pediatr.* **2018**, *61*, 135–149. [CrossRef]
24. Sait, S.; Alqassas, M.; Othman, S.; Shihon, S.; Alqalayta, L.; Alqusair, S. Obesity correlates with neutrophilia. *Hematol. Transfus. Int. J.* **2016**, *3*, 159–162. [CrossRef]
25. Gu, Y.; Hu, K.; Huang, Y.; Zhang, Q.; Liu, L.; Meng, G.; Wu, H.; Xia, Y.; Bao, X.; Shi, H.; et al. White blood cells count as an indicator to identify whether obesity leads to increased risk of type 2 diabetes. *Diabetes Res. Clin. Pr.* **2018**, *141*, 140–147. [CrossRef]
26. Alrubaie, A.; Majid, S.; Alrubaie, R.; Al-Zahraa, B.; Kadhim, F. Effects of Body Mass Index (BMI) on complete blood count parameters. *Prensa Med. Argent* **2019**, *105*, 164–171.
27. Veronelli, A.; Laneri, M.; Ranieri, R.; Koprivec, D.; Vardaro, D.; Paganelli, M.; Folli, F.; Pontiroli, A.E. White blood cells in obesity and diabetes: Effects of weight loss and normalization of glucose metabolism. *Diabetes Care* **2004**, *27*, 2501–2502. [CrossRef] [PubMed]
28. Ohshita, K.; Yamane, K.; Hanafusa, M.; Mori, H.; Mito, K.; Okubo, M.; Hara, H.; Kohno, N. Elevated white blood cell count in subjects with impaired glucose tolerance. *Diabetes Care* **2004**, *27*, 491–496. [CrossRef] [PubMed]

29. Tong, P.C.; Lee, K.-F.; So, W.-Y.; Ng, M.H.; Chan, W.-B.; Lo, M.K.; Chan, N.N.; Chan, J.C. White blood cell count is associated with macro-and microvascular complications in Chinese patients with type 2 diabetes. *Diabetes care* **2004**, *27*, 216–222. [CrossRef] [PubMed]
30. Wärnberg, J.; Marcos, A. Low-grade inflammation and the metabolic syndrome in children and adolescents. *Curr. Opin. Lipidol.* **2008**, *19*, 11–15. [CrossRef]
31. Di Bonito, P.; Pacifico, L.; Chiesa, C.; Invitti, C.; Del Giudice, E.M.; Baroni, M.G.; Moio, N.; Pellegrin, M.; Tomat, M.; Licenziati, M.; et al. White blood cell count may identify abnormal cardiometabolic phenotype and preclinical organ damage in overweight/obese children. *Nutr. Metab. Cardiovasc. Dis.* **2016**, *26*, 502–509. [CrossRef]
32. Park, J.-M.; Lee, D.-C.; Lee, Y.-J. Relationship between high white blood cell count and insulin resistance (HOMA-IR) in Korean children and adolescents: Korean national health and nutrition examination survey 2008–2010. *Nutr. Metab. Cardiovasc. Dis.* **2017**, *27*, 456–461. [CrossRef]
33. Mărginean, C.O.; Meliţ, L.E.; Ghiga, D.V.; Mărginean, M.O. Early inflammatory status related to pediatric obesity. *Front. Pediatr.* **2019**, *7*, 241. [CrossRef] [PubMed]
34. Zhou, P.; Meng, Z.; Liu, M.; Ren, X.; Zhu, M.; He, Q.; Zhang, Q.; Liu, L.; Song, K.; Jia, Q.; et al. The associations between leukocyte, erythrocyte or platelet, and metabolic syndrome in different genders of Chinese. *Medicine* **2016**, *95*, e5189. [CrossRef] [PubMed]
35. Barbieri, M.; Ragno, E.; Benvenuti, E.; Zito, G.A.; Corsi, A.; Ferrucci, L.; Paolisso, G. New aspects of the insulin resistance syndrome: Impact on haematological parameters. *Diabetologia* **2001**, *44*, 1232–1237. [CrossRef] [PubMed]
36. Chen, X.T.; Yang, S.; Yang, Y.M.; Zhao, H.L.; Chen, Y.C.; Zhao, X.H.; Wen, J.B.; Tian, Y.R.; Yan, W.L.; Shen, C. Exploring the relationship of peripheral total bilirubin, red blood cell, and hemoglobin with blood pressure during childhood and adolescence. *J. Pediatr.* **2018**, *94*, 532–538. [CrossRef] [PubMed]
37. Wenzel, B.J.; Stults, H.B.; Mayer, J. Hypoferræmia in obese adolescents. *Lancet* **1962**, *2*, 327–328. [CrossRef]
38. Bekri, S.; Gual, P.; Anty, R.; Luciani, N.; Dahman, M.; Ramesh, B.; Iannelli, A.; Staccini-Myx, A.; Casanova, D.; Ben Amor, I.; et al. Increased adipose tissue expression of hepcidin in severe obesity is independent from diabetes and NASH. *Gastroenterology* **2006**, *131*, 788–796. [CrossRef]
39. Ausk, K.J.; Ioannou, G.N. Is obesity associated with anemia of chronic disease? A population-based study. *Obesity* **2008**, *16*, 2356–2361. [CrossRef]
40. Kim, J.A.; Choi, Y.S.; Hong, J.I.; Kim, S.H.; Jung, H.H.; Kim, S.M. Association of metabolic syndrome with white blood cell subtype and red blood cells. *Endocr. J.* **2006**, *53*, 133–139. [CrossRef]
41. Farhangi, M.A.; Keshavarz, S.A.; Eshraghian, M.; Ostadrahimi, A.; Saboor-Yaraghi, A.-A. White blood cell count in women: Relation to inflammatory biomarkers, haematological profiles, visceral adiposity, and other cardiovascular risk factors. *J. Health Popul. Nutr.* **2013**, *31*, 58–64. [CrossRef]
42. Lim, H.-J.; Seo, M.-S.; Shim, J.-Y.; Kim, K.-E.; Shin, Y.H.; Lee, Y.J. The association between platelet count and metabolic syndrome in children and adolescents. *Platelets* **2015**, *26*, 758–763. [CrossRef] [PubMed]
43. Charles, L.E.; Fekedulegn, D.; McCall, T.; Burchfiel, C.M.; Andrew, M.E.; Violanti, J.M. Obesity, white blood cell counts, and platelet counts among police officers. *Obesity* **2007**, *15*, 2846–2854. [CrossRef]
44. Nuttall, F.Q. Body mass index: Obesity, BMI, and health: A critical review. *Nutr. Today* **2015**, *50*, 117–128. [CrossRef]
45. Nead, K.G.; Halterman, J.S.; Kaczorowski, J.M.; Auinger, P.; Weitzman, M. Overweight children and adolescents: A risk group for iron deficiency. *Pediatrics* **2004**, *114*, 104–108. [CrossRef] [PubMed]
46. Pinhas-Hamiel, O.; Newfield, R.S.; Koren, I.; Agmon, A.; Lilos, P.; Phillip, M. Greater prevalence of iron deficiency in overweight and obese children and adolescents. *Int. J. Obes.* **2003**, *27*, 416–418. [CrossRef] [PubMed]

children

MDPI

Article

Fat Mass Index Associated with Blood Pressure Abnormalities in Children with Chronic Kidney Disease

Chien-Ning Hsu [1,2], Pei-Chen Lu [3] and You-Lin Tain [3,*]

1 Department of Pharmacy, Kaohsiung Chang Gung Memorial Hospital, Kaohsiung 833, Taiwan; cnhsu@cgmh.org.tw
2 School of Pharmacy, Kaohsiung Medical University, Kaohsiung 807, Taiwan
3 Department of Pediatrics, Kaohsiung Chang Gung Memorial Hospital, Chang Gung University College of Medicine, Kaohsiung 833, Taiwan; latina@cgmh.org.tw
* Correspondence: tainyl@cgmh.org.tw; Tel.: +886-975-056-995; Fax: +886-7733-8009

Abstract: Cardiovascular disease (CVD) risk factors are present early in life in children with chronic kidney disease (CKD), consequently cardiovascular morbidity presents in early adulthood. However, risk factors of CVD have been rarely addressed in children with early stage of CKD. This study included 63 children and adolescents aged 8- to 18 years-old with CKD stage G1–G4. Cardiovascular assessments consisted of 24-h ambulatory blood pressure monitoring (ABPM), arterial stiffness index, and echocardiography. We also applied dual-energy x-ray absorptiometry (DXA) scanning to analyze percentage body fat (PBF), lean body mass index (LBMI), fat mass index (FMI), and the android to gynoid fat ratio (A/G ratio). Up to 63.5% of CKD children had abnormal changes in BP detected by ABPM. CKD children with abnormal ABPM were older, had higher numbers of CKD stage G2 to G4, hyperuricemia, obesity, and higher FMI z-score and A/G ratio compared to individuals with normal ABPM (all $p < 0.05$). Among these factors, only FMI z-score showed an independent association with abnormal ABPM using multivariate logistic regression analysis ($p = 0.037$). Our data highlight that body fat plays a key role for an abnormal ABPM in CKD children. The assessment of FMI may have clinical utility in discriminating CV risk in children and adolescents with early stages of CKD.

Keywords: cardiovascular disease; ambulatory blood pressure monitoring; fat mass index; children; dual-energy X-ray absorptiometry; chronic kidney disease; hypertension

Citation: Hsu, C.-N.; Lu, P.-C.; Tain, Y.-L. Fat Mass Index Associated with Blood Pressure Abnormalities in Children with Chronic Kidney Disease. *Children* **2021**, *8*, 621. https://doi.org/10.3390/children8080621

Academic Editors: Odysseas Androutsos and Antonis Zampelas

Received: 20 May 2021
Accepted: 22 July 2021
Published: 22 July 2021

1. Introduction

Patients with chronic kidney disease (CKD) exhibit a marked risk for cardiovascular (CV) comorbidities and mortality [1], whereas major CV events are uncommon in children with CKD [2]. Hypertension is an early sign of CVD and is the most common complication of childhood CKD [2]. Using 24-h ambulatory blood pressure monitoring (ABPM), we and others have shown that hypertension is present in more than one-half of children with CKD, which occurs even in early stages [3,4]. However, subclinical CVD presenting in children is hardly detectable by conventional methods [5]. Left ventricle mass index (LVMI), ambulatory arterial stiffness index (AASI), and carotid intima-media thickness (cIMT) have been considered as surrogate markers for CVD in children with CKD [6–8].

Similar to adults, risk factors of CVD among children with CKD are also related to the traditional risks [2]. Childhood obesity has become a public issue as its prevalence increases at an alarming rate [9]. Obesity is associated with increased risk of developing CVD in people without CKD, but the effect of obesity in people with CKD is inconclusive [10–12]. Despite obesity in children being defined as a body mass index (BMI) at or above the 95th percentile for age and sex [13], BMI does not measure body fat directly. Currently, dual-energy x-ray absorptiometry (DXA) is increasingly used to measure body composition in terms of fat and fat-free mass [14]. Fat mass index measured by DXA has been shown to correlate with CV risk factors in CKD adults [14] and overweight/obese children [15].

Nevertheless, it remains unclear whether these measures are associated with CVD risk in pediatric patients with CKD.

The purpose of our study, therefore, was to assess body composition parameters with DXA and find out their associations with ABPM and other surrogate markers (i.e., LVMI, cIMT, AASI) in pediatric CKD.

2. Materials and Methods

2.1. *Patients and Study Design*

All study procedures were performed in accordance with the ethical standards laid down in the 1964 Declaration of Helsinki and its later amendments. The prospective cohort study was approved by the Institutional Review Boards (IRB) of the Chang Gung Medical Foundation at Taoyuan, Taiwan (201601181A3). A total of 101 children and adolescents aged 3 to 18 years with CKD stage G1–G4 attending the pediatric clinic at Kaohsiung Chang Gung Memorial Hospital was enrolled between December 2016 and October 2018. All of the participants parents gave their written informed consent. CKD was defined by kidney damage or reduced kidney function over at least three months [16]. Kidney damage included functional abnormalities or structural abnormalities identified by kidney biopsy, imaging studies, or urinary sediment [16]. Renal function was determined by estimated glomerular filtration rate (eGFR) using the Schwartz formula on the basis of body height and blood creatinine (Cr) level [17]. Participants were classified into five CKD stages according to their eGFR [16]. Patients were excluded, if they (1) were pregnant; (2) had history of congenital heart disease; (3) had eGFR < 15 mL/min/1.73 m^2, on dialysis maintenance, or ever received renal transplantation; or (4) were incapable of consenting to the follow-up protocol or cooperated with the procedures. All recruited patients were followed up to progression to end-stage kidney disease.

2.2. *Data and Specimens*

The following assessments were performed on each participant at the same clinic visit: (1) history taking and physical examination; (2) anthropometry; (3) office BP and ABPM measurements; and (4) laboratory investigations. The congenital anomalies of the kidney and urinary tract (CAKUT) includes a wide range of structural anomalies like renal agenesis, posterior urethral valves, horseshoe kidney, kidney hypo-/dysplasia, duplex collecting system, multicystic kidney dysplasia, and ureter abnormalities [18]. The etiologies of kidney diseases were divided into two types, namely CAKUT and non-CAKUT. Renal anemia was defined as receiving erythropoietin to correct anemia. Mineral bone disease (MBD) was defined as serum calcium (Ca)-phosphate (P) product $\geq$65 for age $\leq$12 years, serum Ca-P $\geq$55 for age >12 years, or elevated parathyroid hormone according to the CKD stage [18]. Hyperuricemia was defined as serum uric acid level $\geq$5.9 mg/dL for 6- to 8-year-olds, or serum uric acid $\geq$6.1 mg/dL for 9- to 11-year-olds, or serum uric acid $\geq$7.0 mg/dL for age $\geq$12 years in male, or serum uric acid $\geq$6.2 mg/d for age $\geq$12 years in female [19]. Hyperlipidemia was defined as serum levels of total cholesterol $\geq$200 mg/dL, or low-density lipoprotein (LDL) $\geq$130 mg/dL, or triglyceride (TG) $\geq$100 mg/dL for 0- to 9-year-olds, and TG $\geq$130 mg/dL for 10- to 19-year-olds [20].

2.3. *Anthropometric Measurements*

Body height was measured (in cm to the nearest 0.1 cm) with a rigid stadiometer. Body weight was measured (in kg to the nearest 0.1 kg) with an electronic body weight scale. The BMI was calculated as body weight in kilograms divided by the square of body height in meters. BMI percentile for age-and-gender and z-score were determined based on Health Promotion Administration, Ministry of Health and Welfare in Taiwan. Obesity was defined as BMI percentile higher than and equal to the 95th percentile. The fat mass index (FMI), lean body mass index (LBMI), percent body fat (PBF), and the android to gynoid fat ratio (A/G ratio) were measured by DXA (Lunar Prodigy Advance; GE Healthcare, Madison, WI, USA). FMI and LBMI were calculated as both fat mass and lean body mass in kilograms divided by the height in meters squared. FMI for age-and-gender z-scores and LBMI for age-and-gender were based on the reference values underlying the calculator courtesy of the American Journal of Clinical Nutrition [21].

2.4. Office BP and ABPM Measurements

Office BP was recorded using a validated and electronic sphygmomanometer after a subject sat for five minutes. High office BP was defined as systolic blood pressure or diastolic blood pressure >95th percentile. The cuff size had a bladder length by a 1:2 to 2:3 width-to-length ratios based on arm circumference. Data from 24-h ABPM were collected for subjects aged 6–18 years using an Oscar II monitoring device (SunTech Medical, Morrisville, NC, USA) as previously reported [4]. The participants and their parents were asked to record activities that may influence BP measurements and awake and sleep periods. The ABPM was programmed to measure at 30 min intervals from 10 pm to 7 am and at 20 min intervals from 7 am to 10 pm. Outlier readings that could be due to artifacts (e.g., systolic BP > 00 mm Hg) were excluded. The criteria of an abnormal ABPM consisted of (1) daytime, nighttime, systolic, or diastolic BPs ≥95th percentile according to gender and height [22]; (2) 25% or greater of SBP or DBP load; and (3) nocturnal BP load dipping <10% compared with average awake BP load. The AASI was calculated by one minus regression slope of diastolic over systolic blood pressure readings from ABPM [23].

2.5. Cardiovascular Assessments

Carotid ultrasound and echocardiography were done at the same visit. Carotid ultrasound assessment was performed by an experienced pediatric nephrologist (Pei-Chen Lu) using a ProSound α7 ultrasound device (Aloka Co., Tokyo, Japan). This system consisted of computer assisted analysis software (e-TRACKING system, Aloka Co.) [4]. Pediatric cardiologists performed echocardiographic examination using a Philips IE33 system device (Philips, Bothell, WA, USA). Left ventricular mass index (LVMI) was calculated by Devereux′s formula, as reported previously [24].

2.6. Statistical Analysis

All analyses were performed using SPSS version 22.0 (SPSS, Inc, Chicago, IL, USA). Continuous variables with non-normal distribution were presented as medians and interquartile ranges while categorical variables were reported as number (%). The Chi-square test or nonparametric test examined the differences in variables between children with an abnormal ABPM and with a normal ABPM. Multivariate logistic analysis was performed to investigate the associations between risk factors (including age, CKD stage, obesity, hyperuricemia, FMI z-score, and A/G ratio) and an abnormal ABPM. A probability level of $p < 0.05$ indicated statistical significance.

3. Results

A total of 101 subjects were enrolled in this study. Table 1 summarizes the clinical, anthropometric, and blood characteristics of all subjects. The median age of study subjects was 10 years and 59.4% of them were boys. The median eGFR at enrollment was 104 mL/1.73 m^2/min including 70 CKD stage G1 subjects (69.3%), 19 G2 subjects (18.8%), ten G3 subjects (9.9%), and two G4 subjects (2%). CAKUT accounted for 64.4% of the etiologies (n = 65). Since 38 subjects were too young to receive the DXA scan and CV assessments, 63 children were grouped from this CKD cohort and defined as the DXA group for subsequent analysis. Participants in this DXA group were older and had higher creatinine levels compared to the whole study group. However, other characteristics were comparable between these two groups.

Table 1. Clinical, biomedical, and anthropometric characteristics of study population.

Characteristics	DXA Group	Total
	$n = 63$	$n = 101$
Age, years	13.4 (10.4–15.7)	10 (5.9–14.9) *
Male	37 (58.7)	60 (59.4%)
CAKUT	36 (57.1%)	65 (64.4%)
CKD stage		
G1	40 (63.5%)	70 (69.3%)
G2	16 (25.4%)	19 (18.8%)
G3	5 (7.9)	10 (9.9%)
G4	2 (3.2)	2 (2%)
Body height, percentile	50 (15–75)	25 (15–50)
Body weight, percentile	64 (35–60)	50 (15–75)
Body mass index, percentile	64 (28–87)	56 (30–80)
Body mass index z-score	0.37 (−0.57–1.11)	0.15 (−0.54–0.84)
Systolic blood pressure, percentile	50 (50–95)	50 (50–93)
Diastolic blood pressure, percentile	50 (50–90)	50 (50–90)
Creatinine, mg/dL	0.6 (0.5–0.86)	0.51 (0.42–0.76) *
eGFR, mL/min/1.73 m^2	101.4 (80.2–117.1)	104 (83–124)
Urine total protein-to-creatinine ratio, mg/g	64.2 (37.3–410.3)	63.2 (38.4–207.5)
Hemoglobin, g/dL	13.7 (13–14.9)	13.4 (12.7–14.3)
Uric acid, mg/dL	5.7 (4.5–6.7)	5.3 (4.2–6.4)
Sodium, mEq/L	141 (140–142)	141 (140–142)
Potassium, mEq/L	4.3 (4.1–4.5)	4.4 (4.2–4.5)
Calcium, mg/dL	9.5 (9.1–9.9)	9.7 (9.3–9.9)
Phosphate, mg/dL	4.7 (4.3–5.1)	4.9 (4.5–5.3)
LDL-cholesterol, mg/dL	83 (67–102)	87 (70–105)
Total cholesterol, mg/dL	162 (141–185)	163 (146–191)
Triglyceride, mg/dL	67 (53–102)	67 (52–102)

Data given as medians (25th, 75th percentile) or *n* (%). CAKUT = Congenital anomalies of the kidneys and urinary tract. eGFR = Estimated glomerular filtration rate. * $p < 0.05$ by the Chi-square test or the Mann–Whitney *U*-test.

In the DXA group, 18 subjects (28.6%) were obese. Regarding the complications of CKD, our data showed anemia in one subject (1.6%) with CKD stage G4, mineral bone disease in three subjects (4.8%) with CKD stage G3 or G4, hyperlipidemia in 11 subjects (17.5%), hyperuricemia in 19 subjects (30.2%), and hypertension (by office BP) in 20 subjects (31.7%). These data indicate that almost all participants in this group showed the nature of early or mild-to-moderate CKD.

Unlike office BP, up to 63.5% of subjects (40/63) had an abnormal ABPM. ABPM identified ten patients (15.9%) with 24-h hypertension, 11 patients (17.5%) with daytime hypertension, 17 patients (27%) with nighttime hypertension, 32 patients (50.8%) with increased BP load, and 26 patients (41.3%) with non-dipping nocturnal BP. These subjects were classified by ABPM abnormalities in Table 2. CKD children with abnormal ABPM were older ($p = 0.006$), and showed a larger proportion of CKD stages G2–G4 ($p = 0.017$), obesity ($p = 0.039$), and hyperuricemia ($p = 0.005$).

Then, we determined body composition by DXA. As shown in Table 3, FMI z-score and A/G ratio were higher in children with abnormal ABPM (both $p = 0.001$). However, there was no difference in PBF and LBMI z-score between two groups. Additionally, body composition parameters did not differ statistically between subjects with CKD stage G2–G4 and subjects with G1 (all $p > 0.05$, data not shown), indicating that body composition was not affected by CKD stage in this study. Moreover, we observed that AASI, LVMI, and cIMT in CKD children with abnormal ABPM were comparable with those with normal ABPM.

Table 2. Comparison of the clinical characteristics and CKD complications in children with CKD stratified for abnormal and normal ABPM profile.

Characteristics	Abnormal ABPM	Normal ABPM	*p* Value
	n = 40	*n* = 23	
Age, years	14.4 (11.4–16.4)	11.7 (9.6–14.5)	0.006
Male	27 (67.5%)	10 (43.5%)	0.062
CKD G2–G4	19 (47.5%)	4 (17.4%)	0.017
CAKUT	24 (60%)	12 (52.2%)	0.546
Obesity	15 (37.5%)	3 (13%)	0.039
Mineral bone disease	2 (5%)	1 (4.3%)	0.907
Hyperlipidemia	8 (20%)	3 (13%)	0.484
Proteinuria	13 (32.5%)	6 (26.1%)	0.055
Hyperuricemia	17 (42.5%)	2 (8.7%)	0.005

Data are medians (25th, 75th percentile) or *n* (%). ABPM = 24-h ambulatory blood pressure monitoring. CAKUT = Congenital anomalies of the kidneys and urinary tract.

Table 3. Comparison of body composition and CV assessments in children with CKD stratified for abnormal and normal ABPM profile.

	Abnormal ABPM	Normal ABPM	*p* Value
	n = 40	*n* = 23	
Body composition			
PBF, %	32.4 (23.7–37.8)	27.5 (23.7–31.7)	0.077
FMI z-score	0.45 (−0.32–1.22)	−0.41 (−0.78–−0.09)	0.001
LBMI z-score	−0.3 (−1.02–0.6)	−0.59 (−1.35–−0.28)	0.084
A/G ratio	0.9 (0.81–1.04)	0.76 (0.72–0.84)	0.001
Cardiovascular assessment			
AASI	0.397 (0.246–0.492)	0.329 (0.243–0.412)	0.084
LVMI, g/m^2	32.8 (28.6–41.2)	29.8 (26.1–38)	0.253
cIMT, mm	0.4 (0.3–0.4)	0.3 (0.3–0.4)	0.139

Data are medians (25th, 75th percentile) or *n* (%). PBF = Percent body fat. FMI = Fat mass index. LBMI = Lean body mass index. A/G ratio = Android to gynoid fat ratio. AASI = Ambulatory arterial stiffness index. LVMI = Left ventricle mass index. cIMT = Carotid intima-media thickness.

Using multivariate linear regression analyses, we specified the specific role of risk factors of abnormal ABPM (Table 4). A multivariate linear regression model using the stepwise selection was applied for age, CKD stage, obesity, hyperuricemia, FMI z-score, and A/G ratio. Table 4 shows that only the FMI z-score was independently associated with increased risk of abnormal ABPM (adjusted odds ratio [aOR] 3.00, 95% confidence interval [CI] [1.07–8.46], $p = 0.037$).

Table 4. Adjusted regression model estimates of the association between risk factors and abnormal ABPM.

Factor	*p* Value	aOR	95% CI	
			Lower	Upper
Age	0.313	1.16	0.87	1.53
CKD stage (G2–4 vs. G1)	0.099	3.69	0.78	17.39
Obesity	0.523	0.5	0.06	4.2
Hyperuricemia	0.145	4.24	0.61	29.49
FMI z-score	0.037	3.0	1.07	8.46
A/G ratio	0.296	38.8	0.04	37,000.2

aOR = Adjusted odds ratio. 95% CI = 95% confidence interval. FMI = Fat mass index. A/G ratio = Android to gynoid fat ratio.

4. Discussion

In the current study, we provide novel insights into factors involved in subclinical CVD in children with CKD. Our data indicate that body fat plays a key role for an abnormal

ABPM in these children. FMI looks like a more trustworthy index than BMI to assess obesity in relation to abnormal ABPM in CKD children as it considers the body composition.

In line with previous reports [3,4,8], our study found that abnormalities on ABPM were highly common in children with CKD. More than 60% of CKD children with early stages had abnormal ABPM profile, while most of them were not found to have hypertension in the out-patient clinic. Our data support the notion that ABPM is superior to office BP readings in detecting hypertension and in predicting cardiovascular events in pediatric CKD [25].

Our study revealed children with an abnormal ABPM were associated with several factors including age, CKD stage, hyperuricemia, obesity, FMI z-score, and A/G ratio. Age-related increases in BP have been observed in adults but not in children [26]. Our regression analysis indicated that age is a confounding factor as two groups being compared have different age distributions. In support of our findings showing that children with CKD stage G2–G4 are prone to exhibit BP abnormalities, existing evidence indicates that the prevalence of hypertension increases as the CKD stage advances in adults as well as in children [2–4,27]. Uric acid is not just regarded as a risk factor for CVD, but a valuable therapeutic target also in CKD [28]. Our findings might extend the impact of uric acid on adults to children, even with early stages of CKD. Accordingly, there will be a growing need to better understand whether early targeting on uric acid can prevent the development of CVD in children with CKD, even at an early stage.

Although obesity is a risk factor for the development of CKD, it is paradoxically linked with greater survival in patients with advanced CKD [29]. This obesity paradox may be related to the discrepancy between the uses of BMI and other measures in the definition of obesity. BMI is the most commonly used tool to determine obesity in clinical settings. However, BMI cannot directly measure body fat, by which the discrepancy observed between BMI and other adiposity indices reflects the main limitation of BMI [30]. Accordingly, FMI appears to provide higher accuracy than BMI to assess obesity [30]. Our data demonstrated that the association between high FMI z-score and abnormal ABPM remained significant after adjusting for covariates, while obesity did not. This finding suggests that BMI-defined obesity cannot explain an abnormal ABPM sufficiently, in which body fat is of vital importance. Additionally, our study confirms previous research showing that fat measurements are superior to BMI for predicting CV risk, not just for adults but also for children [31–33]. Moreover, examining the relative amount of regional body fat in proportion to other fat regions like A/G ratio is another valid method to test for associations with CV risk [34]. Our study found an association between high A/G ratio and abnormal ABPM in children with CKD, whereas this association did not reach significance after adjusting for other factors. To our knowledge, our results provide the first evidence to show that FMI is associated with CV risk represented by ABPM abnormalities in children with CKD in early stages.

No comparable associations were identified for cardiovascular assessments in our study, despite previous studies reporting that they may increase CV risks in children with CKD [3,6,8]. Given that currently no reference values exist with regard to AASI and cIMT to determine a cut-off value between children with and without CKD, further studies are needed to reveal the relationships using larger populations.

Our study was not without its limitations. First, we acknowledge that the study results may not be generalizable for the entire CKD pediatric population because sampling was performed in one hospital. A multi-center population-based cohort recruiting a large number of patients may be warranted to elucidate the true relationship in the future. Second, considering the long-term nature of childhood CKD, more assessments of body composition with longer follow-ups are required. Third, most participants in our cohort had early stages of CKD. There is a need for further validation of our results in other pediatric cohorts with more advanced CKD. Additionally, dietary intake of salt was not recorded, which comprises a limitation of the present study. Finally, an ethnic difference may be considered because we applied reference values for ABPM, FMI, and LBMI from studies in Europe [21,22].

5. Conclusions

Our data provide evidence that the assessment of fat distribution may be useful in determining CV risk in children with CKD in early stages. Considering the exploding obesity epidemic in children, further research is warranted to replicate our findings in other cohorts, monitor body composition parameters with longer-term CV outcomes, and develop effective control at reducing body fat in children with CKD.

Author Contributions: C.-N.H.: contributed to generation, data interpretation, drafting of the manuscript, critical revision of the manuscript, and approval of the article; P.-C.L.: contributed to concept generation, methodology drafting of the manuscript, critical revision of the manuscript, and approval of the article; Y.-L.T.: contributed to concept generation, data interpretation, methodology, drafting of the manuscript, critical revision of the manuscript, and approval of the article. All authors have read and agreed to the published version of the manuscript.

Funding: This work was supported by Grants CMRPG8K0721, CMRPG8K0722, CORPG8F1031, CORPG8F1032, and CORPG8F1033 from Chang Gung Memorial Hospital, Kaohsiung, Taiwan.

Institutional Review Board Statement: The study was conducted according to the guidelines of the Declaration of Helsinki, and approved by the Institution Review Board and Ethics Committee of Chang Gung Medical Foundation, Taoyuan, Taiwan (Permit numbers: 201601181A3).

Informed Consent Statement: Informed consent was obtained from all subjects involved in the study.

Data Availability Statement: Data will be available upon request.

Conflicts of Interest: The authors declare no conflict of interest. The funders had no role in the design of the study; in the collection, analyses, or interpretation of data; in the writing of the manuscript, or in the decision to publish the results.

References

1. Ruiz-Hurtado, G.; Sarafidis, P.; Fernández-Alfonso, M.S.; Waeber, B.; Ruilope, L.M. Global cardiovascular protection in chronic kidney disease. *Nat. Rev. Cardiol.* **2016**, *13*, 603–608. [CrossRef] [PubMed]
2. Weaver, D.J.; Mitsnefes, M. Cardiovascular Disease in Children and Adolescents with Chronic Kidney Disease. *Semin. Nephrol.* **2018**, *38*, 559–569. [CrossRef] [PubMed]
3. Mitsnefes, M.; Flynn, J.; Cohn, S.; Samuels, J.; Blydt-Hansen, T.; Saland, J.; Kimball, T.; Furth, S.; Warady, B.; CKiD Study Group. Masked hypertension associates with left ventricular hypertrophy in children with CKD. *J. Am. Soc. Nephrol.* **2010**, *21*, 137–144. [CrossRef]
4. Hsu, C.N.; Lu, P.C.; Lo, M.H.; Lin, I.C.; Tain, Y.L. The association between nitric oxide pathway, blood pressure abnormalities, and cardiovascular risk profile in pediatric chronic kidney disease. *Int. J. Mol. Sci.* **2019**, *20*, 5301. [CrossRef] [PubMed]
5. Urbina, E.M.; Williams, R.V.; Alpert, B.S.; Collins, R.T.; Daniels, S.R.; Hayman, L.; Jacobson, M.; Mahoney, L.; Mietus-Snyder, M.; Rocchini, A.; et al. Noninvasive assessment of subclinical atherosclerosis in children and adolescents: Recommendations for standard assessment for clinical research: A scientific statement from the American Heart Association. *Hypertension* **2009**, *54*, 919–950. [CrossRef] [PubMed]
6. Brady, T.M.; Schneider, M.F.; Flynn, J.T.; Cox, C.; Samuels, J.; Saland, J.; White, C.T.; Furth, S.; Warady, B.A.; Mitsnefes, M. Carotid intima-media thickness in children with CKD: Results from the CKiD study. *Clin. J. Am. Soc. Nephrol.* **2012**, *7*, 1930–1937. [CrossRef]
7. Taal, M.W. Arterial stiffness in chronic kidney disease: An update. *Curr. Opin. Nephrol. Hypertens.* **2014**, *23*, 169–173. [CrossRef] [PubMed]
8. Hsu, C.N.; Chang-Chien, G.P.; Lin, S.; Hou, C.Y.; Lu, P.C.; Tain, Y.L. Association of Trimethylamine, Trimethylamine N-oxide, and Dimethylamine with Cardiovascular Risk in Children with Chronic Kidney Disease. *J. Clin. Med.* **2020**, *9*, 336. [CrossRef]
9. Güngör, N.K. Overweight and Obesity in Children and Adolescents. *J. Clin. Res. Pediatr. Endocrinol.* **2014**, *6*, 129–143. [CrossRef]
10. Ortega, F.B.; Lavie, C.J.; Blair, S.N. Obesity and Cardiovascular Disease. *Circ. Res.* **2016**, *118*, 1752–1770. [CrossRef]
11. Ladhani, M.; Craig, J.C.; Irving, M.; Clayton, P.A.; Wong, G. Obesity and the risk of cardiovascular and all-cause mortality in chronic kidney disease: A systematic review and meta-analysis. *Nephrol. Dial. Transplant.* **2017**, *32*, 439–449. [CrossRef] [PubMed]
12. Madero, M.; Sarnak, M.J.; Wang, X.; Sceppa, C.C.; Greene, T.; Beck, G.J.; Kusek, J.W.; Collins, A.J.; Levey, A.S.; Menon, V. Body mass index and mortality in CKD. *Am. J. Kidney Dis.* **2007**, *50*, 404–411. [CrossRef]

13. Barlow, S.E. Expert committee recommendations regarding the prevention, assessment, and treatment of child and adolescent overweight and obesity: Summary report. *Pediatrics* **2007**, *120*, 164–192. [CrossRef] [PubMed]
14. Zoccali, C.; Torino, C.; Tripepi, G.; Mallamaci, F. Assessment of obesity in chronic kidney disease: What is the best measure? *Curr. Opin. Nephrol. Hypertens.* **2012**, *21*, 641–646. [CrossRef] [PubMed]
15. Samsell, L.; Regier, M.; Walton, C.; Cottrell, L. Importance of android/gynoid fat ratio in predicting metabolic and cardiovascular disease risk in normal weight as well as overweight and obese children. *J. Obes.* **2014**, *2014*, 846578. [CrossRef]
16. Kidney Disease: Improving Global Outcomes (KDIGO) CKD Work Group. KDIGO 2012 clinical practice guideline for the evaluation and management of chronic kidney disease. *Kidney Int. Suppl.* **2013**, *3*, 1–150.
17. Schwartz, G.J.; Muñoz, A.; Schneider, M.F.; Mak, R.H.; Kaskel, F.; Warady, B.A.; Furth, S.L. New equations to estimate GFR in children with CKD. *J. Am. Soc. Nephrol.* **2009**, *20*, 629–637. [CrossRef]
18. National Kidney Foundation. K/DOQI clinical practice guidelines for bone metabolism and disease in chronic kidney disease. *Am. J. Kidney Dis.* **2003**, *42*, S1–S201. [CrossRef]
19. Tang, L.; Kubota, M.; Nagai, A.; Mamemoto, K.; Tokuda, M. Hyperuricemia in obese children and adolescents: The relationship with metabolic syndrome. *Pediatr. Rep.* **2010**, *2*, e12.
20. Yoon, J.M. Dyslipidemia in children and adolescents: When and how to diagnose and treat? *Pediatr. Gastroenterol. Hepatol. Nutr.* **2014**, *17*, 85–92. [CrossRef]
21. Pediatric Z-Score Calculator. Available online: https://zscore.research.chop.edu (accessed on 12 January 2021).
22. Soergel, M.; Kirschstein, M.; Busch, C.; Danne, T.; Gellermann, J.; Holl, R.; Krull, F.; Reichert, H.; Reusz, G.S.; Rascher, W. Oscillometric twenty-four-hour ambulatory blood pressure values in healthy children and adolescents: A multicenter trial including 1141 subjects. *J. Pediatr.* **1997**, *130*, 178–184. [CrossRef]
23. Kollias, A.; Stergiou, G.S.; Dolan, E.; O'Brien, E. Ambulatory arterial stiffness index: A systematic review and meta-analysis. *Atherosclerosis* **2012**, *224*, 291–301. [CrossRef]
24. Lo, M.H.; Lin, I.C.; Lu, P.C.; Huang, C.F.; Chien, S.J.; Hsieh, K.S.; Tain, Y.L. Evaluation of endothelial dysfunction, endothelial plasma markers, and traditional metabolic parameters in children with adiposity. *J. Formos. Med Assoc.* **2019**, *118*, 83–91. [CrossRef] [PubMed]
25. Graves, J.W.; Althaf, M.M. Utility of ambulatory blood pressure monitoring in children and adolescents. *Pediatr. Nephrol.* **2006**, *21*, 1640–1652. [CrossRef]
26. Fathallah-Shaykh, S.A.; Cramer, M.T. Uric acid and the kidney. *Pediatr. Nephrol.* **2004**, *29*, 999–1008. [CrossRef]
27. Ku, E.; Lee, B.J.; Wei, J.; Weir, M.R. Hypertension in CKD: Core Curriculum 2019. *Am. J. Kidney Dis.* **2019**, *74*, 120–131. [CrossRef]
28. Gurven, M.; Blackwell, A.D.; Rodríguez, D.E.; Stieglitz, J.; Kaplan, H. Does blood pressure inevitably rise with age? Longitudinal evidence among forager-horticulturalists. *Hypertension* **2012**, *60*, 25–33. [CrossRef]
29. Rhee, C.M.; Ahmadi, S.F.; Kalantar-Zadeh, K. The dual roles of obesity in chronic kidney disease: A review of the current literature. *Curr. Opin. Nephrol. Hypertens.* **2016**, *25*, 208–216. [CrossRef]
30. Peltz, G.; Aguirre, M.T.; Sanderson, M.; Fadden, M.K. The role of fat mass index in determining obesity. *Am. J. Hum. Biol.* **2010**, *22*, 639–647. [CrossRef]
31. Lichtash, C.T.; Cui, J.; Guo, X.; Chen, Y.D.; Hsueh, W.A.; Rotter, J.I.; Goodarzi, M.O. Body adiposity index versus body mass index and other anthropometric traits as correlates of cardiometabolic risk factors. *PLoS ONE* **2013**, *8*, e65954. [CrossRef] [PubMed]
32. Zeng, Q.; Dong, S.Y.; Sun, X.N.; Xie, J.; Cui, Y. Percent body fat is a better predictor of cardiovascular risk factors than body mass index. *Braz. J. Med. Biol Res.* **2012**, *45*, 591–600. [CrossRef] [PubMed]
33. Daniels, S.R.; Morrison, J.A.; Sprecher, D.L.; Khoury, P.; Kimball, T.R. Association of body fat distribution and cardiovascular risk factors in children and adolescents. *Circulation* **1999**, *99*, 541–545. [CrossRef] [PubMed]
34. Staiano, A.E.; Gupta, A.K.; Katzmarzyk, P.T. Cardiometabolic risk factors and fat distribution in children and adolescents. *J. Pediatr.* **2014**, *164*, 560–565. [CrossRef] [PubMed]

children

MDPI

Article

Using Body Composition Groups to Identify Children and Adolescents at Risk of Dyslipidemia

Alina Ofenheimer [1,2,*], Robab Breyer-Kohansal [1,3], Sylvia Hartl [1,3,4], Otto C. Burghuber [1,4], Florian Krach [5], Frits M. E. Franssen [2,6], Emiel F. M. Wouters [1,2] and Marie-Kathrin Breyer [1,3]

1 Ludwig Boltzmann Institute for Lung Health, 1140 Vienna, Austria; robab.breyer-kohansal@leadstudy.at (R.B.-K.); sylvia.hartl@leadstudy.at (S.H.); otto.burghuber@leadstudy.at (O.C.B.); emiel.wouters@leadstudy.at (E.F.M.W.); marie-kathrin.breyer@leadstudy.at (M.-K.B.)
2 NUTRIM, Maastricht University Medical Center, 6200 MD Maastricht, The Netherlands; fritsfranssen@ciro-horn.nl
3 Department of Respiratory and Critical Care Medicine, Clinic Penzing, 1140 Vienna, Austria
4 Medical School, Sigmund Freud University, 1020 Vienna, Austria
5 Department of Mathematics, ETH Zürich, 8092 Zurich, Switzerland; florian.krach@math.ethz.ch
6 Department of Research and Education, CIRO, 6085 NM Horn, The Netherlands
* Correspondence: alina.ofenheimer@leadstudy.at

Abstract: The impact of body composition on the early origin of chronic diseases is an increasingly appreciated phenomenon. Little is known about the characteristics of children with varying body composition. The aim of this study was to investigate serum lipid profiles and other characteristics in relation to body composition. The data of 1394 participants (aged 6 to <18 years) of the observational general population-based Austrian LEAD Study have been analyzed. Body composition groups were defined by appendicular lean mass (ALMI) and fat mass (FMI) indices assessed by DXA. Serum lipid profiles (triglycerides, LDL-c, HDL-c) and other characteristics (e.g., prematurity, smoke exposure, physical activity, nutrition) were investigated in these body composition groups. Different body composition groups, which are not distinguishable by BMI, exist. Children with high ALMI and high FMI showed higher triglycerides and LDL-c, but lower HDL-c levels. In contrast, levels did not differ between those with high FMI but low (or normal) ALMI, and other body composition groups. BMI should be interpreted cautiously, and body composition should be measured by more precise techniques. In particular, children and adolescents with high FMI who have concomitantly high ALMI should be followed closely in future studies to investigate whether they are at increased risk of cardiovascular problems.

Keywords: dyslipidemia; cardiovascular risk; physical activity; body composition; muscle mass; body compartments

Citation: Ofenheimer, A.; Breyer-Kohansal, R.; Hartl, S.; Burghuber, O.C.; Krach, F.; Franssen, F.M.E.; Wouters, E.F.M.; Breyer, M.-K. Using Body Composition Groups to Identify Children and Adolescents at Risk of Dyslipidemia. *Children* **2021**, *8*, 1047. https://doi.org/10.3390/children8111047

Academic Editors: Odysseas Androutsos and Antonis Zampelas

Received: 30 September 2021
Accepted: 11 November 2021
Published: 13 November 2021

Publisher's Note: MDPI stays neutral with regard to jurisdictional claims in published maps and institutional affiliations.

1. Introduction

The increasing prevalence of increased weight and obesity in children and adults worldwide imposes a tremendous burden on global health and economies [1]. The large increase in their prevalence in children is of great concern, since individuals being overweight or obese at a young age carry a higher risk for the early development of chronic diseases, including cardiovascular, metabolic, and respiratory diseases [1–4]. Commonly, body mass index (BMI, kg/m^2) is used to classify individuals into weight categories to detect these conditions of overweight and obesity. However, body composition can vary widely within a group of subjects sharing the same BMI [5]. In 1964, Forbes described two groups of obesity in childhood: one group with increased lean mass (LM) additionally to high fat mass (FM); and a second group without an increase in lean mass [6]. Recent evidence suggests that the body composition group with high FM and low muscle mass is associated with greater health risks than either compartment alone [7,8]. Therefore, the amount of appendicular lean mass (ALM), a marker for skeletal muscle mass, might be important to consider when screening for early diseases.

The lack of differentiation between various components of body weight might be the reason for the conflicting results of studies investigating the impact of BMI alone on health conditions [9]. There is increasing awareness that prenatal and early-life represent important periods for development, and several traits contributing to the development of body composition have been proposed [10–12]. The origins of atherosclerosis, diabetes mellitus type 2, and chronic respiratory diseases are also believed to be partly situated in utero or early infancy [13–16]. It has been shown that incidence of dyslipidemia at young age, especially in overweight and obese adolescents, is associated with increased carotid intima–media thickness in adulthood, an early marker of atherosclerosis [17,18]. Despite the increasing awareness of its importance, little is known about the various body constitutions in relation to risk profiles in children and adolescents.

The aim of this study was to define body composition groups by appendicular lean mass and fat mass amounts assessed by dual-energy X-ray absorptiometry (DXA) in different BMI categories and to investigate their serum lipid profiles and early life risk factors (preterm birth, no breastfeeding), smoke exposure (second-hand, maternal), and lifestyle characteristics (physical activity, nutrition, socio-economic status).

2. Materials and Methods

2.1. Study Design and Population

The data were sampled by the Austrian LEAD (Lung, hEart, sociAl, boDy) Study (ClinicalTrials.gov; NCT01727518, https://clinicaltrials.gov/, accessed on 12 November 2021), a longitudinal, observational, population-based cohort study, representative of the general Austrian population. Inhabitants were randomly recruited based on the inhabitants' register. More details about the study design, methodology, and the representativeness of the LEAD study cohort can be found elsewhere [19]. For the present analysis, cross-sectional data of participants aged 6 to <18 years, examined between 2011 and 2019, were included. Further inclusion criteria were valid DXA scans (verified by a trained study nurse) and availability of blood samples. The LEAD study was approved by the local ethics committee in Vienna (protocol number: EK-11-117-0711), and the study participants signed for their informed consent. For all minors, it was signed by their parents or legal representatives.

2.2. Measurements

All study participants were examined in the study center after a fasting period of 8 h. A team of trained study nurses ensured the quality of data assessment. Body height was measured by a stadiometer and body weight by a high precision scale (exacta CLASSIC by SOEHNLE™). BMI was calculated as weight (kg) divided by height^2 (m^2). Waist circumference was measured in the standardized fashion. Handgrip strength was measured on the dominant hand with a hand dynamometer (Trailite, TL-LSC100™, LiteXpress GmbH, Ahaus, Deutschland); three successive measurements were performed and the highest value was kept. Whole-body scans were obtained by a Lunar Prodigy™ (GE Lunar Corp.; Madison, WI, USA) DXA and analyzed with enCORE™ (version 17, 2016). The coefficient of variation has been reported previously and found to be similar to other studies [20]. Body compartments should be normalized for total body height, hence it was essential to use the right index considering that height squared may not work best in children [21–23]. We recently confirmed that the exponents 2.5 and 3.5 account best for body height in FM and ALM, respectively [24]. Therefore, ALMI was calculated as the sum of the lean mass of all four limbs (kg) divided by height$^{3.5}$ ($m^{3.5}$), and FMI as fat mass (kg) divided by height$^{2.5}$ ($m^{2.5}$).

Venous blood samples were collected, and total cholesterol, high-density lipoprotein cholesterol (HDL-c), and triglycerides were measured by photometric enzymatic method (Siemens Dimension Vista 150™, Diamond Diagnostics Inc, Holliston MA USA). Low-density lipoprotein cholesterol (LDL-c) was calculated with the Friedewald formula.

2.3. *Definition of Variables*

BMI categories. Reference values of the WHO were applied to calculate BMI z-scores. Study participants were classified into the following BMI categories according to their BMI z-scores: extreme thinness was defined as z-scores <-2, thinness as z-scores ≥-2 but <-1, overweight as z-scores >1 but ≤ 2, obesity as z-scores >2 and normal as z-scores ≥-1 but ≤ 1 [25].

2.4. *Body Composition Groups*

The cut-offs for the FMI and ALMI group categories low, normal, and high were set to the 25th and 75th age- and sex-specific percentiles. Low FMI and ALMI were defined as ≤25th percentile, normal as >25th but <75th percentile, and high as ≥75th percentile. The following combined body composition groups were assigned: high ALMI–FMI, low ALMI–FMI, low ALMI–high FMI, and high ALMI–low FMI. The normal FMI–ALMI group was defined as FMI >25th but <75th percentile and/or ALMI >25th but <75th percentile.

2.5. *Blood Samples*

Altered levels of total values were defined by using the cut-offs recommended by the American Academy of Pediatrics (AAP) guidelines: ≥100 mg/dl for triglycerides in children aged 6 to ≤9 years; ≥130 mg/dl for triglycerides in children >9 years; ≥130 mg/dl for LDL-c. A decreased HDL-c was defined as <40 mg/dl [26]. For each of the analyzed blood markers, an LMS model was further applied to construct smoothed percentiles accounting for age- and sex-specific differences. For analysis based on z-scores, values ≥95th percentile were defined as elevated, and for HDL-c values ≤5th percentile as decreased.

2.6. *Questionnaires*

All participants performed validated interviewer-based questionnaires, providing the following information. Socio-economic status (SES) was defined as a score based on education level, household income, and the occupational status of the parents or legal representatives. Three different categories of SES were defined: low as the 1st quintile (20th percentile); normal as quintiles 2 to 4; and high as the 5th quintile (80th percentile). Preterm birth was defined as birthweight <2500 g or gestational age <260 days. Second-hand smoking was defined by participants reporting to be or have been exposed to cigarette smoke during most days or nights. For maternal smoking, three categories were analyzed (ever smoking during, prior, or after pregnancy). Physical activity was calculated by the individuals' reported number of minutes spent performing moderate- to high-intensity physical activity each day. Healthy nutrition was defined as drinking sugar-sweetened beverages less than daily and eating ≥2 portions of fruits and/or vegetables per day.

2.7. *Statistics*

The sample was split into ALMI and FMI groups, and combined body composition groups, similar to the approach by Prado et al., were constructed [27]. The prevalence of each group was calculated for the total sample, as well as for males and females. The whole study cohort was divided into BMI categories, and the prevalence of the body composition groups was analyzed in each BMI category. Different characteristics were compared between all body composition groups. First, all parameters were evaluated if they corresponded to a normal distribution via the Kolmogorov–Smirnov test. For continuous, non-normally distributed variables, medians and quantiles 25 (q25) and 75 (q75) were calculated. For all continuous, normally distributed variables, means and standard deviations were calculated. To test for differences between the groups, Wilcoxon-rank sum test was performed. For all dichotomous variables, the prevalence was calculated, and group differences were tested by Fisher's exact test. Multiple testing correction (Bonferroni–Holm) was applied to all p-values, and the significance level was set to 5%.

3. Results

Of the 1573 children and adolescents with valid DXA measurements, 1394 study participants had blood samples available. The sample size and distribution of ALMI and FMI groups are shown in Table 1.

Table 1. Sample size and prevalence of body composition groups and BMI categories.

	Males *n* (%)	Females *n* (%)	Overall *n* (%)
FMI (kg/m$^{2.5}$)			
normal	348 (46.8%)	307 (54.1%)	655 (47.0%)
low	206 (27.7%)	177 (27.2%)	383 (27.5%)
high	190 (25.5%)	166 (25.5%)	356 (25.5%)
ALMI (kg/m$^{3.5}$)			
normal	361 (48.5%)	333 (51.2%)	694 (49.8%)
low	185 (24.9%)	159 (24.5%)	344 (24.7%)
high	198 (26.6%)	158 (24.3%)	356 (25.5%)
ALMI-FMI groups			
normal ALMI-FMI	535 (71.9%)	459 (70.6%)	994 (71.3%)
low ALMI-FMI	72 (9.7%)	74 (11.4%)	146 (10.5%)
high ALMI-FMI	79 (10.6%)	84 (12.9%)	163 (11.7%)
low ALMI-high FMI	21 (2.8%)	14 (2.2%)	35 (2.5%)
high ALMI-low FMI	37 (5.0%)	19 (2.9%)	56 (4.0%)
BMI category			
extreme thinness	16 (2.2%)	9 (1.4%)	25 (1.8%)
thinness	83 (11.2%)	93 (14.3%)	176 (12.6%)
normal	443 (59.5%)	397 (61.1%)	840 (60.3%)
overweight	131 (17.6%)	111 (17.1%)	242 (17.4%)
obesity	71 (9.5%)	40 (6.2%)	111 (8.0%)
total	744	650	1394

Table shows sample size (*n*) and prevalence (%) of ALMI, FMI, combined body composition groups, and BMI categories. FMI = fat mass index, ALMI = appendicular lean mass index, BMI = Body mass index.

Among the body composition groups, the prevalence of the normal ALMI–FMI group was highest (71.3% overall), followed by the high ALMI–FMI (11.7% overall) and the low ALMI–FMI groups (10.5% overall). The low ALMI–high FMI group was prevalent in only 2.5%, and the high ALMI–low FMI group showed a prevalence of 4.0%. The prevalence was similar across all age groups (Table S1). Body composition data for the different BMI categories are illustrated in Figure 1. In the extreme thinness and thinness groups, the majority had low FMI and low ALMI values. Changes in the ALMI and FMI were masked in the normal BMI category. The prevalence of high ALMI and high FMI increased in the overweight and obese categories (Figure 1). In contrast to the sharp increase of high FMI from the normal to the overweight category (from 7.6% to 75.2%), the increase in ALMI was more gradual (from 18.2% to 48.8%). Figure 2 illustrates the prevalence of the body composition groups in the different BMI categories. Almost all extremely thin participants had low ALMI–FMI, while only half of the thinness group showed this body composition combination. In children and adolescents classified as normal according to their BMI, all different body composition groups were present. In participants classified as overweight, 30.2% had high ALMI–FMI values, which increased to 75.7% in the subjects with obesity.

Isolated high FMI was present in a minority of these children and adolescents (3.3% and 1.8%, respectively).

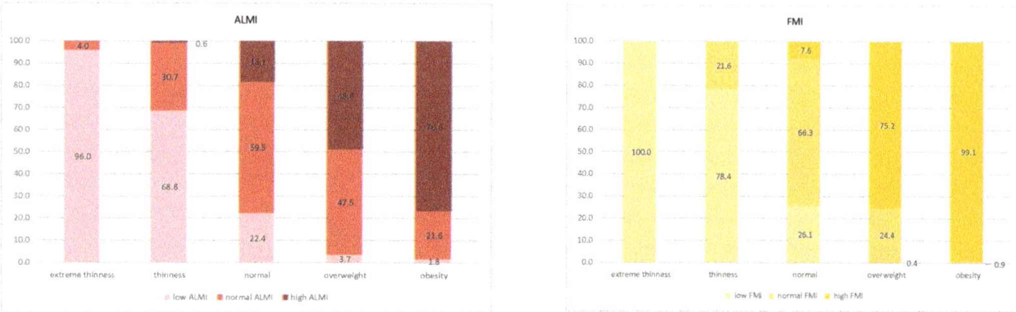

(a) (b)

Figure 1. Prevalence of body compartments in BMI categories. Figure shows the prevalence of the different groups (low, normal, high) for FMI (**a**) and ALMI (**b**), respectively, in different BMI categories. BMI categories (extreme thinness, thinness, normal, overweight, obesity) were defined according to WHO. FMI and ALMI groups were defined by age- and sex-specific groups (low: ≤25th percentile, normal: >25th but <75th percentile, high: ≥75th percentile).

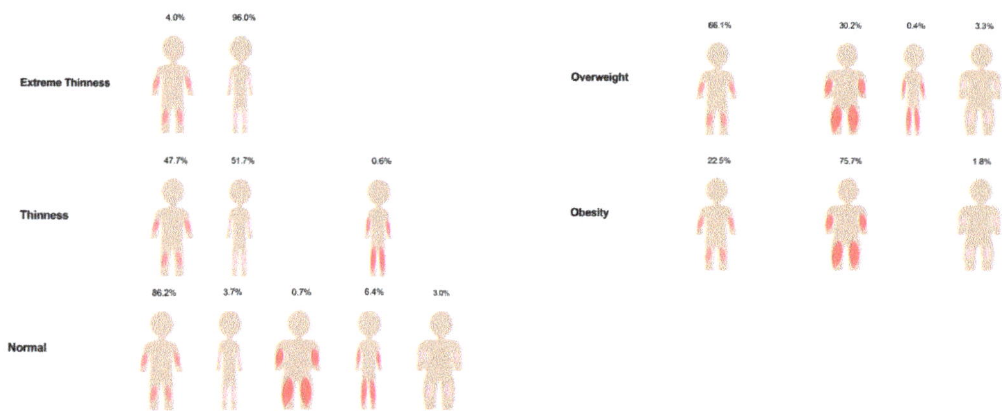

Figure 2. Prevalence of body composition groups in BMI categories. Figure shows the prevalence of the different body composition groups in BMI categories. The different body composition groups are represented by different figures. Varying amount of FMI are indicated by body shape (thin, middle, broad), ALMI amount by color shade (light, middle, dark) of displayed skeletal muscles mass in arms and legs region.

3.1. Serum Lipid Profiles

HDL-c was higher in the low ALMI and the low FMI groups, while levels were lower in the high FMI group (Table 2). In contrast, LDL-c and triglycerides were higher in children and adolescents with high FMI. The high ALMI–FMI body composition group showed marked differences in serum lipid levels. LDL-c and triglyceride levels were significantly higher in the high ALMI–FMI group. Contrary, HDL-c levels were significantly lower in this group. These differences persisted considering z-scores (which account for age and sex) of the serum lipids. Prevalence of decreased HDL-c and elevated triglycerides was higher in the high ALMI–FMI group compared to the normal ALMI–FMI group (Figure 3). Prevalence of altered serum lipids in all body composition groups are provided in the supplemental material (Table S2).

Table 2. Characteristics of serum lipid profiles in body composition groups.

	HDL-c [mg/dL]	HDL-c z-Scores	LDL-c [mg/dL]	LDL-c z-Scores	Triglycerides [mg/dL]	Triglycerides z-Scores
FMI [kg/m$^{2.5}$]						
normal	61.0 (53.0, 70.0)	0.0 (−0.6, 0.6)	86.6 (70.4, 102.5)	0.0 (−0.7, 0.6)	63.0 (48.0, 86.0)	0.0 (−0.7, 0.6)
low	65.0 (54.0, 75.0) ■	0.3 (−0.3, 1.0) ■	82.4 (68.3, 97.3)	−0.2 (−0.8, 0.4)	57.0 (47.0, 77.0)	−0.3 (−0.8, 0.4)
high	54.5 (46.0, 64.0) ■ †	−0.4 (−1.2, 0.2) ■ †	92.8 (77.3, 111.3) ■ †	0.3 (−0.4, 0.9) ■ †	74.0 (56.0, 104.3) ■ †	0.3 (−0.3, 1.0) ■ †
ALMI [kg/m$^{3.5}$]						
normal	59.0 (51.0, 70.0)	0.0 (−0.7, 0.6)	86.6 (71.9, 102.0)	0.0 (−0.6, 0.6)	62.0 (49.0, 86.0)	0.0 (−0.7, 0.6)
low	64.0 (54.8, 72.0) ■	0.2 (−0.4. 0.8) ■	84.2 (69.9, 102.9)	−0.1 (−0.8, 0.6)	62.0 (48.0, 86.3)	−0.1 (−0.7, 0.6)
high	58.0 (48.8, 69.0) †	−0.1 (−0.9, 0.6) †	90.2 (74.0, 108.4)	0.1 (−0.6, 0.8)	66.0 (51.0, 93.3)	0.1 (−0.6, 0.9)
ALMI-FMI groups						
normal ALMI-FMI	60.0 (52.0, 70.0)	0.0 (−0.6, 0.6)	86.7 (71.4, 102.0)	−0.0 (−0.6, 0.6)	62.0 (49.0, 87.8)	−0.0 (−0.7, 0.6)
low ALMI-FMI	66.0 (57.0, 75.8) *	0.4 (−0.3, 1.0)*	80.8 (66.9, 97.8)	−0.2 (−0.8, 0.5)	58.5 (47.0, 78.0)	−0.3 (−0.8, 0.4)
high ALMI-FMI	53.0 (45.0, 62.0) *▲●	−0.5 (−1.2, 0.0) *▲●	97.4 (77.6, 113.7) *▲	0.4 (−0.5, 1.0)*▲	76.0 (57.0, 106.0) *▲●	0.4 (−0.3, 1.1) *▲●
low ALMI-high FMI	60.0 (50.0, 64.5)	−0.2 (−0.8, 0.5)	92.0 (76.1, 112.1)	0.4 (−0.4, 1.0)	70.0 (54.0, 91.0)	0.3 (−0.4, 0.7)
high ALMI-low FMI	66.0 (53.8, 77.0)	0.4 (−0.4, 1.0)	84.3 (70.8, 96.2)	-0.1 (−0.7, 0.5)	57.5 (46.8, 69.2)	−0.4 (−0.7, 0.3)

Table shows medians and quartiles (q25, q75) for each continuous parameter not corresponding to a normal distribution (according to Kolmogorov–Smirnov test), and means ± standard deviation for normally distributed parameters (according to Kolmogorov–Smirnov test). p-values were calculated by Wilcoxon-rang sum test to test for differences between each group, and Bonferroni–Holm correction was applied. *Comparison between separate low, normal, high ALMI (FMI, respectively) groups*: † p-value < 0.05 vs. low ALMI (FMI, respectively), ■ p-value < 0.05 vs. normal ALMI (FMI, respectively). *Comparison between combined ALMI-FMI body composition groups*: * p-value < 0.05 vs. normal ALMI-FMI group. ▲ p-value < 0.05 vs. low ALMI-FMI group, ● p-value < 0.05 vs. high ALMI-low FMI group. HDL-c = high-density lipoprotein cholesterol, LDL-c = low-density lipoprotein cholesterol. FMI = fat mass index, ALMI = appendicular lean mass index.

3.2. Body Composition Group Characteristics

While half of the children and adolescents with low FMI had a high SES, only 22.9% and 31.3% with high FMI had a high SES. Birthweight, preterm birth, and breast feeding did not differ significantly between body composition groups (Table 3). Only 8% of the high ALMI–FMI group were born preterm, while it was 15.1% in the low ALMI–FMI group. A quarter of those with high FMI reported second-hand smoking. This was found in both body composition groups with high FMI, independent from ALMI amount. The prevalence of children exposed to maternal smoking was also highest in the low ALMI–high FMI group, followed by the high ALMI–FMI group. Physical activity was higher in the high ALMI–low FMI group compared to the low ALMI–FMI, and the high ALMI–FMI group. Moreover, the prevalence of physical activity ≥60 min per day was significantly higher in the high ALMI–low FMI group compared to the low ALMI–FMI group (91.1% and 58.9%, respectively) and to the high ALMI–FMI group (91.1% and 54.0%, respectively). Between all other groups physical activity was similar. All other analyzed parameters did not show significant differences, and although hand grip strength was highest in the high ALMI–FMI group, it did not differ significantly.

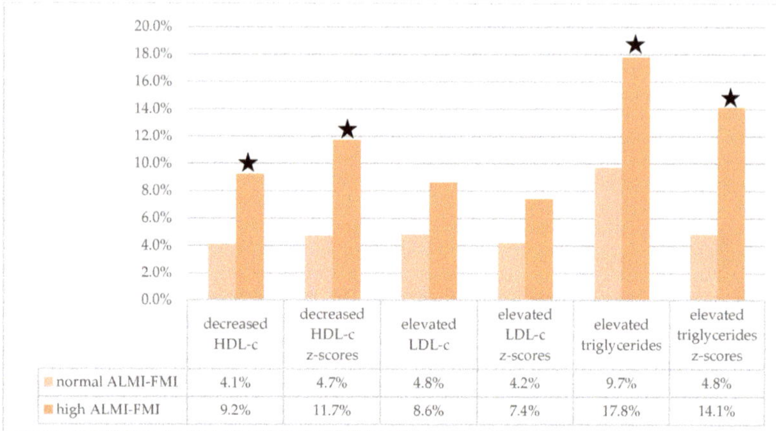

Figure 3. Prevalence of altered lipid profiles in the high ALMI–FMI and the normal ALMI–FMI groups. Figure shows the prevalence of elevated (for HDL-c decreased, respectively) serum lipid levels in the high. ALMI–FMI and the normal ALMI–FMI groups. Star indicates a p-value < 0.05 according to Fisher's Exact test after applying Bonferroni–Holm multiple testing correction.

Table 3. Characteristics of body composition groups.

	Normal ALMI FMI	Low ALMI FMI	High ALMI FMI	Low ALMI High FMI	High ALMI Low FMI
			Demographics		
age [years]	10.8 (8.3, 14.6)	10.9 (8.7, 14.5)	10.8 (8.7, 14.5)	9.9 (8.5, 15.1)	10.4 (8.2, 16.1)
sex [%females]	46.2%	50.7%	51.5%	40.0%	33.9%
height [cm]	146.0 (132.0, 164.0)	151.5 (133.0,168.0)	148.0 (135.0,163.0)	146.0 (136.0,166.5)	140.5 (127.0,168.3)
weight [kg]	39.0 (28.0, 55.0)	33.5 (24.0, 46.8) *	54.0 (38.0, 71.0) *	43.0 (32.5, 60.0)	33.5 (25.8, 56.5)
waist circumference [cm]	65.5 (59.0, 74.5)	60.5 (55.6, 68.5) *	80.5 (70.8, 90.3) *	76.0 (66.3, 80.5) *	60.8 (56.0, 68.6)
hand grip strength [kg]	17.5 (12.9, 26.2)	18.2 (12.5, 26.5)	19.5 (14.2, 27.2)	13.2 (11.4, 21.9)	19.6 (13.9, 32.2)
Socio-economic status					
low	11.8%	8.2%	21.5%	8.6%	10.7%
normal	44.3%	39.7%	47.2%	68.6%	39.3%
high	43.9%	50.7%	31.3%	22.9%	50.0%
			Early life risk factors		
birthweight [kg]	3.3 (2.8, 3.7)	3.2 (2.4, 3.6)	3.3 (2.7, 3.7)	3.3 (1.0, 3.9)	3.2 (2.3, 3.5)
preterm birth [%]	9.2%	15.1%	8.0%	11.4%	14.3%
low birthweight	20.7%	26.0%	22.1%	34.3%	28.6%
breast feeding ever	89.5%	91.1%	82.8%	82.9%	96.4%

Table 3. *Cont.*

	Normal ALMI FMI	Low ALMI FMI	High ALMI FMI	Low ALMI High FMI	High ALMI Low FMI
			Smoke exposure		
second-hand smoking	17.5%	15.1%	25.3%	25.7%	3.6%
maternal smoking					
prior pregnancy	38.2%	34.2%	44.2%	65.7%	26.8%
during pregnancy	9.3%	4.1%	14.7%	22.9%	3.6%
after pregnancy	11.9%	8.9%	16.6%	22.9%	5.4%
			Lifestyle factors		
physical activity [minutes/day]	79.3 (49.8, 110.7)	70.7 (43.2, 103.6)	62.1 (41.4, 99.3)	70.7 (44.3, 98.3)	95.0 (77.9, 120.5)
physical activity ≥60 minutes/day	67.8%	58.9%	54.0% *	68.6%	91.1% *
healthy nutrition	26.9%	26.0%	28.2%	25.7%	26.8%

Table shows medians and quartiles (q25, q75) for continuous parameters, and prevalence (%) for dichotomous parameters for each body composition group. * p-value < 0.05 after Bonferroni–Holm correction compared to normal ALMI–FMI group.

4. Discussion

The present study found that BMI poorly reflects differences in body composition existing in childhood. It shows that low ALMI–FMI and high ALMI–FMI are the most frequent aberrant body composition groups. Early dyslipidemia was found in children and adolescents who have high FMI with simultaneously high ALMI values.

Those within the normal BMI category showed an especially huge variety of ALMI and FMI combinations. This has been previously shown in specific groups [28,29], but the present study extends this finding on a population-based level with data obtained by DXA. Even though BMI is widely used as a surrogate measure of fat mass mainly, we show that a close relationship exists also with ALMI, since the prevalence of high ALMI increased similarly with the BMI category, and was also highest in the obesity category (76.6%). Equally, FMI and ALMI both decreased by decreasing the BMI category. The present analysis shows that already at young age, different body composition groups exist within the general population, with similar distribution throughout childhood and adolescence. Longitudinal assessment in birth cohorts has shown that most children usually track along with stable BMI percentiles [30–33]. Our cross-sectional data suggest a comparable persistence for body composition patterns.

Fetal and early postnatal programming is considered as an important factor for skeletal muscle development [34]. Aside from the normal ALMI–FMI group, the groups with high ALMI–FMI and with low ALMI–FMI showed the highest frequency. These results confirm that the body compartments LM and FM correlate positively with each other [20,35]. They further show that an increase in ALMI and FMI is dominant in overweight and obese children, and that an isolated increase in FMI only is rarely found.

Low muscle mass contributes to several adverse health outcomes, and, together with low muscle strength, is consistently associated with reduced bone parameters during growth, increasing the risk of osteoporosis at older age [8,36–39]. Sarcopenic obesity, which is the combination of high FM and low muscle mass, is associated with greater health risks [7,8]. In our cohort, this group with high FMI–low ALMI had a prevalence of 2.5% only. Our data stress the importance of body composition assessment in childhood and indicate that individual body constitution is probably determined early and often persists throughout life. Longitudinal data on body composition are needed to identify and specify body composition changes and to track low muscle mass.

Which traits determine body composition is not yet fully understood. In the present analysis, suggested early-life risk factors have been investigated, but no significant difference in any of these factors existed between the body composition groups analyzed. Besides genetics and early risk factors, nutrition and physical activity may also actively influence body composition [10,40]. While the prevalence of healthy nutrition was similar across all groups, reported physical activity differed. The high ALMI–low FMI group reported to spend more time doing physical activity. Considering that these parameters were assessed by questionnaire, they should be interpreted cautiously. Still, other studies, which measured physical activity objectively by accelerometer, support this result [40–42]. Future studies are needed to explore the role of protein intake in relation to the body composition groups analyzed. Indeed, higher protein intake is associated with higher fat-free mass (FFM), increased FM, and obesity risk [43–45]. Furthermore, it is not only the protein quantity but also the protein source which seems to influence muscle mass in childhood [43,46].

Intriguingly, the group of high ALMI–FMI showed marked differences in lipids compared to other groups. HDL-c levels were lower, while triglyceride and LDL-c levels were higher. A previous observational study in 660 adolescents (aged 16–17 years) reported that those with low muscle mass had significantly higher values of fasting triglycerides and atherogenic index, defined as the ratio triglycerides/HDL-c [47]. In contrast, study subjects with high ALMI of the present cohort values showed higher values of triglycerides, also after adjusting for age and sex. Another paper reported that in school children a greater hand grip strength was associated with healthier triglyceride and HDL-c concentrations, although these relationships were not independent of BMI [39]. In a cohort of 3320 participants (aged 5–18 years), body fat (assessed by skinfold thickness) $\geq$25% in males and $\geq$30% in females, was associated with cardiovascular risk factors [48]. Others confirmed that excess body fat (>37%) is associated with increased triglycerides, total cholesterol, LDL-c, and decreased HDL-c levels [49]. The prevalence of dyslipidemia was also reported to be significantly higher in U.S. youths with high adiposity. Prevalence varied from 12.4% (LDL-c) to 21.3% (triglycerides) in the high adiposity subgroup [50]. We found decreased HDL-c levels in the high FMI and the high ALMI–FMI groups. A negative association between HDL-c and ALMI has been previously reported [51,52]. Our study does not only support this finding, but confirms it by age- and sex-adjusted lipid values. Taking these factors into account is important, since FFM and muscle mass growth vary greatly by development stage [8]. Our data confirm that, total body weight and FMI amount in particular are associated with dyslipidemia in youth, but suggest that ALMI amount seems to additionally alter lipid levels to varying degrees. Hence, these findings stress the importance of early cardiovascular risk assessment, in particular for those with excess body fat. However, further longitudinal studies are required to demonstrate a link between this body composition group and cardiovascular risks as effects on pulse wave velocity. In that case, in-depth metabolic profiling is needed to explore possible underlying biological mechanisms.

A strength of the present study is that body compartments were assessed by DXA with very good reported accuracy and repeatability [53,54]. Additionally, FM and ALM were accurately normalized for body size; thus, biases related to body height were minimized as much as possible. Many studies did not consider that height squared often does not account for height correctly in children [21–23]. Another important strength is that our results were confirmed by age- and sex-adjusted lipid levels. To our knowledge, combined body composition groups have not been analyzed previously in such a large cohort of children and adolescents recruited from the general population. Our data underscore the need to use this approach in future studies and to consider the relationship between body compartments. Tracking these body composition groups longitudinally and investigating related traits could contribute fundamentally to understanding body composition development. However, some limitations have to be considered as well. First of all, cut-offs for ALMI and FMI groups were chosen arbitrarily due to the lack of validated cut-offs. Secondly, the reported data were collected in a single centre, recruiting Caucasian Middle-European youths. Thirdly, body composition data from birth to the age of 6 years were not available.

5. Conclusions

The results confirm that BMI does not reflect body constitution and we urge that the use of more precise assessment techniques should be emphasized for investigating body composition. In addition, children and adolescents with high ALMI–FMI showed an adverse cardiovascular risk profile. In the future, the prevalence of low muscle mass in these youths should be focused on. Prospective longitudinal data in the presented body composition groups are needed to define the impact of body composition on overall health across the lifecycle and to identify novel preventive strategies.

Supplementary Materials: The following are available online at https://www.mdpi.com/article/10.3390/hildren8111047/s1, Table S1: Sample size and prevalence of body composition groups in different age groups. Table S2: Prevalence of elevated levels of serum lipids in body composition groups.

Author Contributions: A.O. carried out the statistical analyses and wrote the initial manuscript; M.-K.B. and R.B.-K. designed the study, coordinated and supervised data collection and reviewed and revised the manuscript; S.H. and O.C.B. designed the study, coordinated and supervised data collection; F.K. designed the statistical analyses and supported them; F.M.E.F. reviewed and revised the manuscript; E.F.M.W. reviewed and revised the manuscript and contributed to the discussion with important intellectual content. All authors have read and agreed to the published version of the manuscript.

Funding: The Austrian LEAD Study is supported by the Ludwig Boltzmann Society, the Municipal Department of Health and Environment of Vienna, the Federal State Governmental Department of Health of Lower Austria, and unrestricted scientific grants from AstraZeneca, Böhringer Ingelheim, Chiesi Pharma, Glaxo Smith Kline, Menarini Pharma, MSD, and Novartis Pharma. None of the supporting parties had any participation in the data, nor did they contribute to the design or the content of the present manuscript.

Institutional Review Board Statement: The LEAD study was conducted according to the guidelines of the Declaration of Helsinki. It was approved by the ethics committee of Vienna on the 21st of November 2011 (protocol number: EK-11-117-0711).

Informed Consent Statement: Informed consent was obtained from all subjects involved in the study.

Data Availability Statement: Data is contained within the article or supplementary material.

Conflicts of Interest: All authors declare no conflict of interest related to the manuscript.

References

1. Abarca-Gómez, L.; Abdeen, Z.A.; Hamid, Z.A.; Abu-Rmeileh, N.M.; Acosta-Cazares, B.; Acuin, C.; Adams, R.J.; Aekplakorn, W.; Afsana, K.; Aguilar-Salinas, C.A.; et al. Worldwide trends in body-mass index, underweight, overweight, and obesity from 1975 to 2016: A pooled analysis of 2416 population-based measurement studies in 128 million children, adolescents, and adults. *Lancet* **2017**, *390*, 2627–2642. [CrossRef]
2. Singh, G.M.; Danaei, G.; Farzadfar, F.; Stevens, G.A.; Woodward, M.; Wormser, D.; Kaptoge, S.; Whitlock, G.; Qiao, Q.; Lewington, S. The age-specific quantitative effects of metabolic risk factors on cardiovascular diseases and diabetes: A pooled analysis. *PLoS ONE* **2013**, *8*, e65174.
3. Anandacoomarasamy, A.; Caterson, I.; Sambrook, P.; Fransen, M.; March, L. The impact of obesity on the musculoskeletal system. *Int. J. Obes.* **2008**, *32*, 211–222. [CrossRef]
4. Forno, E.; Han, Y.-Y.; Libman, I.M.; Muzumdar, R.H.; Celedón, J.C. Adiposity and Asthma in a Nationwide Study of Children and Adults in the United States. *Ann. Am. Thorac. Soc.* **2017**, *15*, 322–330. [CrossRef]
5. Prentice, A.M.; Jebb, S.A. Beyond body mass index. *Obes. Rev.* **2001**, *2*, 141–147. [CrossRef] [PubMed]
6. Forbes, G.B. Lean Body mass and fat in obese children. *Pediatrics* **1964**, *34*, 308–314. [PubMed]
7. Van Aller, C.; Lara, J.; Stephan, B.C.M.; Donini, L.M.; Heymsfield, S.; Katzmarzyk, P.T.; Wells, J.C.K.; Prado, C.M.; Siervo, M. Sarcopenic obesity and overall mortality: Results from the application of novel models of body composition phenotypes to the National Health and Nutrition Examination Survey 1999. *Clin. Nutr.* **2019**, *38*, 264–270. [CrossRef] [PubMed]
8. Orsso, C.E.; Tibaes, J.R.B.; Oliveira, C.L.P.; Rubin, D.A.; Field, C.J.; Heymsfield, S.B.; Prado, C.M.; Haqq, A.M. Low muscle mass and strength in pediatrics patients: Why should we care? *Clin. Nutr.* **2019**, *38*, 2002–2015. [CrossRef] [PubMed]
9. Donini, L.M.; Pinto, A.; Giusti, A.M.; Lenzi, A.; Poggiogalle, E. Obesity or BMI Paradox? *Beneath the Tip of the Iceberg. Front. Nutr.* **2020**, *7*, 53. [PubMed]
10. Emmett, P.M.; Jones, L.R. Diet, growth, and obesity development throughout childhood in the Avon Longitudinal Study of Parents and Children. *Nutr. Rev.* **2015**, *73*, 175–206. [CrossRef]

11. Fleming, T.P.; Watkins, A.J.; Velazquez, M.A.; Mathers, J.C.; Prentice, A.M.; Stephenson, J.; Barker, M.; Saffery, R.; Yajnik, C.S.; Eckert, J.J.; et al. Origins of lifetime health around the time of conception: Causes and consequences. *Lancet* **2018**, *391*, 1842–1852. [CrossRef]
12. Mameli, C.; Mazzantini, S.; Zuccotti, G.V. Nutrition in the first 1000 days: The origin of childhood obesity. *Int. J. Environ. Res. Public Health* **2016**, *13*, 838. [CrossRef] [PubMed]
13. Origin of atherosclerosis in childhood and adolescence. *Am. J. Clin. Nutr.* **2000**, *72*, 1307s–1315s. [CrossRef] [PubMed]
14. Nadeau, K.J.; Anderson, B.J.; Berg, E.G.; Chiang, J.L.; Chou, H.; Copeland, K.C.; Hannon, T.S.; Huang, T.T.-K.; Lynch, J.L.; Powell, J.; et al. Youth-Onset Type 2 Diabetes Consensus Report: Current Status, Challenges, and Priorities. *Diabetes Care* **2016**, *39*, 1635–1642. [CrossRef] [PubMed]
15. Forsén, T.; Eriksson, J.; Tuomilehto, J.; Reunanen, A.; Osmond, C.; Barker, D. The Fetal and Childhood Growth of Persons Who Develop Type 2 Diabetes. *Ann. Intern. Med.* **2000**, *133*, 176–182. [CrossRef] [PubMed]
16. 1Svanes, C.; Sunyer, J.; Plana, E.; Dharmage, S.; Heinrich, J.; Jarvis, D.; de Marco, R.; Norbäck, D.; Raherison, C.; Villani, S.; et al. Early life origins of chronic obstructive pulmonary disease. *Thorax* **2010**, *65*, 14–20. [CrossRef] [PubMed]
17. Juonala, M.; Wu, F.; Sinaiko, A.; Woo, J.G.; Urbina, E.M.; Jacobs, D.; Steinberger, J.; Prineas, R.; Koskinen, J.; Sabin, M.A.; et al. Non-HDL Cholesterol Levels in Childhood and Carotid Intima-Media Thickness in Adulthood. *Pediatrics* **2020**, *145*, e20192114. [CrossRef] [PubMed]
18. Magnussen, C.G.; Venn, A.; Thomson, R.; Juonala, M.; Srinivasan, S.R.; Viikari, J.S.A.; Berenson, G.S.; Dwyer, T.; Raitakari, O.T. The Association of Pediatric Low- and High-Density Lipoprotein Cholesterol Dyslipidemia Classifications and Change in Dyslipidemia Status With Carotid Intima-Media Thickness in Adulthood: Evidence From the Cardiovascular Risk in Young Finns Study, the Bog. *J. Am. Coll. Cardiol.* **2009**, *53*, 860–869. [CrossRef]
19. Breyer-Kohansal, R.; Hartl, S.; Burghuber, O.C.; Urban, M.; Schrott, A.; Agusti, A.; Sigsgaard, T.; Vogelmeier, C.; Wouters, E.; Studnicka, M.; et al. The LEAD (Lung, Heart, Social, Body) Study: Objectives, Methodology, and External Validity of the Population-Based Cohort Study. *J. Epidemiol.* **2019**, *29*, 315–324. [CrossRef] [PubMed]
20. Ofenheimer, A.; Breyer-Kohansal, R.; Hartl, S.; Burghuber, O.C.; Krach, F.; Schrott, A.; Wouters, E.F.M.; Franssen, F.M.E.; Breyer, M.-K. Reference values of body composition parameters and visceral adipose tissue (VAT) by DXA in adults aged 18–81 years—results from the LEAD cohort. *Eur. J. Clin. Nutr.* **2020**, *74*, 1181–1191. [CrossRef] [PubMed]
21. Orsso, C.E.; Tibaes, J.R.B.; Rubin, D.A.; Field, C.J.; Heymsfield, S.B.; Prado, C.M.; Haqq, A.M. Metabolic implications of low muscle mass in the pediatric population: A critical review. *Metab.-Clin. Exp.* **2019**, *99*, 102–112. [CrossRef]
22. Peterson, C.M.; Su, H.; Thomas, D.M.; Heo, M.; Golnabi, A.H.; Pietrobelli, A.; Heymsfield, S.B. Tri-Ponderal Mass Index vs Body Mass Index in Estimating Body Fat During Adolescence. *JAMA Pediatr.* **2017**, *171*, 629–636. [CrossRef] [PubMed]
23. Burton, R.F. Measures of adiposity: The inappropriate use of the fat mass index. *Int. J. Obes.* **2010**, *34*, 213. [CrossRef]
24. Ofenheimer, A.; Breyer-Kohansal, R.; Hartl, S.; Burghuber, O.C.; Krach, F.; Schrott, A.; Franssen, F.M.E.; Wouters, E.F.M.; Breyer, M.-K. Reference charts for body composition parameters by dual-energy X-ray absorptiometry in European children and adolescents aged 6 to 18 years—Results from the Austrian LEAD (Lung, hEart, sociAl, boDy) cohort. *Pediatr. Obes.* **2020**, *16*, e12695. [CrossRef] [PubMed]
25. Onis, M.D.; Onyango, A.W.; Borghi, E.; Siyam, A.; Nishida, C.; Siekmann, J. Development of a WHO growth reference for school-aged children and adolescents. *Bull. World Health Organ.* **2007**, *85*, 660–667. [CrossRef]
26. Expert panel on integrated guidelines for cardiovascular health and risk reduction in children and adolescents: Summary report. *Pediatrics* **2011**, *128*, S213. [CrossRef] [PubMed]
27. Prado, C.M.M.; Siervo, M.; Mire, E.; Heymsfield, S.B.; Stephan, B.C.M.; Broyles, S.; Smith, S.R.; Wells, J.C.K.; Katzmarzyk, P.T. A population-based approach to define body-composition phenotypes. *Am. J. Clin. Nutr.* **2014**, *99*, 1369–1377. [CrossRef] [PubMed]
28. Engelen, M.P.K.J.; Schroder, R.; Van der Hoorn, K.; Deutz, N.E.P.; Com, G. Use of body mass index percentile to identify fat-free mass depletion in children with cystic fibrosis. *Clin. Nutr.* **2012**, *31*, 927–933. [CrossRef]
29. Freedman, D.S.; Wang, J.; Maynard, L.M.; Thornton, J.C.; Mei, Z.; Pierson, R.N.; Dietz, W.H.; Horlick, M. Relation of BMI to fat and fat-free mass among children and adolescents. *Int. J. Obes.* **2005**, *29*, 1–8. [CrossRef] [PubMed]
30. Bichteler, A.; Gershoff, E.T. Identification of Children's BMI Trajectories and Prediction from Weight Gain in Infancy. *Obesity* **2018**, *26*, 1050–1056. [CrossRef] [PubMed]
31. Wibaek, R.; Vistisen, D.; Girma, T.; Admassu, B.; Abera, M.; Abdissa, A.; Mudie, K.; Kæstel, P.; Jørgensen, M.E.; Wells, J.C.K.; et al. Body mass index trajectories in early childhood in relation to cardiometabolic risk profile and body composition at 5 years of age. *Am. J. Clin. Nutr.* **2019**, *110*, 1175–1185. [CrossRef] [PubMed]
32. Aris, I.M.; Oken, E. Childhood Adiposity Trajectories: Discerning Order Amongst the Chaos. *Am. J. Clin. Nutr.* **2019**, *110*, 1049–1050. [CrossRef] [PubMed]
33. Guo, S.S.; Chumlea, W.M.C.; Roche, A.F.; Siervogel, R.M. Age-and maturity-related changes in body composition during adolescence into adulthood: The Fels Longitudinal Study. *Appl. Radiat. Isot.* **1998**, *49*, 581–585. [CrossRef]
34. Chomtho, S.; Wells, J.C.K.; Williams, J.E.; Lucas, A.; Fewtrell, M.S. Associations between birth weight and later body composition: Evidence from the 4-component model. *Am. J. Clin. Nutr.* **2008**, *88*, 1040–1048. [CrossRef]
35. Bosy-Westphal, A.; Müller, M.J. Identification of skeletal muscle mass depletion across age and BMI groups in health and disease—there is need for a unified definition. *Int. J. Obes.* **2015**, *39*, 379–386. [CrossRef]

36. Kim, S.G.; dong Ko, K.; Hwang, I.C.; Suh, H.S.; Kay, S.; Caterson, I.; Kim, K.K. Relationship between indices of obesity obtained by anthropometry and dual-energy X-ray absorptiometry: The Fourth and Fifth Korea National Health and Nutrition Examination Survey (KNHANES IV and V, 2008–2011). *Obes. Res. Clin. Pract.* **2015**, *9*, 487–498. [CrossRef] [PubMed]
37. Peterson, M.D.; Zhang, P.; Saltarelli, W.A.; Visich, P.S.; Gordon, P.M. Low Muscle Strength Thresholds for the Detection of Cardiometabolic Risk in Adolescents. *Am. J. Prev. Med.* **2016**, *50*, 593–599. [CrossRef] [PubMed]
38. Murphy, M.J.; Metcalf, B.S.; Jeffery, A.N.; Voss, L.D.; Wilkin, T.J. Does lean rather than fat mass provide the link between birth weight, BMI, and metabolic risk? EarlyBird 23. *Pediatr. Diabetes* **2006**, *7*, 211–214. [CrossRef] [PubMed]
39. Blakeley, C.E.; Van Rompay, M.I.; Schultz, N.S.; Sacheck, J.M. Relationship between muscle strength and dyslipidemia, serum 25 (OH) D, and weight status among diverse schoolchildren: A cross-sectional analysis. *BMC Pediatr.* **2018**, *18*, 23. [CrossRef] [PubMed]
40. Meredith-Jones, K.; Haszard, J.; Moir, C.; Heath, A.-L.; Lawrence, J.; Galland, B.; Taylor, B.; Gray, A.; Sayers, R.; Taylor, R. Physical activity and inactivity trajectories associated with body composition in pre-schoolers. *Int. J. Obes.* **2018**, *42*, 1621–1630. [CrossRef] [PubMed]
41. Butte, N.F.; Puyau, M.R.; Wilson, T.A.; Liu, Y.; Wong, W.W.; Adolph, A.L.; Zakeri, I.F. Role of physical activity and sleep duration in growth and body composition of preschool-aged children. *Obesity* **2016**, *24*, 1328–1335. [CrossRef] [PubMed]
42. McCormack, L.; Meendering, J.; Specker, B.; Binkley, T. Associations between sedentary time, physical activity, and dual-energy X-ray absorptiometry measures of total body, android, and gynoid fat mass in children. *J. Clin. Densitom.* **2016**, *19*, 368–374. [CrossRef] [PubMed]
43. Jen, V.; Karagounis, L.G.; Jaddoe, V.W.V.; Franco, O.H.; Voortman, T. Dietary protein intake in school-age children and detailed measures of body composition: The Generation R Study. *Int. J. Obes.* **2018**, *42*, 1715–1723. [CrossRef] [PubMed]
44. Voortman, T.; Braun, K.V.E.; Kiefte-de Jong, J.C.; Jaddoe, V.W.V.; Franco, O.H.; van den Hooven, E.H. Protein intake in early childhood and body composition at the age of 6 years: The Generation R Study. *Int. J. Obes.* **2016**, *40*, 1018–1025. [CrossRef] [PubMed]
45. Rolland-Cachera, M.F.; Deheeger, M.; Maillot, M.; Bellisle, F. Early adiposity rebound: Causes and consequences for obesity in children and adults. *Int. J. Obes.* **2006**, *30*, S11–S17. [CrossRef]
46. Assmann, K.E.; Joslowski, G.; Buyken, A.E.; Cheng, G.; Remer, T.; Kroke, A.; Günther, A.L.B. Prospective association of protein intake during puberty with body composition in young adulthood. *Obesity* **2013**, *21*, E782–E789. [CrossRef] [PubMed]
47. Burrows, R.; Correa-Burrows, P.; Reyes, M.; Blanco, E.; Albala, C.; Gahagan, S. Low muscle mass is associated with cardiometabolic risk regardless of nutritional status in adolescents: A cross-sectional study in a Chilean birth cohort. *Pediatr. Diabetes* **2017**, *18*, 895–902. [CrossRef]
48. Williams, D.P.; Going, S.B.; Lohman, T.G.; Harsha, D.W.; Srinivasan, S.R.; Webber, L.S.; Berenson, G.S. Body fatness and risk for elevated blood pressure, total cholesterol, and serum lipoprotein ratios in children and adolescents. *Am. J. Public Health* **1992**, *82*, 358–363. [CrossRef]
49. Choi, J.W.; Pai, S.H.; Kim, S.K. Associations between total body fat and serum lipid concentrations in obese human adolescents. *Ann. Clin. Lab. Sci.* **2002**, *32*, 271–278.
50. Lamb, M.M.; Ogden, C.L.; Carroll, M.D.; Lacher, D.A.; Flegal, K.M. Association of body fat percentage with lipid concentrations in children and adolescents: United States, 1999–20041–4. *Am. J. Clin. Nutr.* **2011**, *94*, 877–883. [CrossRef] [PubMed]
51. Duran, I.; Martakis, K.; Schafmeyer, L.; Jackels, M.; Rehberg, M.; Schoenau, E. Inverse association of high-density lipoprotein cholesterol concentration with muscle mass in children. *Child. Obes.* **2019**, *15*, 476–484. [CrossRef] [PubMed]
52. Pietrobelli, A.; Lee, R.C.; Capristo, E.; Deckelbaum, R.J.; Heymsfield, S.B. An independent, inverse association of high-density-lipoprotein-cholesterol concentration with nonadipose body mass. *Am. J. Clin. Nutr.* **1999**, *69*, 614–620. [CrossRef] [PubMed]
53. Borga, M.; West, J.; Bell, J.D.; Harvey, N.C.; Romu, T.; Heymsfield, S.B.; Leinhard, O.D. Advanced body composition assessment: From body mass index to body composition profiling. *J. Investig. Med.* **2018**, *66*, 1–9. [CrossRef] [PubMed]
54. Heard-Lipsmeyer, M.E.; Hull, H.; Sims, C.R.; Cleves, M.A.; Andres, A. Evaluating body composition in infancy and childhood: A comparison between 4C, QMR, DXA, and ADP. *Pediatr. Obes.* **2020**, *15*, e12617. [CrossRef] [PubMed]

children

MDPI

Article

Percentile Reference Values for the Neck Circumference of Mexican Children

Evelyn Valencia-Sosa [1,2], Clío Chávez-Palencia [1,3,*], Juan R. Vallarta-Robledo [1], Enrique Romero-Velarde [4], Alfredo Larrosa-Haro [4], Edgar Manuel Vásquez-Garibay [4] and César Octavio Ramos-García [3]

1 Instituto Tecnológico de Estudios Superiores de Occidente, Anillo Periférico Sur Manuel Gómez Morín 8585, Santa María Tequepexpan, Tlaquepaque, Jalisco 45604, Mexico; envs_621@hotmail.com (E.V.-S.); cybervallarta@hotmail.com (J.R.V.-R.)

2 Departamento de Salud Pública, Centro Universitario de Ciencias de la Salud (CUCS), Universidad de Guadalajara, 950 Sierra Mojada St. Guadalajara, Jalisco 44340, Mexico

3 División de Ciencias de la Salud, Centro Universitario de Tonalá, Universidad de Guadalajara, Av. Nuevo Periférico 555 Ejido San José Tatepozco, Tonalá, Jalisco 45425, Mexico; cesar.ramos@ciansc.com

4 Instituto de Nutrición Humana, Universidad de Guadalajara, Salvador Quevedo y Zubieta 750, Guadalajara, Jalisco 44340, Mexico; enrique.rvelarde@academicos.udg.mx (E.R.-V.); alfredo.larrosa@academicos.udg.mx (A.L.-H.); edgar.vgaribay@academicos.udg.mx (E.M.V.-G.)

* Correspondence: clio.chavez@academicos.udg.mx; Tel.: +33-35403020 (ext. 64033)

Citation: Valencia-Sosa, E.; Chávez-Palencia, C.; Vallarta-Robledo, J.R.; Romero-Velarde, E.; Larrosa-Haro, A.; Vásquez-Garibay, E.M.; Ramos-García, C.O. Percentile Reference Values for the Neck Circumference of Mexican Children. *Children* **2021**, *8*, 407. https://doi.org/10.3390/children8050407

Academic Editors: Odysseas Androutsos and Antonis Zampelas

Received: 20 April 2021
Accepted: 16 May 2021
Published: 18 May 2021

Abstract: Neck circumference was studied for the first time in a pediatric population in 2010. Since then, various countries have proposed cutoff values to identify overweight, obesity, and metabolic syndrome. However, no reference values have been established for the Mexican child population. The aim of this study is to provide percentile reference values for the neck circumference of Mexican schoolchildren. Only normal-weight schoolchildren aged 6–11 years were included. Percentiles and growth charts were constructed based on the "Generalized Additive Model for Location, Scale and Shape" (GAMLSS). A total of 1059 schoolchildren (52.9% female) was evaluated. Weight, height, and BMI values were higher for males; however, this difference was not statistically significant. The 50th percentile for females was 24.6 cm at six years old and 28.25 cm at 11 years old, and for males, it was 25.75 cm and 28.76 cm, respectively. Both males and females displayed a pronounced increase in neck circumference between 10 and 11 years of age. The greatest variability was found in the 11-year-old group, with an increase of 5.5 cm for males and 5.4 cm for females. This study presents the first reference values for neck circumference for a Mexican child population.

Keywords: childhood; neck circumference; percentile; anthropometry

1. Introduction

The prevalence of childhood obesity has increased worldwide in recent decades. According to the National Health and Nutrition Survey in Mexico (ENSANUT, 2018), three out of ten children are overweight or obese, whereas the prevalence in adolescents surpasses 38% [1]. These numbers place Mexico among the countries with the highest overweight/obesity prevalence in the pediatric population. Furthermore, it has been reported that prominent fat depots in the upper body increase the risk for metabolic disturbances to a greater extent than general adiposity [2]. Therefore, it is important to develop practical and noninvasive indicators to assess body fat distribution.

Body mass index (BMI) is the most practical and utilized index for assessing normal weight ranges; however, it has been proven to be unsuitable for determining fat mass volume and location [3,4]. On the other hand, waist circumference (WC) and waist-to-height ratio have been proposed as reliable tools to identify individuals at metabolic risk, as both reflect central adiposity [5]. Moreover, the fact that a few technical issues may arise when measuring WC has led to the study of novel indicators.

Neck circumference (NC) has been proposed as a simple, minimally invasive, and inexpensive indicator to identify upper-body adiposity. Research on NC began in 2010 and ever since, it has been shown a wide association with central adiposity [6] along the onset of metabolic alterations [7]. Several cutoff points have been proposed to identify overweight and obesity [8–10], non-alcoholic fatty liver disease (NAFLD) [11], hypertension [12] and metabolic syndrome [13]. Nonetheless, the usefulness of these values is limited to the screened population. In Mexico, we have previously demonstrated that NC shows a high correlation with WC, which indicates that it might be utilized for the identification of elevated central adiposity [14]. However, no reference values have been established for the Mexican child population. In this regard, the creation of new reference tables with percentile values distribution for neck circumference might be useful to determine how distant they are from the mean as for sex and age, which could facilitate the identification of individuals at risk in clinical practice. Thus, the aim of this study is to provide percentile reference values of neck circumference of Mexican schoolchildren.

2. Materials and Methods

2.1. Participants

The sample for this study was obtained as part of a broader project entitled "Active intervention to improve feeding habits and physical activity in school children" conducted in six different schools located in Acatlán de Juárez and Villa Corona, Jalisco, Mexico. The sample size of the original project was calculated on data from a similar study by Li et al. [15] obtaining a total sample size of 288 children among the 6 schools (3 control and 3 intervention, selected by convenience sampling), with an extra 30% due to possible dropouts. A type I error of 0.05 and power of 80% were considered. Nevertheless, anthropometric measurements were taken in all children from the six schools at baseline and final stages. Remarkably, the data for this study were obtained from the baseline stage. Invitation to participate was granted to 2070 students; however, the assessments were performed in 1802 children aged 6–11 years (the elementary education in Mexico comprises these ages), Nevertheless, only 1059 were included because data from normal-weight children are required to create reference values.

The inclusion criteria were as follows: children attending six elementary schools in Acatlán and Villa Corona Jalisco, Mexico from November 2015 to January 2016. Exclusion criteria consisted of being under any type of nutritional or medical treatment and/or having a chronic disease.

This study was conducted according to the guidelines of the Declaration of Helsinki. All procedures involving human subjects/patients were approved by the Comité de Ética en Investigación del Centro Universitario de Tonalá (003–2016). Verbal informed consent was obtained and formally recorded from all the subjects and their tutors. Written informed consent was obtained from all the school directors.

2.2. Measurements

All anthropometrical measurements were carried out by two trained researchers according to the Habitch method [16]. Height was measured using a portable stadiometer with a precision of 0.1 cm (SECA, Hamburg, Germany) with the subject shoeless and the child's head held in the Frankfurt horizontal plane. Body weight was measured using a calibrated electronic weighing scale (SECA, Hamburg, Germany) with a precision of 0.05 kg with children shoeless and without heavy extra clothing such as sweaters and jackets. WC and NC were measured to the nearest 0.1 cm using a metallic tape (606PMMX, Apex Tool Group, Lufkin, Queretaro, Mexico). WC was measured at the midway point between the lowest rib and the top of the iliac crest with the subject standing and at the end of a regular expiration. NC was measured at the midpoint of the neck at the level of the thyroid cartilage and perpendicular to the neck axis with the participant's body held erect, eyes facing forward, and breathing normally. The triceps and subscapular skinfold thickness [17] were used to estimate the body fat percentage (BF%) according to Slaughter's equation [18]. BMI was obtained by dividing the weight in kilograms by the height in square meters.

All measurements were taken twice. However, when the height, weight, and circumference differed by 1% or more, or by 5% in the case of skinfold thickness, a third measurement was performed. The mean of these values was used for the analyses.

2.3. Operational Definitions of Terms

Underweight was indicated by a BMI for age < −2 SD, normal weight was a BMI between −2 and +1 SD, overweight was a BMI between +1 and +2 SD, and obesity was indicated by a BMI > +2 SD [19]. Only normal weight participants were included to assure that the reference curves represented *"a standard healthy population"* in accordance with the World Health Organization (WHO).

2.4. Statistical Analysis

To determine the distribution of the quantitative variables, the Kolmogorov–Smirnov normality test was used. The significance level was established at $p < 0.05$ for all hypothesis tests. Numerical variables are reported as the mean and standard deviation (SD). Comparisons were conducted between groups using Student's *t*-test for independent samples. Pearson correlation coefficients were calculated to explore the associations between NC and anthropometric variables. The following classification was used to categorized r values: Low or weak correlations (<0.35), modest or moderate correlations (0.36 to 0.67), and strong or high correlations (0.68–1.0). However, r coefficients > 0.90 were considered as "very high correlations" [20]. A $p < 0.05$ was considered statistically significant.

Percentiles and growth charts for neck circumference were constructed based on the *"Generalized Additive Model for Location, Scale and Shape"* (GAMLSS) [21], which is an extension of the lambda-mu-sigma method (LMS) by Cole and Green (1992). This method allows the construction of smooth curves at different percentile intervals based on age when the distribution is not normal. LMS transforms the age value at a specific exponential; thus, it prevents the tendency for distortion due to the classic rapid growth that occurs at early age stages [22]. Lambda (L) represents the skewness, mu (M) reflects the median, and sigma (S) is equal to the coefficient of variation. In contrast to the LMS method, which is based only on skewness, the GAMLSS method applies two additional submethodologies: Box-Cox power transform (LMSP) and Box-Cox t (LMST). These submethodologies are also adjusted by kurtosis [23,24].

For this study, all three methods (LMS, LMSP, and LMST) were applied for each sex and age transformation. The most suitable model was selected based on the lowest value for the Akaike information criterion (AIC). To validate this model, Q–Q and worm plots were created, and the Filliben plot correlation coefficient was calculated. Then, all the values were verified to confirm that the mean, standard deviation, skewness, and kurtosis were close to 0, 1, 0, and 3, respectively [21].

After the model selection, percentile values were calculated: 3rd, 5th, 10th, 25th, 50th, 75th, 85th, 95th, and 97th by sex and age using the following formula:

$$X = M(1 + LSz)\hat{\ }(1/L)$$

where,

X = percentile value
Z = z score
M = mu
S = sigma
L = lambda

Statistical analyses were performed using R (3.4.4, R Foundation for Stati, Vienna, Austria) and RStudio (1.2.1335, PBC, Boston, MA, USA) software. The GAMLSS package was used for the construction of growth charts [25].

3. Results

For this study, a total of 1059 schoolchildren aged 6 to 11 years were included (52.9% female). Weight, height, and BMI values were higher for males; however, this difference was not statistically significant. Similarly, males showed greater neck and waist circumferences than females ($p < 0.05$). On the other hand, the body fat percentage was higher for females (2.59% higher than males; $p < 0.001$).

Table 1 shows the anthropometric measurements of the studied population by sex and age. It also shows that NC and WC displayed higher values in males, except for the 11-year-old group, in which girls showed higher WC values compared to boys.

Table 1. Anthropometric measurements of schoolchildren aged 6–11 years in Acatlán de Juárez and Villa Corona Jalisco, México.

Gender	Age (years)	*n*	Weight (kg)		Height (cm)		BMI (kg/m^2)		WC (cm)		NC (cm)		BF (%)	
			Mean	SD	Mean	SD	Mean	SD	Mean	SD	Mean	SD	Mean	SD
Female	6	114	20.67	2.81	116.33	5.52	15.16	1.19	52.24 +	3.42	24.61 *	1.08	15.42 *	4.12
	7	99	23.60	3.89	123.16	5.50	15.33	1.16	54.01	3.11	25.25 *	1.14	15.77 *	3.58
	8	94	25.72	3.60	127.75 +	5.51	15.71	1.45	55.54	3.58	25.79 *	1.32	16.91 *	4.50
	9	92	29.54	3.81	134.39	6.40	16.31	1.33	58.00	3.43	26.50 *	1.18	18.43 *	3.91
	10	90	33.04	4.90	141.00	7.32	16.51	1.45	59.66	4.52	27.07 *	1.19	18.62 +	4.15
	11	71	39.42 +	6.95	148.90 +	7.58	17.67	2.06	62.99	5.11	28.40 +	1.59	20.70 *	4.47
Male	6	91	21.12	2.47	117.30	5.14	15.21	0.93	53.32 +	2.69	25.74 *	1.10	13.18 *	2.84
	7	85	23.54	2.62	123.09	5.03	15.51	1.08	54.64	3.12	26.28 *	1.26	13.68 *	3.65
	8	76	26.60	3.06	129.66 +	5.29	15.78	1.08	56.48	3.25	27.06 *	1.05	14.07 *	3.22
	9	92	29.44	3.87	134.64	5.73	16.18	1.38	58.30	4.26	27.40 *	1.28	14.47 *	4.23
	10	85	33.12	5.22	140.33	5.63	16.74	1.91	60.93	5.46	28.10 *	1.50	16.59 +	5.42
	11	70	36.74 +	5.44	145.77 +	6.30	17.18	1.71	62.65	4.69	28.96 +	1.39	17.46 *	5.75

BMI: body mass index; WC: waist circumference; NC: neck circumference; BF: body fat; SD: standard deviation. * represents the statistical differences ($p < 0.001$) for one anthropometric parameter between males and females of the same age. + represents the statistical differences ($p < 0.05$) for one anthropometric parameter between males and females of the same age.

The correlations between NC and all the anthropometric variables were statistically significant, regardless of sex and age. For WC and BMI, the correlations ranged from r = 0.5 to r = 0.8, whereas for BF% and skinfold thickness, the correlations had lower values, ranging from r = 0.2 to r = 0.7 (Table 2).

Table 2. Correlation coefficients between neck circumference and adiposity anthropometric indicators by sex and age.

Sex	Age	*n*	BMI	WC	BF (%)	TSF	SSF	*p*
Female	6	114	0.51	0.60	0.35	0.37	0.34	<0.001
	7	99	0.62	0.65	0.43	0.45	0.30	<0.001
	8	94	0.63	0.62	0.49	0.49	0.44	<0.001
	9	92	0.53	0.59	0.32	0.34	0.24	<0.001
	10	90	0.63	0.65	0.36	0.31	0.32	<0.001
	11	71	0.65	0.70	0.52	0.42	0.55	<0.001
Male	6	91	0.59	0.67	0.53	0.48	0.47	<0.001
	7	85	0.53	0.54	0.42	0.38	0.41	<0.001
	8	76	0.56	0.60	0.33	0.29	0.33	<0.001
	9	92	0.67	0.70	0.57	0.49	0.57	<0.001
	10	85	0.84	0.84	0.73	0.68	0.77	<0.001
	11	70	0.65	0.72	0.24	0.26	0.22	<0.001

BMI: body mass index; WC: waist circumference; BF: body fat; TSF: tricipital skinfold thickness; SSF: subscapular skinfold thickness.

Regarding the percentile distribution for neck circumference, both males and females displayed a pronounced increase between 10 and 11 years of age. The 6-to-7-year-old group had the lowest increase. The greatest variability (97th percentile minus the 3rd percentile) was found for the 11-year-old group, with an increase of 5.45 cm for males and 5.39 cm for females.

The least variability was observed for the six-year-old children, with increments of 4.18 and 4.03 cm for the males and females, respectively. At the 50th percentile, the yearly increment of neck circumference ranged from 0.5 to 1.0 cm (Table 3).

Table 3. Percentile distribution of neck circumference (cm) of schoolchildren aged 6–11 years in Acatlán de Juarez and Villa Corona Jalisco, México.

				Percentiles									
Age (years)	**L**	**M**	**S**	**3**	**5**	**10**	**25**	**50**	**75**	**85**	**90**	**95**	**97**
						Female							
6	0.1292	24.6009	0.0435	22.66	22.89	23.26	23.89	24.60	25.33	25.73	26.01	26.42	26.69
7	−1.4136	25.1297	0.0447	23.21	23.43	23.78	24.40	25.13	25.92	26.36	26.68	27.16	27.48
8	−2.4229	25.7153	0.0460	23.77	23.99	24.34	24.96	25.72	26.56	27.05	27.40	27.96	28.34
9	−2.0373	26.3952	0.0474	24.32	24.55	24.93	25.59	26.40	27.28	27.80	28.16	28.73	29.12
10	−0.5193	27.2256	0.0490	24.88	25.16	25.59	26.35	27.23	28.15	28.66	29.02	29.56	29.92
11	1.0549	28.2507	0.0507	25.55	25.89	26.41	27.28	28.25	29.22	29.73	30.08	30.60	30.94
						Male							
6	−0.1859	25.7458	0.0431	23.76	23.99	24.37	25.01	25.75	26.51	26.93	27.22	27.65	27.94
7	−0.7149	26.2513	0.0439	24.23	24.47	24.84	25.49	26.25	27.05	27.49	27.80	28.27	28.58
8	−1.2831	26.8052	0.0449	24.74	24.98	25.36	26.02	26.81	27.65	28.12	28.46	28.97	29.31
9	−1.8881	27.4073	0.0459	25.30	25.54	25.92	26.60	27.41	28.30	28.81	29.17	29.73	30.12
10	−2.5277	28.0587	0.0470	25.91	26.15	26.53	27.22	28.06	29.00	29.56	29.95	30.58	31.01
11	−3.2002	28.7605	0.0482	26.56	26.80	27.18	27.89	28.76	29.76	30.37	30.81	31.51	32.01

L: lambda, M: mean, S: variation coefficient.

Growth charts for neck circumference showed a linear and constant tendency for both sexes starting with the 10-year-old group. Similarly, neck circumference was greater for males than for females (Figure 1).

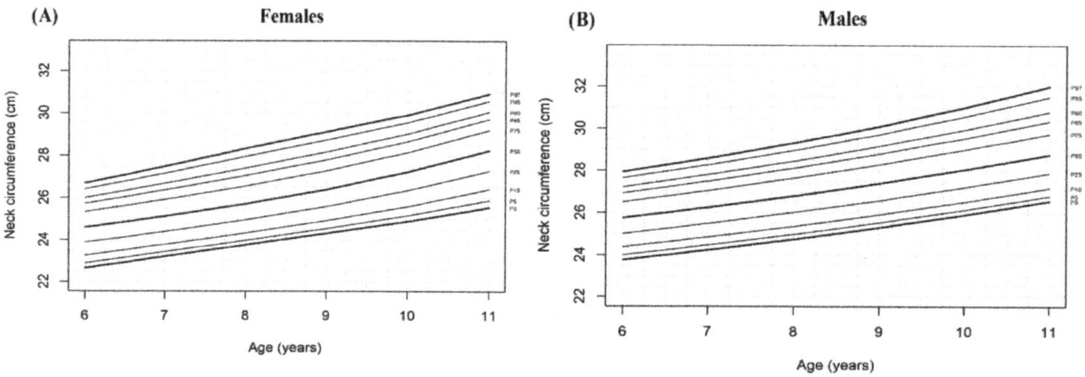

Figure 1. Percentiles of neck circumference of females (**A**) and males (**B**) aged 6–11 years in Acatlán de Juarez and Villa Corona Jalisco, México.

4. Discussion

This study provides reference percentile values for neck circumference from a Mexican schoolchildren cohort. Remarkably, as of the submission of this report, there was no other similar study for the Mexican population. Given their correlation with upper body adiposity, our results may be applied as references in future research and in clinical practice to identify individuals at risk for overweight and obesity [14,26,27].

The correlation values we presented here were statistically significant but lower than those we previously reported (where overweight and obese children were included) [14]. This reinforces the hypothesis that the correlation values tend to be higher for these children. In general, BMI and WC were the variables with the highest correlation values with NC, which was consistent with other studies that included children, regardless of body weight [26,28–30].

Neck circumference was greater for the male than for the female participants, and it showed an age-dependent increase. The age group with the smallest increment was for children between 6 and 7 years old, whereas the most pronounced increment was for children between 10 and 11 years old for both sexes. This increment might be related to the onset of puberty, which normally begins at 10–11 years and 13–14 years for girls and boys, respectively. Most importantly, female participants had the greatest yearly increment, regardless of age, for neck circumference, waist circumference, and body fat percentage. This finding represents relevant evidence of the association between neck circumference and adiposity indicators in this gender.

The percentile values presented herein were similar to those reported by Katz [31] for Canadian children. These results can be explained by the fact that both samples included children of normal weight. Although the Canadian study presented data from 6-17-year-old participants, the values appear to be consistent with those in our study of Mexican children up to 11 years of age. The 50th percentile values for girls were slightly lower in our study, with differences ranging from 0.0 to 0.3 cm. Nevertheless, the values for boys were higher than those in Katz's report, differing from 0.0 to 0.5. This was not the case for the report of European children by Nagy [32], who included normal-weight subjects, with NC values for both males and females 6–10 years of age displaying greater differences, ranging from 0.2 to 0.9 cm, than in our study.

Our data were also similar to those reported by Mazicioglu [26] in Turkey. The values of female subjects at the 50th percentile varied between 24.9 and 28.5 cm, close to the data reported herein (24.6 to 28.3 cm). Male participants showed a similar pattern, ranging from 25.6 to 28.8 cm versus 25.8 to 28.8 cm for the Turkish data and Mexican data, respectively. Notably, the values represent Turkish children between the 3rd and 97th percentiles for weight. Similarly, the percentiles reported by Hosseini for Iranian children showed values at the 50th percentile ranging from 26.3 to 28.3 cm for males and 25.4 to 28.2 cm for females [33]. Nonetheless, these percentiles were based on children who were 7 years of age and older, regardless of weight.

On the other hand, Coutinho's report [34] on Brazilian children showed higher values at the 50th percentile compared to those in our study. This difference was more notable for girls 6–8 years of age (0.3–1.2 cm). For boys, the difference was less (0.1–0.6 cm). However, these percentiles were calculated based on children with NC values within $\pm$3 SD. A graphical comparison between the 25th, 50th and 75th percentiles for NC data from the aforementioned countries is shown in Figure 2.

According to the WHO, to create reference growth charts, it is necessary to include normal weight subjects only to represent the ideal increment. It is evident that the data we present showed lower values compared to most of the studies we cited, except for those studies that included only normal-weight children.

It has been previously mentioned in other research reports that neck circumference represents certain advantages over other indicators, such as BMI or waist circumference [13,28,35]. Neck circumference is an adiposity indicator for the upper body segment, and it is not necessarily repeated to obtain a reliable measurement, as it can be taken at any time of the day without variation.

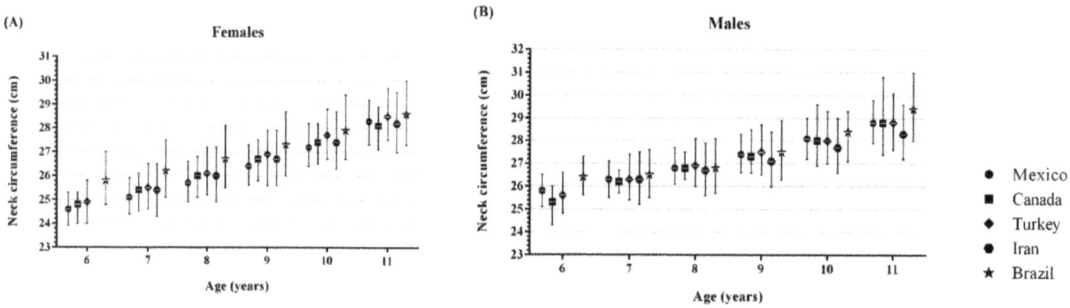

Figure 2. Comparison of 25th, 50th and 75th percentiles values for neck circumference (NC) among females (**A**) and males (**B**) from different countries. Note: Mexico and Canada [31] included normal weight children. Turkey [26] removed the extreme values by determining the 3rd–97th percentiles for weight. Iran [33] included the entire sample. Brazil [34] included children with NC values within ±3 SD.

One limitation of this study might be the fact that the sample size was selected by convenience. Furthermore, it is important to point out that although this study included a considerable sample size, due to the differences in body composition related to the diversity of the Mexican population, the data do not represent the entire country's population. Notably, although we adopted the WHO recommendations to create a pattern reference, we included only the BMI/age standard to classify normal weight children.

5. Conclusions

This study presents the first reference percentile values for the neck circumference of Mexican children, which may be applied to the identification of subjects far from the mean as well as for clinical follow-up.

Author Contributions: Conceptualization, E.V.-S. and C.C.-P.; methodology, J.R.V.-R. and C.O.R.-G.; formal analysis, J.R.V.-R.; investigation, E.V.-S., C.C.-P. and C.O.R.-G.; writing—original draft preparation, E.V.-S., C.C.-P. and J.R.V.-R.; writing—review and editing, E.R.-V., E.M.V.-G. and A.L.-H.; supervision, E.R.-V., E.M.V.-G. and A.L.-H. All authors have read and agreed to the published version of the manuscript.

Funding: This research was funded by the Tresmontes Lucchetti Company, Universidad de Guadalajara, and the Instituto Tecnológico de Estudios Superiores de Occidente. They had no role in the design, analysis, or writing of this article.

Institutional Review Board Statement: The study was conducted according to the guidelines of the Declaration of Helsinki and approved by the Comité de Ética en Investigación del Centro Universitario de Tonalá (003–2016).

Informed Consent Statement: Verbal informed consent was obtained and formally recorded from all the subjects and their tutors. Written informed consent was obtained from all the school directors.

Data Availability Statement: Data supporting reported results are available upon request to the corresponding author.

Acknowledgments: We acknowledge the financial funding from the Tresmontes Lucchetti Company and the support of the National Institute of Public Health. The authors thank Roberto Rodriguez-Echevarria, for manuscript review.

Conflicts of Interest: The funders had no role in the design of the study; in the collection, analyses, or interpretation of data; in the writing of the manuscript, or in the decision to publish the results.

References

1. Instituto Nacional de Salud Pública. I.N. de E. y G. Encuesta Nacional de Salud y Nutrición. 2018. Available online: https://ensanut.insp.mx/encuestas/ensanut2018/informes.php (accessed on 13 July 2020).
2. Lurbe, E.; Alvarez, V.; Redon, J. Obesity, body fat distribution, and ambulatory blood pressure in children and adolescents. *J. Clin. Hypertens. (Greenwich)* **2001**, *3*, 362–367. [CrossRef]
3. Javed, A.; Jumean, M.; Murad, M.H.; Okorodudu, D.; Kumar, S.; Somers, V.K.; Sochor, O.; Lopez-Jimenez, F. Diagnostic performance of body mass index to identify obesity as defined by body adiposity in children and adolescents: A systematic review and meta-analysis. *Pediatr. Obes.* **2015**, *10*, 234–244. [CrossRef]
4. Kaufer-Horwitz, M.; Toussaint, G. Indicadores antropométricos para evaluar sobrepeso y obesidad en pediatría. *Bol. Med. Hosp. Infant. Mex.* **2008**, *65*, 502–518.
5. Vorwieger, E.; Kelso, A.; Steinacker, J.M.; Kesztyüs, D. Cardio-metabolic and socio-environmental correlates of waist-to-height ratio in German primary schoolchildren: A cross-sectional exploration. *BMC Public Health* **2018**, *18*, 1–10. [CrossRef]
6. de Filgueiras, M.S.; de Albuquerque, F.M.; Castro, A.P.P.; Rocha, N.P.; Milagres, L.C.; de Novaes, J.F. Neck circumference cutoff points to identify excess android fat. *J. Pediatr. Rio. J.* **2020**, *96*, 356–363. [CrossRef]
7. Valencia-Sosa, E.; González-Pérez, G.J.; Martinez-Lopez, E.; Rodriguez-Echevarria, R. Neck and Wrist Circumferences as Indicators of Metabolic Alterations in the Pediatric Population: A Scoping Review. *Children* **2021**, *8*, 297. [CrossRef] [PubMed]
8. Mucelin, E.; Traebert, J.; Zaidan, M.A.; Piovezan, A.P.; Nunes, R.D.; Traebert, E. Accuracy of neck circumference for diagnosing overweight in six- and seven-year-old children. *J. Pediatr. Rio. J.* **2021**. [CrossRef]
9. Ma, C.; Wang, R.; Liu, Y.; Lu, Q.; Liu, X.; Yin, F. Diagnostic performance of neck circumference to identify overweight and obesity as defined by body mass index in children and adolescents: Systematic review and meta-analysis. *Ann. Hum. Biol.* **2017**, *44*, 223–229. [CrossRef] [PubMed]
10. Asif, M.; Aslam, M.; Wyszyńska, J.; Altaf, S.; Ahmad, S. Diagnostic performance of neck circumference and cut-off values for identifying overweight and obese pakistani children: A receiver operating characteristic analysis. *JCRPE J. Clin. Res. Pediatr. Endocrinol.* **2020**, *12*, 366–376. [CrossRef] [PubMed]
11. Hatipoğlu, N.; Doğan, S.; Mazıcıoğlu, M.M.; Kurtoğlu, S. Relationship between Neck Circumference and Non-Alcoholic Fatty Liver Disease in Childhood Obesity. *J. Clin. Res. Pediatr. Endocrinol.* **2016**, *8*, 32–39. [CrossRef] [PubMed]
12. Nafiu, O.O.; Zepeda, A.; Curcio, C.; Prasad, Y. Association of neck circumference and obesity status with elevated blood pressure in children. *J. Hum. Hypertens.* **2014**, *28*, 263–268. [CrossRef] [PubMed]
13. Androutsos, O.; Grammatikaki, E.; Moschonis, G.; Roma-Giannikou, E.; Chrousos, G.P.; Manios, Y.; Kanaka-Gantenbein, C. Neck circumference: A useful screening tool of cardiovascular risk in children. *Pediatr. Obes.* **2012**, *7*, 187–195. [CrossRef]
14. Valencia-Sosa, E.; Chávez-Palencia, C.; Romero-Velarde, E.; Larrosa-Haro, A.; Vásquez-Garibay, E.M.; Ramos-García, C.O. Neck circumference as an indicator of elevated central adiposity in children. *Public Health Nutr.* **2019**, *22*, 1755–1761. [CrossRef] [PubMed]
15. Li, X.H.; Lin, S.; Guo, H.; Huang, Y.; Wu, L.; Zhang, Z.; Ma, J.; Wang, H.J. Effectiveness of a school-based physical activity intervention on obesity in school children: A nonrandomized controlled trial. *BMC Public Health* **2014**, *14*. [CrossRef] [PubMed]
16. Habicht, J.P. Standardization of quantitative epidemiological methods in the field. *Bol. Oficina Sanit. Panam.* **1974**, *76*, 375–384. [PubMed]
17. Stewart, A.A.; Marfell-Jones, M.; Olds, T.; Al., E. *International Standards for Anthropometric Assessment*; International Society for the Advancement of Kinanthropometry: Lower Hutt, New Zeland, 2011; ISBN 0620362073.
18. Slaughter, A.M.H.; Lohman, T.G.; Boileau, R.A.; Horswill, C.A.; Stillman, R.J.; Loan, M.D.V.A.N.; Bemben, D.A. Skinfold Equations for Estimation of Body Fatness in Children and Youth. *Hum. Biol.* **2013**, *60*, 709–723. Available online: http://www.jstor.org/stable/41464064 (accessed on 15 January 2019).
19. De Onis, M.; Onyango, A.W.; Borghi, E.; Siyam, A.; Nishida, C.; Siekmann, J. Development of a WHO growth reference for school-aged children and adolescents. *Bull. World Health Organ.* **2007**, *85*, 660–667. [CrossRef]
20. Taylor, R. Interpretation of the Correlation Coefficient: A Basic Review. *J. Diagn. Med. Sonogr.* **1990**, *6*, 35–39. [CrossRef]
21. Stasinopoulos, M.D.; Rigby, R.A.; Heller, G.Z.; Voudouris, V.; De Bastiani, F. Flexible regression and smoothing: Using GAMLSS in R. *Flex. Regres. Smoothing Using GAMLSS R* **2017**, 1–549. [CrossRef]
22. Indrayan, A. Demystifying LMS and BCPE methods of centile estimation for growth and other health parameters. *Indian Pediatr.* **2014**, *51*, 37–43. [CrossRef]
23. Rigby, R.A.; Stasinopoulos, D.M. Smooth centile curves for skew and kurtotic data modelled using the Box-Cox power exponential distribution. *Stat. Med.* **2004**, *23*, 3053–3076. [CrossRef]
24. Rigby, R.A.; Stasinopoulos, D.M. Using the Box-Cox t distribution in GAMLSS to model skewness and kurtosis. *Stat. Model.* **2006**, *6*, 209–229. [CrossRef]
25. Stasinopoulos, D.M.; Rigby, R.A. Generalized Additive Models for Location Scale and Shape (GAMLSS) in R. *J. Stat. Softw.* **2007**, *23*, 1–46. [CrossRef]
26. Mazicioglu, M.M.; Kurtoglu, S.; Ozturk, A.; Hatipoglu, N.; Cicek, B.; Ustunbas, H.B. Percentiles and mean values for neck circumference in Turkish children aged 6–18 years. *Acta Paediatr. Int. J. Paediatr.* **2010**, *99*, 1847–1853. [CrossRef] [PubMed]

27. Da Silva, C.C.; Zambon, M.P.; Vasques, A.C.; Rodrigues, A.M.; Camilo, D.F.; Antonio, M.A.; Cassani, R.S.; Geloneze, B. Neck circumference as a new anthropometric indicator for prediction of insulin resistance and components of metabolic syndrome in adolescents: Brazilian Metabolic Syndrome Study. *Rev. Paul. Pediatr.* **2014**, *32*, 221–229. [CrossRef]
28. Lou, D.-H.; Yin, F.-Z.; Wang, R.; Ma, C.-M.; Liu, X.-L.; Lu, Q. Neck circumference is an accurate and simple index for evaluating overweight and obesity in Han children. *Ann. Hum. Biol.* **2012**, *39*, 161–165. [CrossRef] [PubMed]
29. Taheri, M.; Kajbaf, T.Z.; Taheri, M.R.; Aminzadeh, M. Neck circumference as a useful marker for screening overweight and obesity in children and adolescents. *Oman Med. J.* **2016**, *31*, 170–175. [CrossRef] [PubMed]
30. Kelishadi, R.; Djalalinia, S.; Motlagh, M.E.; Rahimi, A.; Bahreynian, M.; Arefirad, T.; Ardalan, G.; Safiri, S.; Hasani, M.; Asayesh, H.; et al. Association of neck circumference with general and abdominal obesity in children and adolescents: The weight disorders survey of the CASPIAN-IV study. *BMJ Open* **2016**, *6*, 1–10. [CrossRef]
31. Katz, S.L.; Vaccani, J.-P.; Clarke, J.; Hoey, L.; Colley, R.C.; Barrowman, N.J. Creation of a reference dataset of neck sizes in children: Standardizing a potential new tool for prediction of obesity-associated diseases? *BMC Pediatr.* **2014**, *14*, 159. [CrossRef] [PubMed]
32. Nagy, P.; Kovacs, E.; Moreno, L.A.; Veidebaum, T.; Tornaritis, M.; Kourides, Y.; Siani, A.; Lauria, F.; Sioen, I.; Claessens, M.; et al. Percentile reference values for anthropometric body composition indices in European children from the IDEFICS study. *Int. J. Obes.* **2014**, *38*, 15–25. [CrossRef]
33. Hosseini, M.; Motlagh, M.E.; Yousefifard, M.; Qorbani, M.; Ataei, N.; Asayesh, H.; Yaseri, M.; Mohammadi, R.; Baikpour, M.; Abbasi, A.; et al. Neck circumference percentiles of Iranian children and adolescents: The weight disorders survey of CASPIAN IV study. *Int. J. Endocrinol. Metab.* **2017**, *15*. [CrossRef] [PubMed]
34. Coutinho, C.A.; Longui, C.A.; Monte, O.; Conde, W.; Kochi, C. Measurement of neck circumference and its correlation with body composition in a sample of students in Sao Paulo, Brazil. *Horm. Res. Paediatr.* **2014**, *82*, 179–186. [CrossRef] [PubMed]
35. Nafiu, O.O.; Burke, C.; Lee, J.; Voepel-Lewis, T.; Malviya, S.; Tremper, K.K. Neck circumference as a screening measure for identifying children with high body mass index. *Pediatrics* **2010**, *126*, e306–e310. [CrossRef] [PubMed]

children

MDPI

Article

Comparison of Bioelectrical Impedance-Based Methods on Body Composition in Young Patients with Obesity

Alexandra Thajer [1,*], Gabriele Skacel [1], Katharina Truschner [1], Anselm Jorda [1], Martin Vasek [1], Brian Horsak [2], Johanna Strempfl [3], Alexandra Kautzky-Willer [4], Franz Kainberger [5] and Susanne Greber-Platzer [1]

1 Division of Pediatric Pulmonology, Allergology and Endocrinology, Department of Pediatrics and Adolescent Medicine, Medical University of Vienna, Waehringer Guertel 18–20, 1090 Vienna, Austria; gabriele.skacel@akhwien.at (G.S.); kathianna568@gmail.com (K.T.); anselm.jorda@meduniwien.ac.at (A.J.); martin.vasek@meduniwien.ac.at (M.V.); susanne.greber-platzer@meduniwien.ac.at (S.G.-P.)

2 Institute of Health Sciences, St. Pölten University of Applied Sciences, Matthias-Corvinus-Straße 15, 3100 St. Pölten, Austria; brian.horsak@fhstp.ac.at

3 Department of Physiotherapy, St. Pölten University of Applied Sciences, Matthias-Corvinus-Straße 15, 3100 St. Pölten, Austria; Johanna.Strempfl@fhstp.ac.at

4 Gender Medicine Unit, Division of Endocrinology and Metabolism, Department of Internal Medicine III, Medical University of Vienna, Waehringer Guertel 18–20, 1090 Vienna, Austria; alexandra.kautzky-willer@meduniwien.ac.at

5 Division of Neuro- and Musculoskeletal Radiology, Department of Biomedical Imaging and Image-Guided Therapy, Medical University of Vienna, Waehringer Guertel 18–20, 1090 Vienna, Austria; franz.kainberger@meduniwien.ac.at

* Correspondence: alexandra.thajer@meduniwien.ac.at

Citation: Thajer, A.; Skacel, G.; Truschner, K.; Jorda, A.; Vasek, M.; Horsak, B.; Strempfl, J.; Kautzky-Willer, A.; Kainberger, F.; Greber-Platzer, S. Comparison of Bioelectrical Impedance-Based Methods on Body Composition in Young Patients with Obesity. *Children* **2021**, *8*, 295. https://doi.org/10.3390/children8040295

Academic Editor: Odysseas Androutsos

Received: 3 March 2021
Accepted: 7 April 2021
Published: 11 April 2021

Abstract: (1) Background: The determination of body composition is an important method to investigate patients with obesity and to evaluate the efficacy of individualized medical interventions. Bioelectrical impedance-based methods are non-invasive and widely applied but need to be validated for their use in young patients with obesity. (2) Methods: We compiled data from three independent studies on children and adolescents with obesity, measuring body composition with two bioelectrical impedance-based devices (TANITA and BIACORPUS). For a small patient group, additional data were collected with air displacement plethysmography (BOD POD) and dual-energy X-ray absorptiometry (DXA). (3) Results: Our combined data on 123 patients (age: 6–18 years, body mass index (BMI): 21–59 kg/m^2) and the individual studies showed that TANITA and BIACORPUS yield significantly different results on body composition, TANITA overestimating body fat percentage and fat mass relative to BIACORPUS and underestimating fat-free mass ($p < 0.001$ for all three parameters). A Bland–Altman plot indicated little agreement between methods, which produce clinically relevant differences for all three parameters. We detected gender-specific differences with both methods, with body fat percentage being lower ($p < 0.01$) and fat-free mass higher ($p < 0.001$) in males than females. (4) Conclusions: Both bioelectrical impedance-based methods provide significantly different results on body composition in young patients with obesity and thus cannot be used interchangeably, requiring adherence to a specific device for repetitive measurements to ascertain comparability of data.

Keywords: pediatric obesity; body composition analysis; bioelectrical impedance analysis; air displacement plethysmography; DXA; body fat percentage; fat mass; fat free mass

1. Introduction

Obesity is a health issue of world-wide relevance and of increasing prevalence, considered to have reached the level of a pandemic disease over the past 50 years [1]. Besides the often-perceived general reduction of quality of life, obesity is known to profoundly increase the risk for numerous diseases such as diabetes [2], cardiovascular diseases [3], cancer [4], and, as has recently become evident, the severity of viral diseases such as COVID-19 [5,6].

The enhanced prevalence of obesity is already detectable in children and adolescents and a recent study found that 90% of children with obesity at the age of 3 years have overweight at adolescence [7], and almost 60% with obesity at 2 to 19 years were predicted to be obese at an age of 35 [8]. Available longitudinal data suggest that early obesity also enhances the risk of developing metabolic syndrome [9] among numerous other complications [10], and long-term observational studies even identified childhood adiposity as a predictor for morbidity and mortality independent of the weight status in adulthood [11]. The early detection of signs indicating enhanced risk of developing obesity such as unusually high body fat or body fat percentage during childhood and adolescence thus appears of critical importance to allow one to take preventive measures or to initiate medical interventions before the development of obesity-related diseases is triggered.

Among the different methods to diagnose overweight and obesity, the determination of body fat and of total body composition is of particular relevance. Different approaches for correct classification include measurements with bioelectrical impedance analysis (BIA) [12], air displacement plethysmography [13], and dual X-ray absorptiometry (DXA) [14], all of which are reportedly used in clinical practice. DXA is described as a reference method for body composition determination [11,15] but is an expensive tool therefore with limited availability in centers and patients are exposed to radiation. Bioelectrical impedance-based determination of body composition is a relatively simple method, not very time-consuming, non-invasive, and does not expose the patient to radiation [12]. It is widely applied in clinical practice and appears ideally suited for the use with children and adolescents [16]. However, there are different devices available using bioelectrical impedance measuring principles and only limited data on young patients with obesity are available. It is crucial to fill this gap in the literature concerning different bioelectrical impedance devices in children and adolescents with obesity.

Although these methods are well established for the application with adult patients, data on their use in children and adolescents with obesity are associated with considerable uncertainty, since several assumptions underlying these methods may be valid in adults but not in children. For example, changes in protein, water and mineral content occur during growth and maturation and affect the estimates of percent fat and fat-free mass [17]. Additionally, the share of different body compartments is not a constant from childhood to adulthood and shows considerable developmental differences between boys and girls [18]. Observations in 5–7 year old school kids even indicated changes in fat mass and distribution that occur totally independently from changes in body mass index (BMI) [15]. Thus, methods applied to determine body composition in adults need to be validated for their use in children [17,19]. In addition, as different methods may produce variable results, it is important to know how these methods compare with each other to be able to set into perspective data generated using a particular methodology and to evaluate the impact of individualized treatment regimes.

Recently, we could publish differences related to body composition in children with obesity examined by two methods based on bioelectrical impedance, TANITA (Type BC-418MA, TANITA Corporation, Tokyo, Japan), and BIACORPUS (BIACORPUS RX 4000, Medical Healthcare GmbH, Karlsruhe, Germany) [20]. These data were based on a small study population, but still indicated that TANITA overestimated body fat percentage and fat mass and underestimated fat free mass compared to BIACORPUS [20]. In the present study, the main objective was to reevaluate these results and to specifically test in a larger study population whether the use of different devices for the determination of body composition in children and adolescents yields different results. For this purpose, we thus evaluated different devices and compared them with each other to obtain insight into the accuracy and reliability of the methods for the determination of body composition in patients with obesity during childhood and adolescence and to assess their usability in routine clinical practice.

Data on body composition assessments of children with obesity were obtained with TANITA and BIACORPUS. In addition, in order to expand this comparison and to obtain evidence as to which further comparison may be important for future research, these bioelectrical impedance-based data were compared to air displacement plethysmography BOD POD (BOD POD COSMED Inc., Concord, CA, USA) and DXA.

2. Materials and Methods

2.1. Patient Recruitment

Data from patients of three different studies performed at the Medical University Vienna were analyzed. All patients were recruited at the outpatient clinic of obesity, lipometabolic disorder and nutritional medicine at the Department of Pediatrics and Adolescent Medicine, Medical University of Vienna. The obesity outpatient clinic provides care by an interdisciplinary team of doctors, dieticians, nurses, psychologists and social workers.

Study #1, the BODCOP study—a retrospective study on data obtained between August 2015 and May 2016, included 67 children and adolescents who were patients at the above-named institution for whom body composition with TANITA and BIACORPUS, and anthropometric data were routinely assessed during their treatment schedule. All patients were aged under 18 years and had a body mass index (BMI) exceeding the 97th percentile according to the recommendations given by Kromeyer-Hauschild and colleagues [21]. The primary end point of this study was the comparison of both devices for measuring body composition in obese and non-obese individuals. The study was approved by the local Ethics Committee of the Medical University Vienna (EC Nr: 1357/2016).

Study #2, Children's KNEEs study—a randomized controlled trial performed between September 2015 and May 2017, with data from 44 patients which were collected during a 12-week strength and neuromuscular exercise program for the lower extremity on knee load, pain and function in pediatric patients with obesity [22]. Patients were aged between 10 and 18 years and exceeded the 97th percentile described by Kromeyer-Hauschild [21]. The exclusion criteria were: (i) syndromes associated with obesity (e.g., Prader–Willi syndrome); (ii) chronic joint diseases; (iii) neuro-motor diseases; and (iv) any history of a lower extremity joint surgery. The bioelectrical impedance-based methods, TANITA and BIACORPUS, were performed in all patients. The primary end point of this study was to assess the impact of an exercise program for the lower extremity on knee load, pain and function in young patients with obesity. The study was approved by the local Ethics Committee of the Medical University Vienna (EC Nr: 1445/2013) and was registered at clinicaltrials.gov (accessed on 4 March 2021) NCT02545764.

Study #3, the MotiMove Study—a pilot for an intervention study on movement in adolescents with obesity conducted between October and December 2019, included 12 patients aged between 14 and 18 years with a BMI exceeding the 90th percentile according to Kromeyer-Hauschild [21]. Despite the slightly different approach to include patients with a BMI above the 90th percentile, all patients in fact exceeded the 97th percentile as in the other two studies. The exclusion criteria were: (i) syndromes associated with obesity (e.g., Prader–Willi syndrome); (ii) chronic joint diseases; (iii) neuro-motor diseases; and (iv) any history of a lower extremity joint surgery. The primary end point of this study was to determine the effect of a physiotherapeutic group training on the motivation to perform physical activity in adolescents with obesity. In all patients, TANITA and BIACORPUS measurements were performed. In a subgroup, BOD POD and DXA were used to measure the body composition.

The Ethical Committee of the Medical University of Vienna approved the study (EC Nr: 1572/2019). In study #2 and #3, written informed consent was obtained from all parents and children.

2.2. Anthropometric Measurements

Anthropometric parameters which were recorded included body height and weight, and circumference at the waist, hip abdomen, and mid upper arm and were measured under the same standardized procedures in all patients. These latter measures of circumferences were not included in the present analysis. Procedures were as outlined previously [20], but a calibrated TANITA scale was used instead of a SECA scale (SECA 959, Seca Gmbh & Co., Hamburg, Germany). Body weight was measured in patients wearing light underwear, in standing position and within 0.1 kg precision. These values were also applied for the calculation of BIACORPUS-based body composition determination. Body height was measured in an upright facing position with feet together and the back against the wall to the nearest 0.1 cm as the maximum distance between the floor and the highest point on the head. Body weight and body height of each patient were assessed as a prerequisite to allow for the determination of the body composition with the methods used here.

2.3. *Measurements of Body Composition*

The present study applied four different methods for the assessment of body composition, two of these based on bioelectrical impedance analysis (TANITA, BIACORPUS), one on air displacement plethysmography (BOD POD), and one on dual X-ray absorptiometry (DXA). The main parameters determined and analyzed were body fat percentage (BFP, %), fat mass (FM, kg) and fat-free mass (FFM, kg). The two devices based on bioelectrical impedance analysis, TANITA and BIACORPUS, were performed in all patients. BOD POD and DXA were only conducted in patients from study #3. All body composition measurements were performed under the same standardized procedures. To reduce measurement variability, all measurements in each patient were conducted by the same trained team member, and under the same conditions such as overnight fast of the patient, with an empty bladder, wearing only light clothes, at the same daytime, and measurements were performed at room temperature by calibrated devices.

2.4. *TANITA and BIACORPUS*

The assessment of body composition with the devices followed standard procedures as outlined before [20]. Patients were measured with the TANITA scale (Type BC-418MA, TANITA Corporation, Tokyo, Japan) and immediately afterwards with the BIACORPUS device (BIACORPUS RX 4000, Medical Healthcare GmbH, Karlsruhe, Germany), using a frequency setting of 50 kHz and applying 8 electrodes for both devices, assuring maximum comparability. Patients were measured in the morning, had fasted overnight and had emptied their bladder prior to the measurement.

2.5. *BOD POD*

The determination of body composition with air displacement plethysmography was conducted with the BOD POD Gold Standard Body Composition Tracking System (BOD POD COSMED Inc., Concord, CA, USA), following the instructions of the manufacturer. This type of measurement applies the principles of whole-body densitometry to determine body composition. For this, a high-precision weight assessment is combined with a body volume measurement and determination of the thoracic gas volume, and then, based on densitometric equations BFP, FM and FFM are calculated.

2.6. *DXA*

The DXA-based method for the determination of body composition applies the fact that fat and non-fat tissue attenuate X-rays to a different extent, allowing one to calculate the proportion of each compartment following a whole-body scan.

Whole-body DXA was performed with a Hologic Horizon system A (Hologic Inc., Marlborough, MA, USA), applying the latest available Hologic Horizon A software release, for estimating body composition by the same trained technicians according to the prescriptions by the manufacturer and after daily calibration. Patients were scanned according to the most recent version of the official position for pediatric patients of the International Society for Clinical Densitometry (ISCD 2019). All patients were placed in supine position with the limbs in a standardized way according to the guidelines. Previous fractures or systemic diseases with a potential to influence the results had been excluded prior to the investigation and were excluded during reporting by checking the shape of the bones and the symmetry of the density values on the scout image. The precision and accuracy of DXA is reported as acceptable to define body composition in children [23]. Fat mass, lean tissue mass, and bone mineral content of the whole body were measured. The radiation dose, expressed as dose area product, was 7.9 cGy·cm^2.

2.7. Statistics

Statistical analyses were performed with the SPSS software package (SPSS Inc., Chicago, IL, version 24.0). Data were analyzed for normal distribution allowing the application of parametric tests. The results are expressed as means ± SD unless otherwise indicated. Assumptions were checked before conducting parametric tests. Data on body composition methods were analyzed with paired *t*-tests and a *p*-value of < 0.05 was considered as statistically significant. Further, data obtained with TANITA and BIACORPUS were compared by the Pearson correlation coefficient, and the Bland–Altman plot was used to assess the extent of agreement between the two methods [24]. The 95% limits of agreement were calculated as the mean difference ± 1.96 SD [25].

3. Results

3.1. General Characteristics of the Study Population

The data obtained from three independent studies were either separately evaluated for each study or presented as a pooled study population. As shown in Table 1, the common patient population consisted of 123 children and adolescents with 63% male participants and a mean age of 13.6 years. Study #3 included only adolescents aged 14 to 18 years. The overall BMI was 35.2 kg/m^2 and slightly higher in study group 3 with an average BMI of 37 kg/m^2.

Table 1. Demographic patient characteristics and anthropometric parameters.

Characteristics	All Patients (n = 123)	Study 1 (n = 67)	Study 2 (n = 44)	Study 3 (n = 12)
Male (absolute; %)	78 (63)	43 (64)	28 (64)	7 (58)
Age (years)	13.6 ± 2.6	13.4 ± 2.6	13.3 ± 2.4	15.9 ± 1.4
Height (cm)	165.2 ± 12.1	165.0 ± 12.7	164.2 ± 11.4	169.8 ± 10.4
Weight (kg)	97.5 ± 27.6	98.0 ± 28.7	94.0 ± 27.1	107.4 ± 21.8
BMI (kg/m^2)	35.2 ± 7.0	35.4 ± 7.5	34.3 ± 6.8	37.0 ± 4.7

Results are presented as means ± SD or as numbers of subjects (%); BMI body mass index.

3.2. Comparison of Bioelectric Impedance-Based Methods

Values on body composition determined with the bioelectrical impedance-based methods, TANITA and BIACORPUS are summarized in Table 2. In the overall study population with 123 patients, BFP amounted to 41.4% when determined with TANITA and 39.1% when assessed with the BIACORPUS device, a statistically highly significant difference ($p < 0.001$). Similar effects were shown for the FM with an average of 41.6 kg determined with TANITA and a significantly lower value of 38.6 kg determined with BIACORPUS ($p < 0.001$). In contrast, the average FFM of 56.2 kg seen with TANITA was significantly lower than that of 58.9 kg seen with BIACORPUS ($p < 0.001$). These differences were highly significant in patients from study #1 and #2, while in study #3 the differences were not significant, presumably due to the lower patient frequencies.

Table 2. Comparison of body composition parameters determined with bioelectrical impedance-based methods.

	All Patients (n = 123)			Study 1 (n = 67)			Study 2 (n = 44)			Study 3 (n = 12)		
	TANITA	BIACORPUS	p-Value	TANITA	BIACORPUS	p-Value	TANITA	BIACORPUS	p-Value	TANITA	BIACORPUS	p-Value
BFP (%)	41.4 ± 7.8	39.1 ± 6.7	0.000	41.6 ± 8.1	39.4 ± 6.8	0.000	40.7 ± 7.7	37.9 ± 6.0	0.000	43.0 ± 7.0	42.1 ± 8.2	0.481
FM (kg)	41.6 ± 17.8	38.6 ± 14.3	0.000	41.8 ± 18.5	39.0 ± 14.6	0.000	39.3 ± 17.4	36.3 ± 13.8	0.001	48.6 ± 14.5	45.2 ± 12.7	0.252
FFM (kg)	56.2 ± 14.2	58.9 ± 16.2	0.000	56.2 ± 14.3	59.1 ± 16.9	0.000	54.7 ± 13.6	57.7 ± 15.4	0.001	61.3 ± 15.9	62.1 ± 15.6	0.520

Results are presented as mean ± SD; comparison of the body composition methods was made using paired samples t-test; BFP body fat percentage, FM fat mass, FFM fat free mass. TANITA (Type BC-418MA, TANITA Corporation, Tokyo, Japan), BIACORPUS (BIACORPUS RX 4000, Medical Healthcare GmbH, Karlsruhe, Germany).

TANITA and BIACORPUS showed a significant difference in BFP ($p < 0.01$) and FFM ($p < 0.001$) between male and female participants, but not in FM (Table 3). As a result, BFP was lower in males compared to females. FFM was considerably higher in males than in females. Most of these differences were also observed and remained significant in study #1 and #2, except BFP in the TANITA of study #2, and in the BIACORPUS of study #3. The total FM measured with TANITA was higher in males than females, although this was not significant. In contrast, FM measured with BIACORPUS was equal in both genders and even slightly increased in females.

Table 3. Gender-specific values of body composition parameters determined with TANITA and BIACORPUS.

	All Patients (n = 123)			All Patients (n = 123)		
Parameter	Male (n = 78)	Female (n = 45)	p-value	Male (n = 78)	Female (n = 45)	p-value
		TANITA			BIACORPUS	
BFP (%)	39.9 ± 8.1	43.9 ± 6.9	0.007	36.5 ± 5.3	43.6 ± 6.6	0.000
FM (kg)	42.6 ± 18.8	39.8 ± 16.1	0.408	38.2 ± 13.7	39.4 ± 15.3	0.638
FFM (kg)	60.8 ± 14.3	48.1 ± 9.7	0.000	64.9 ± 16.0	48.5 ± 10.1	0.000
	Study 1 (n = 67)			Study 1 (n = 67)		
Parameter	Male (n = 43)	Female (n = 24) TANITA	p-value	Male (n = 43)	Female (n = 24) BIACORPUS	p-value
BFP (%)	39.9 ± 8.4	44.5 ± 6.8	0.024	36.8 ± 5.6	44.0 ± 6.2	0.000
FM (kg)	42.5 ± 19.4	40.5 ± 17.1	0.682	38.5 ± 14.1	39.9 ± 15.8	0.700
FFM (kg)	60.9 ± 13.9	48.8 ± 10.8	0.000	65.0 ± 16.7	48.5 ± 10.3	0.000
	Study 2 (n = 44)			Study 2 (n = 44)		
Parameter	Male (n = 28)	Female (n = 16) TANITA	p-value	Male (n = 28)	Female (n = 16) BIACORPUS	p-value
BFP (%)	39.9 ± 8.2	42.1 ± 6.6	0.362	36.8 ± 4.9	41.4 ± 6.3	0.002
FM (kg)	40.8 ± 18.8	36.6 ± 14.7	0.443	36.4 ± 13.8	36.0 ± 14.3	0.912
FFM (kg)	58.8 ± 14.0	47.5 ± 9.3	0.006	63.3 ± 15.5	48.1 ± 9.5	0.001
	Study 3 (n = 12)			Study 3 (n = 12)		
Parameter	Male (n = 7)	Female (n = 5) TANITA	p-value	Male (n = 7)	Female (n = 5) BIACORPUS	p-value
BFP (%)	40.6 ± 5.4	46.5 ± 8.0	0.158	37.7 ± 5.1	48.3 ± 8.0	0.017
FM (kg)	50.0 ± 15.0	46.5 ± 15.2	0.698	43.2 ± 11.7	48.1 ± 14.8	0.528
FFM (kg)	68.2 ± 17.6	51.6 ± 5.5	0.072	70.8 ± 14.0	50.0 ± 7.8	0.014

Results are presented as mean ± SD; comparison of the body composition methods was made using independent samples t-test; BFP, body fat percentage, FM, fat mass, FFM, fat free mass.

For further evaluation of differences in body composition determined with TANITA or BIACORPUS, Pearson correlation coefficients were used, and Bland–Altman plots generated. As shown in Figure 1, the Pearson correlation was highly significant between the TANITA and BIACORPUS-based determination of BFP (Pearson $r = 0.817$; $p < 0.01$, $n = 123$), of FM (Pearson $r = 0.953$; $p < 0.01$, $n = 123$), and of FFM (Pearson $r = 0.941$; $p < 0.01$, $n = 123$). This contrasts with the result obtained with the Bland–Altman plot, which displays the mean difference between the methods ± 1.96 SD. This analysis indicated important, potentially clinically relevant differences between both methods for BFP, FM and FFM. Thus, the average difference and the 95% limits of agreement was amounted to −2.27% (average) and from −11.17% to 6.62% (range) for BFP, to −2.93 kg (average) and from −14.81 kg to 8.95 kg (range) for FM, and to 2.69 kg (average) and from −8.17 kg to 13.56 kg (range) for FFM. The discrepancy between the methods increased at higher values of BFP, resulting in little agreement in the upper range of the data. Similar discrepancies were determined for FM, which differed by ± 12 kg at a mean of around 40 kg and displayed considerable skewness in the upper data range, and for FFM with a ± 11 kg range at a mean of around 57 kg.

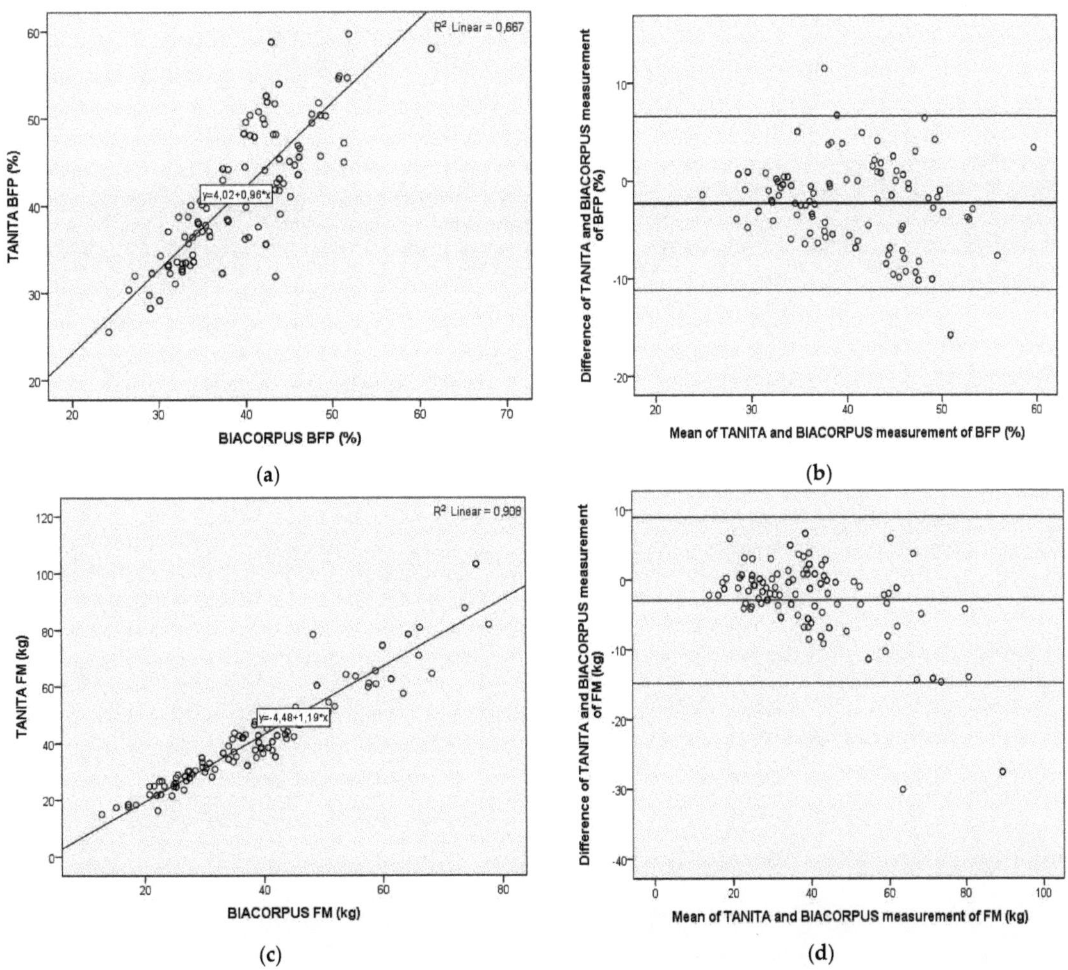

Figure 1. *Cont.*

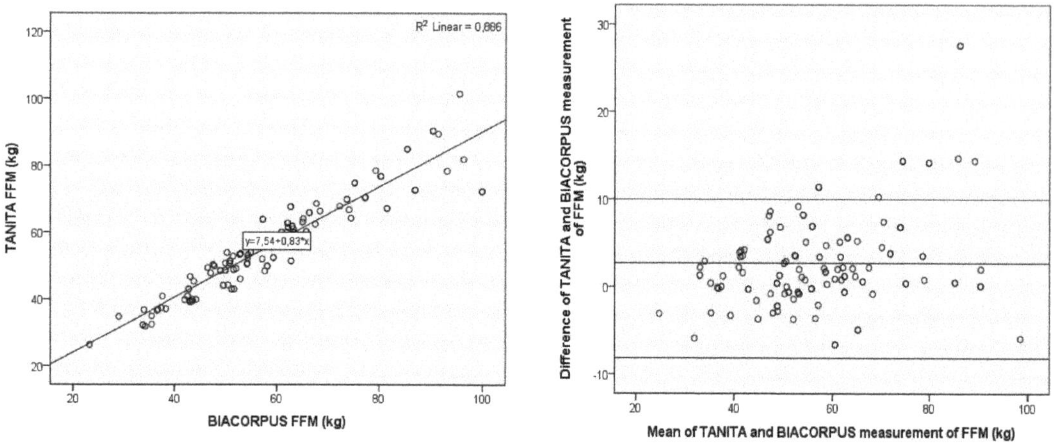

(e) (f)

Figure 1. Pearson correlation analysis and Bland–Altman plot of body composition data obtained with two bioelectrical impedance-based methods, TANITA (Type BC-418MA, TANITA Corporation, Tokyo, Japan) and BIACORPUS (BIACORPUS RX 4000, Medical Healthcare GmbH, Karlsruhe, Germany). Left panel: Pearson correlation for body fat percentage (**a**), fat mass (**c**), and fat free mass (**e**), with R^2 and the equation of the best linear fit given for each correlation plot. Right panel: Bland–Altman plots for body fat percentage (**b**), fat mass (**d**), fat free mass (**f**), with the middle line indicating the mean difference and the upper- and lower-line representing limits of agreement. BFP = body fat percentage, FM = fat mass, FFM = fat-free mass.

3.3. Comparison of BOD POD- and DXA-Based Determination of Body Composition

In study #3, the body composition from 12 individuals was obtained with TANITA and BIACORPUS. In addition, seven patients were also subjected to comparative measurements using air displacement plethysmography and DXA-based assessments. In this sub-group, three persons were male (43%), the average age was 16.1 $\pm$ 1.5 years, and mean ($\pm$SD) values for height, weight and BMI were 166.9 $\pm$ 10.2 cm, 105.1 $\pm$ 17.3 kg, and 37.0 $\pm$ 4.7 kg/m^2, respectively.

In spite of apparent differences, the most pronounced related to the determination of FM (ranging from 48.6 to 53.2 kg); due to the small sample size, we could not detect significant differences between methods (Table 4). Overall, there was a satisfying agreement for the BFP values, only BOD POD determined somewhat higher values, and a greater variability in for FM, which was lowest with BIACORPUS, and a reasonably good agreement for values on FFM.

Table 4. Comparison of body composition data obtained with TANITA, BIACORPUS, BOD POD and DXA.

Parameter	TANITA	BIACORPUS	BOD POD	DXA
BFP (%)	46.3 ± 7.3	46.0 ± 8.0	48.5 ± 4.8	46.8 ± 3.4
FM (kg)	53.2 ± 16.3	48.6 ± 12.1	50.6 ± 8.0	51.5 ± 7.7
FFM (kg)	56.1 ± 11.2	56.6 ± 12.5	54.5 ± 12.2	56.6 ± 10.7

Results are presented as mean $\pm$ SD, n = 7; DXA: dual-energy X-ray absorptiometry, BFP: body fat percentage; FM: fat mass; FFM: fat free mass.

4. Discussion

In line with a previous study [20], our current assessment, using a much larger sample size (n = 123 as compared to n = 38), confirmed that body composition determination in young patients with obesity yields significantly different results depending on whether it is conducted using the TANITA or the BIACORPUS device. The TANITA-based determination of body composition overestimated BFP and FFM relative to BIACORPUS, and significantly underestimated FFM. Pearson correlation indicated an expected association for all three parameters between both methods, and Bland–Altmann plot tests established clinically relevant differences. At an average BFP of around 40% measured with both methods, a difference of −2.3% may not seem excessive. However, these differences ranged from −11% to 6.6%, which appears rather large. In addition, we observed that the discrepancy between both methods increased at higher values of BFP. This showed considerable skewness for FM in the upper data range and was also unsatisfying for FFM. It seems obvious that body composition should be measured with the same bioelectrical impedance-based method. Hence, we recommend using exclusively the same BIA-device, either TANITA or BIACORPUS, for both clinical and research baseline and follow-up measurements of body composition within a study population group. This is supported by findings on adolescent patients with obesity [26] and on adult women with obesity, the latter study comparing TANITA with yet another device based on the same measuring principle [27]. Similarly, in a comparison of two overall rather similar bioelectrical impedance-based method with DXA in Taiwanese children, the impedance devices were found to overestimate FFM and to underestimate FM and BFP [16] and to agree with DXA to a different extent.

We could also confirm previously indicated gender-related differences in body composition [20]. Independent from the bioelectrical impedance-based method, male patients had significantly lower BFP and significantly higher FFM, whereas FM did not differ between genders. Physiologically, these differences may reflect the larger muscle mass in males and the enhanced fat deposition in developing females [28]. Overall, these data underline that gender differences need to be taken into account when comparing different groups [29].

For the same patient group, the four methods for body composition, TANITA, BIACORPUS, BOD POD and DXA showed a good agreement for BFP and FFM, but a higher variability for FM. It appears that TANITA and BIACORPUS, which are non-invasive and easily applied devices, could be appropriate to measure body composition in young patients with obesity. Previous studies found that bioelectrical impedance-based methods may be accurate enough to provide reliable measurements [30,31], and in other studies a more limited reliability was defined [26,32,33], except for individuals with obesity, but not for patients with morbid obesity [34], or a clear need of optimization for young patients with obesity [35], highlighting the need to be cautious when applying it. It thus appears important that future research conducts a thorough comparison of TANITA and BIACORPUS with DXA. As reported before, bioelectrical impedance-based methods may in some regards show excellent agreement with DXA but in others display considerable discrepancies [16,36]. It thus appears that for each specific device, a comparison is required to provide ultimate insight as to which bioelectrical impedance-based method is best comparable to the reference method.

Some limitations have to be mentioned in interpreting our results. Most obviously, the sample size of the group with the BODPOD and DXA method comparison is rather small and therefore the results have to be considered as preliminary. An increased and consistent sample size in the comparison of all four body composition methods is the next step to apply a more robust methodology and to make sure that the outcomes are generalizable to the pediatric population affected by obesity. Another research focus should be especially the comparison of the different bioelectrical impedance devices with the gold standard DXA measurement to identify whether TANITA or BIACORPUS is more accurate in detecting the true body composition. The discrepancies between devices increases proportional to body fat percentage and therefore a follow-up study with lower body mass index percentile should be conducted to complement this study.

A major strength of the study is that it fills a gap concerning different bioelectrical impedance analysis methods in pediatric individuals with obesity in a young western population. Other strengths are that the same devices, the same trained study team members as well as standardized procedures were used in all body composition measurements assuring that factors affecting measuring variability other than the devices applied were kept to a minimum.

5. Conclusions

The evaluation of data from three independent studies in young patients with obesity could show that the body composition differed significantly for body fat percentage, fat mass and fat free mass depending on the bioelectrical impedance-based devices. TANITA and BIACORPUS, which are appropriate for clinical and research body composition measurements, cannot be used interchangeably and need to be strictly adhered to in order to receive reliable repetitive data. We recommend consistently using one of the easily applicable and non-invasive bioelectrical impedance-based devices, either TANITA or BIACORPUS, in pediatric patients with obesity. Further research on the comparison of the different bioelectrical impedance-based methods with DXA in young patients with obesity is warranted.

Author Contributions: Conceptualization, A.T.; methodology, A.T.; formal analysis, A.T.; investigation, A.T., G.S., K.T., A.J. and M.V.; resources, A.K.-W., F.K. and S.G.-P.; data curation, A.T.; writing—original draft preparation, A.T.; writing—review and editing, G.S., K.T., A.J., M.V., B.H., J.S., A.K.-W., F.K., S.G.-P.; supervision, S.G.-P.; project administration, A.T., B.H. and J.S.; funding acquisition, B.H. All authors have read and agreed to the published version of the manuscript.

Funding: The Children's KNEEs study was funded by NFB—Lower Austrian Research and Education Company and Provincial Government of Lower Austria, Department of Science and Research (LSC13–009).

Institutional Review Board Statement: The three independent studies were approved by the Institutional Review Board of the Medical University Vienna (study #1: EC Nr: 1357/2016; study #2: EC Nr: 1445/2013; and study #3: EC Nr: 1572/2019).

Informed Consent Statement: Written informed consent has been obtained from the patients in study #2 and #3. In study #1, a retrospective study, a written informed consent was not applicable.

Data Availability Statement: Data and intervention materials are available upon request to the corresponding author.

Acknowledgments: We would like to thank all children and adolescents who have participated in these three studies. The authors also thank the BODCOP, Children's KNEEs, the MotiMove study team and Thomas Thajer for his great assistance in visualization of the data.

Conflicts of Interest: The authors declare no conflict of interest.

Abbreviations

BFP: Body fat percentage; BIA: Bioelectrical impedance analysis; BMI: Body mass index; DXA: Dual energy X-ray absorptiometry; FFM: Fat free mass; FM: Fat mass

References

1. Blüher, M. Obesity: Global Epidemiology and Pathogenesis. *Nat. Rev. Endocrinol.* **2019**, *15*, 288–298. [CrossRef]
2. Bhupathiraju, S.N.; Hu, F.B. Epidemiology of Obesity and Diabetes and Their Cardiovascular Complications. *Circ. Res.* **2016**, *118*, 1723–1735. [CrossRef]
3. Cercato, C.; Fonseca, F.A. Cardiovascular Risk and Obesity. Diabetol. *Metab. Syndr.* **2019**, *11*, 1–15.
4. Allott, E.H.; Hursting, S.D. Obesity and Cancer: Mechanistic Insights from Transdisciplinary Studies. *Endocr. Relat. Cancer* **2015**, *22*, R365–R386. [CrossRef]
5. de Siqueira, J.V.V.; Almeida, L.G.; Zica, B.O.; Brum, I.B.; Barceló, A.; de Siqueira Galil, A.G. Impact of Obesity on Hospitalizations and Mortality, Due to COVID-19: A Systematic Review. *Obes. Res. Clin. Pract.* **2020**. [CrossRef]
6. Kass, D.A.; Duggal, P.; Cingolani, O. Obesity Could Shift Severe COVID-19 Disease to Younger Ages. *Lancet* **2020**, *395*, 1544–1545. [CrossRef]
7. Geserick, M.; Vogel, M.; Gausche, R.; Lipek, T.; Spielau, U.; Keller, E.; Pfäffle, R.; Kiess, W.; Körner, A. Acceleration of BMI in Early Childhood and Risk of Sustained Obesity. *N. Engl. J. Med.* **2018**, *379*, 1303–1312. [CrossRef] [PubMed]

8. Ward, Z.J.; Long, M.W.; Resch, S.C.; Giles, C.M.; Cradock, A.L.; Gortmaker, S.L. Simulation of Growth Trajectories of Childhood Obesity into Adulthood. *N. Engl. J. Med.* **2017**, *377*, 2145–2153. [CrossRef] [PubMed]
9. Kelsey, M.M.; Zaepfel, A.; Bjornstad, P.; Nadeau, K.J. Age-Related Consequences of Childhood Obesity. *Gerontology* **2014**, *60*, 222–228. [CrossRef] [PubMed]
10. Lee, Y.S. Consequences of Childhood Obesity. Ann. Acad. *Med. Singap.* **2009**, *38*, 75–81.
11. Must, A.; Jacques, P.F.; Dallal, G.E.; Bajema, C.J.; Dietz, W.H. Long-Term Morbidity and Mortality of Overweight Adolescents. *N. Engl. J. Med.* **1992**, *327*, 1350–1355. [CrossRef]
12. Ward, L.C. Bioelectrical Impedance Analysis for Body Composition Assessment: Reflections on Accuracy, Clinical Utility, and Standardisation. *Eur. J. Clin. Nutr.* **2019**, *73*, 194–199. [CrossRef]
13. Fields, D.A.; Gunatilake, R.; Kalaitzoglou, E. Air Displacement Plethysmography: Cradle to Grave. *Nutr. Clin. Pract.* **2015**, *30*, 219–226. [CrossRef]
14. Shepherd, J.A.; Ng, B.K.; Sommer, M.J.; Heymsfield, S.B. Body Composition by DXA. *Bone* **2017**, *104*, 101–105. [CrossRef]
15. Mast, M.; Körtzinger, I.; König, E.; Müller, M.J. Gender Differences in Fat Mass of 5-7-Year Old Children. *Int. J. Obes.* **1998**, *22*, 878–884. [CrossRef] [PubMed]
16. Lee, L.W.; Liao, Y.S.; Lu, H.K.; Hsiao, P.L.; Chen, Y.Y.; Chi, C.C.; Hsieh, K.C. Validation of Two Portable Bioelectrical Impedance Analyses for the Assessment of Body Composition in School Age Children. *PLoS ONE* **2017**, *12*. [CrossRef]
17. Lohman, T.G.; Going, S.B. Body Composition Assessment for Development of an International Growth Standard for Preadolescent and Adolescent Children. *Food Nutr. Bull.* **2006**, *27*. [CrossRef] [PubMed]
18. Weber, D.R.; Leonard, M.B.; Zemel, B.S. Body Composition Analysis in the Pediatric Population. Pediatr. *Endocrinol. Rev.* **2012**, *10*, 130–139.
19. Wells, J.C.K.; Williams, J.E.; Chomtho, S.; Darch, T.; Grijalva-Eternod, C.; Kennedy, K.; Haroun, D.; Wilson, C.; Cole, T.J.; Fewtrell, M.S. Body-Composition Reference Data for Simple and Reference Techniques and a 4-Component Model: A New UK Reference Child. *Am. J. Clin. Nutr.* **2012**, *96*, 1316–1326. [CrossRef]
20. Kreissl, A.; Jorda, A.; Truschner, K.; Skacel, G.; Greber-Platzer, S. Clinically Relevant Body Composition Methods for Obese Pediatric Patients. *BMC Pediatr.* **2019**, *19*. [CrossRef] [PubMed]
21. Kromeyer-Hauschild, K.; Wabitsch, M.; Kunze, D.; Geller, F.; Geiß, H.C.; Hesse, V.; Von Hippel, A.; Jaeger, U.; Johnsen, D.; Korte, W.; et al. Percentiles of Body Mass Index in Children and Adolescents Evaluated from Different Regional German Studies. *Monatsschr. Kinderheilkd.* **2001**, *149*, 807–818. [CrossRef]
22. Horsak, B.; Artner, D.; Baca, A.; Pobatschnig, B.; Greber-Platzer, S.; Nehrer, S.; Wondrasch, B. The Effects of a Strength and Neuromuscular Exercise Programme for the Lower Extremity on Knee Load, Pain and Function in Obese Children and Adolescents: Study Protocol for a Randomised Controlled Trial. *Trials* **2015**, *16*, 1–10. [CrossRef] [PubMed]
23. Wohlfahrt-Veje, C.; Tinggaard, J.; Winther, K.; Mouritsen, A.; Hagen, C.P.; Mieritz, M.G.; De Renzy-Martin, K.T.; Boas, M.; Petersen, J.H.; Main, K.M. Body Fat throughout Childhood in 2647 Healthy Danish Children: Agreement of BMI, Waist Circumference, Skinfolds with Dual X-Ray Absorptiometry. *Eur. J. Clin. Nutr.* **2014**, *68*, 664–670. [CrossRef] [PubMed]
24. Martin Bland, J.; Altman, D.G. Statistical Methods for Assessing Agreenebt between Two Methods of Clinical Measurement. *Lancet* **1986**, *327*, 307–310. [CrossRef]
25. Bland, J.M.; Altman, D.G. Comparing Methods of Measurement: Why Plotting Difference against Standard Method Is Misleading. *Lancet* **1995**, *346*, 1085–1087. [CrossRef]
26. Lazzer, S.; Boirie, Y.; Meyer, M.; Vermorel, M. Evaluation of Two Foot-to-Foot Bioelectrical Impedance Analysers to Assess Body Composition in Overweight and Obese Adolescents. *Br. J. Nutr.* **2003**, *90*, 987–992. [CrossRef] [PubMed]
27. Minderico, C.S.; Silva, A.M.; Keller, K.; Branco, T.L.; Martins, S.S.; Palmeira, A.L.; Barata, J.T.; Carnero, E.A.; Rocha, P.M.; Teixeira, P.J.; et al. Usefulness of Different Techniques for Measuring Body Composition Changes during Weight Loss in Overweight and Obese Women. *Br. J. Nutr.* **2008**, *99*, 432–441. [CrossRef]
28. Leonard, M.B.; Elmi, A.; Mostoufi-Moab, S.; Shults, J.; Burnham, J.M.; Thayu, M.; Kibe, L.; Wetzsteon, R.J.; Zemel, B.S. Effects of Sex, Race, and Puberty on Cortical Bone and the Functional Muscle Bone Unit in Children, Adolescents, and Young Adults. *J. Clin. Endocrinol. Metab.* **2010**, *95*, 1681–1689. [CrossRef]
29. Kirchengast, S. Gender Differences in Body Composition from Childhood to Old Age: An Evolutionary Point of View. *J. Life Sci.* **2010**, *2*, 1–10. [CrossRef]
30. Brantlov, S.; Ward, L.C.; Jødal, L.; Rittig, S.; Lange, A. Critical Factors and Their Impact on Bioelectrical Impedance Analysis in Children: A Review. *J. Med. Eng. Technol.* **2017**, *41*, 22–35. [CrossRef]
31. Leskinen, T.; Eloranta, A.M.; Tompuri, T.; Saari, A.; Ollila, H.; Mäkelä, J.; Niinikoski, H.; Lagström, H. Changes in Body Composition by Age and Obesity Status in Preschool-Aged Children: The STEPS Study. *Eur. J. Clin. Nutr.* **2020**, 1–9. [CrossRef] [PubMed]
32. Sim, P.Y.; Su, T.T.; Majid, H.A.; Nahar, A.M.; Jalaludin, M.Y. A Comparison Study of Portable Foot-to-Foot Bioelectrical Impedance Scale to Measure Body Fat Percentage in Asian Adults and Children. *Biomed Res. Int.* **2014**, *2014*. [CrossRef] [PubMed]
33. Eisenmann, J.C.; Heelan, K.A.; Welk, G.J. Assessing Body Composition among 3- to 8-Year-Old Children: Anthropometry, BIA, and DXA. *Obes. Res.* **2004**, *12*, 1633–1640. [CrossRef] [PubMed]
34. Verney, J.; Metz, L.; Chaplais, E.; Cardenoux, C.; Pereira, B.; Thivel, D. Bioelectrical Impedance Is an Accurate Method to Assess Body Composition in Obese but Not Severely Obese Adolescents. *Nutr. Res.* **2016**, *36*, 663–670. [CrossRef]

35. Lyons-Reid, J.; Ward, L.C.; Kenealy, T.; Cutfield, W. Bioelectrical Impedance Analysis—an Easy Tool for Quantifying Body Composition in Infancy? *Nutrients* **2020**, *12*, 920. [CrossRef]
36. Lim, J.S.; Hwang, J.S.; Lee, J.A.; Kim, D.H.; Park, K.D.; Jeong, J.S.; Cheon, G.J. Cross-Calibration of Multi-Frequency Bioelectrical Impedance Analysis with Eight-Point Tactile Electrodes and Dual-Energy X-Ray Absorptiometry for Assessment of Body Composition in Healthy Children Aged 6-18 Years. *Pediatr. Int.* **2009**, *51*, 263–268. [CrossRef]

children

MDPI

Article

Changes in Body Composition and Anthropomorphic Measurements in Children Participating in Swimming and Non-Swimming Activities

Grzegorz Bielec [1,*], Anna Gozdziejewska [1] and Piotr Makar [2]

1 Department of Tourism, Recreation and Ecology, University of Warmia and Mazury in Olsztyn, ul. Oczapowskiego 5, 10-719 Olsztyn, Poland; gozdzik@uwm.edu.pl
2 Department of Swimming, University School of Physical Education and Sport in Gdansk, ul. Gorskiego 1, 80-336 Gdansk, Poland; piotrmakar@wp.pl
* Correspondence: grzegorz.bielec@uwm.edu.pl

Abstract: Background. Physical activity is a well-known means of obesity prevention, but the relationship between exercise frequency and body composition in children has not been thoroughly investigated. Objective: The aim of this study was to compare the body composition of children aged 11–12 who regularly performed swimming and other sports as an organized extra-curricular physical activity for a 12-week period. Methods: The study included 46 students who attended swimming classes and 42 students who participated in training activities in other sports, including, but not limited to, football, basketball and athletics. Body height and body composition were measured using a Tanita BC 418 MA analyzer. The students individually reported their rate of perceived exertion during training using the Pictorial Children's Effort Rating Table PCERT scale. Results: The weekly volume of training was substantially higher in the group of swimmers than in that playing other sports (12.3 h/week vs. 5.2 h/week, $p < 0.01$). After 12 weeks of training, body height and weight significantly increased in both groups ($p < 0.001$). However, the BMI value and adipose tissue content only increased in the group of non-swimmers. Swimmers perceived greater exertion during training than non-swimmers (7.1 vs. 5.8 on the PCERT scale, $p < 0.01$). Conclusions: In early pubescent children, engaging in vigorous exercise such as swimming for at least 10 h a week may restrain the growth of adipose tissue. However, the variety of exercises that are typical of team sports, if performed for no more than 5 h a week, may be insufficient to restrain adipose tissue growth.

Keywords: body composition; body mass index; physical activity; pubescence; training

Citation: Bielec, G.; Gozdziejewska, A.; Makar, P. Changes in Body Composition and Anthropomorphic Measurements in Children Participating in Swimming and Non-Swimming Activities. *Children* **2021**, *8*, 529. https://doi.org/10.3390/children8070529

Academic Editor: Odysseas Androutsos

Received: 23 April 2021
Accepted: 18 June 2021
Published: 22 June 2021

1. Introduction

Regular physical activity, if performed at a sufficient intensity and with adequate frequency, has a positive effect on cardiovascular, mental and musculoskeletal health, and adiposity [1]. In children and adolescents, the most common result of an insufficient amount of physical activity and an improper diet is being overweight or obese, whereas intensified physical activity reduces body fat content in prepubertal children [2–4]. The World Health Organization (WHO) states that children should be encouraged to participate in various kinds of physical activity to improve cardiorespiratory and muscle fitness and reduce symptoms of anxiety and depression [5]. Following WHO suggestions, American, African and European governmental organizations have presented guidelines on the recommended daily amount of physical activity for children according to regional context [6,7]. In most countries, it is recommended that children engage in moderate-to-vigorous intensity exercise for at least 60 min a day. The energy costs of different sports activities performed by children can vary; for example, gymnastics demands 4 METs (metabolic units), basketball 8.2 METs, soccer 8.8 METs, and swimming from 8.4 to 11.6 METs depending on the velocity of swimming [8]. Additionally, children should limit to 2 h a day the amount of time spent using a computer, playing video games and watching TV.

The level of compliance with the recommended amounts of physical activity in schoolchildren differs widely. In Latin America, 2 h of recreational screen-time is frequently exceeded by school-age children [9]. Moreover, only 15% of Latin American adolescents undertake 60 min of moderate-to-vigorous exercise per day [10]. In the United States, children and adolescents aged 6–17 are also insufficiently active, and only 24% of them engage in 60 min of exercise every day [11]. In South Africa, on the other hand, nearly 70% of 8–14-year-old children undertake moderate-to-vigorous physical activity [12]. In this regard, European countries are very diverse. Swiss children dedicate only a small percentage of their daytime to running or jumping, instead spending 75% of the time in a still position (mostly sitting) [13], whereas in a group of Estonian children, 11% of the children met the desired criteria for physical activity, in a group of British children, 70% of the subjects met the criteria [14,15]. A different study involving 9-year-old British children attending ad hoc and scheduled sports activities with various weekly frequencies led researchers to conclude that more frequent participation in sports activities (up to five times a week) substantially contributes to compliance with the recommended amounts of physical activity [16].

Extracurricular sports activities are usually organized by educational institutions, sports clubs and other organizations that are supervised by adults. Children may often participate in these activities for free, but in sports clubs, a membership fee may be required. Participation in an organized extra-curricular sport has a stronger effect on developmental factors, such as proper body mass, BMI and academic achievements, than spontaneous physical activity [17]. Moreover, organized physical activity can be defined more precisely in terms of training loads than spontaneous, self-organized physical activity.

To our knowledge, no research has been undertaken so far to examine anthropometric differences among children who participate in extra-curricular physical activities of various volumes (e.g., 30 min, 60 min or 90 min) and frequencies (e.g., once a week, three times per week or five times per week). Therefore, the objective of this study was to determine the relationships between body composition and regular physical activity of various volumes (5–12 h/week) performed over a 12-week period of time by 11- to 12-year-old children.

2. Materials and Methods

2.1. Participants

We aimed to create a study group with children with similar socioeconomic backgrounds and similar daily routines due to attendance at the same public school. Therefore, we chose a quasi-experimental, two-group design that used non-random sampling. The study involved 100 students attending a state-funded primary school, which serves approximately 600 students from a range of socioeconomic backgrounds in the city of Olsztyn, Poland. Students at this school attend compulsory physical education classes for 3 h a week. The students were divided into two groups: 50 children (including 21 girls) who regularly participated in swimming workouts in addition to compulsory physical education, and 50 children (including 22 girls) who regularly participated in other forms of extra-curricular physical activity (details on duration and frequency of activity are presented in Results). Children participating in non-swimming sports were grouped together for two reasons: (i) the number of children engaged in each non-swimming discipline (e.g., in basketball) was too small to create separate groups; (ii) the children in the group of non-swimmers constitute all of the school's pupils aged 11–12 years who participated in any organized extra-curricular sports activities other than competitive swimming. The examined children of both groups did not differ substantially in terms of socioeconomic status, i.e., they lived in the same residential area, were being raised by two parents of middle-to-higher education background and had 1–2 siblings. The information on the children's participation in these extra-curricular sporting activities was collected by physical education teachers during school classes. The average age in the group of swimmers was 11.7, whereas that in the group of non-swimmers was 11.9 (non-significant difference).

2.2. Measurements

In September 2018, the first anthropometric measurements were taken in both groups. Body height was measured to one decimal place (0.1 cm) by means of a Seca 216 stadiometer (Seca GmbH, Hamburg, Germany). Body mass, BMI and body fat percentage were estimated using a Tanita BC 418 MA analyzer (Tanita Corp.,Tokyo, Japan). The measurements were conducted in the school nurse's office between 9:00 a.m. and 11:00 a.m. The children were dressed in light clothes without shoes. Each measurement of body height and body composition took about 12 min. The measurements were taken twice and the coefficient of variation (CV) was calculated for each pair of measurements. The range of CV values was from 2.2 to 3.2. The children conducted a self-assessment of the stage of their sexual development at home, in the presence of their parents, on the basis of the Tanner scale [18]. In September 2018, children from both groups were acquainted with the Pictorial Children's Effort Rating Table (PCERT) scale of the subjective assessment of physical exertion [19]. The children were asked to regularly record their assessment of intensity of exertion immediately after their training session every week. The children from the group of swimmers continued to record this every Tuesday for 12 weeks. Non-swimmers recorded the data every Tuesday or Wednesday for the same duration, depending on the day their training was scheduled. In October 2018, both groups were surveyed on their eating habits. The reliability and accuracy of the questionnaire used had been validated by the studies conducted in the past on children aged 10–13 [20]. Eating habits did not differ significantly between groups. In December 2018, after 12 weeks from the anthropometric measurements, follow-up measurements were performed according to the same procedure. As some of the children had been absent from the sports activities (20% or more absences within 12 weeks), the results of 46 children (including 19 girls) from the swimmers and 42 children (including 19 girls) from the non-swimmers were taken into consideration.

2.3. Ethical Clearance

The study was conducted in accordance with the Declaration of Helsinki. The head teacher of the school gave permission for conducting the tests, and the parents of each child provided written consent for their children to participate in the study. Informed consent was also obtained from all children participating in the study. Finally, the Ethics Committee of the University of Warmia and Mazury in Olsztyn approved the study (KB-8/17).

2.4. Statistics

The statistical calculations were performed using Statistica 12.0 software (StatSoft, Tulsa, OK, USA). According to the Shapiro–Wilk test, the distribution of the data was normal. The results of the first and second anthropometric measurements were compared using a t-test. Statistical significance was set at $p < 0.05$. The statistical power of the comparisons varied from 0.32 to 0.46.

3. Results

Table 1 presents the weekly volume of training in both groups and their self-assessments of their physical exercise intensity. The children from the group of non-swimmers declared their participation in the following sports activities: football (9), basketball (6), swimming (6), athletics (4), martial arts (4), dance classes (4), fencing (3) and table tennis (3). To clarify, swimming activities by children in the non-swimmers group differed from the training sessions of those in the swimmers group. Non-swimmers participated in three 45 min classes per week, focusing on technical aspects and usually covering 700–800 m during each class. Swimmers participated in eight to ten 90 min workouts per week, covering 3.6–4.2 km per workout. In individual cases among the non-swimmers, the children stated that they participated in other sports on a regular basis, e.g., acrobatics, horse riding, wrestling and roller skating. The students from the group of swimmers participated in training sessions considerably more frequently. In addition, they perceived their exertion in these sessions as substantially more difficult than that of their peers from the group of non-swimmers.

Swimmers mostly perceived their exertion as "hard", whereas non-swimmers usually expressed their effort as "starting to get hard" and "getting quite hard".

Table 1. Training loads, puberty stage and rate of perceived exertion in the examined groups.

	Number of Training Sessions per Week Mean (SD)	Number of Training Hours per Week Mean (SD)	Puberty Stage (Tanner Scale) Mean (SD)	Rate of Perceived Exertion (PCERT Scale) Mean (SD)
Swimmers (n = 46)	8.2 (1.4) **	12.3 (2.2) **	2.0 (0.6)	7.1 (0.7) **
Non-swimmers (n = 42)	3.5 (0.9)	5.2 (1.8)	2.0 (0.7)	5.8 (0.8)

** $p < 0.01$ compared with non-swimmers group; SD—standard deviation; PCERT—Pictorial Children's Effort Rating Table.

Figures 1–4 present the results of the anthropometric measurements taken in September 2018 and December 2018. After 12 weeks, a noticeable increase in body height and weight was observed in both groups. However, a statistically significant increase in BMI value and body fat percentage was noted only in the group of non-swimmers. The correlation between BMI and body fat percentage was strong both in girls ($r = 0.99$) and in boys ($r = 0.95$).

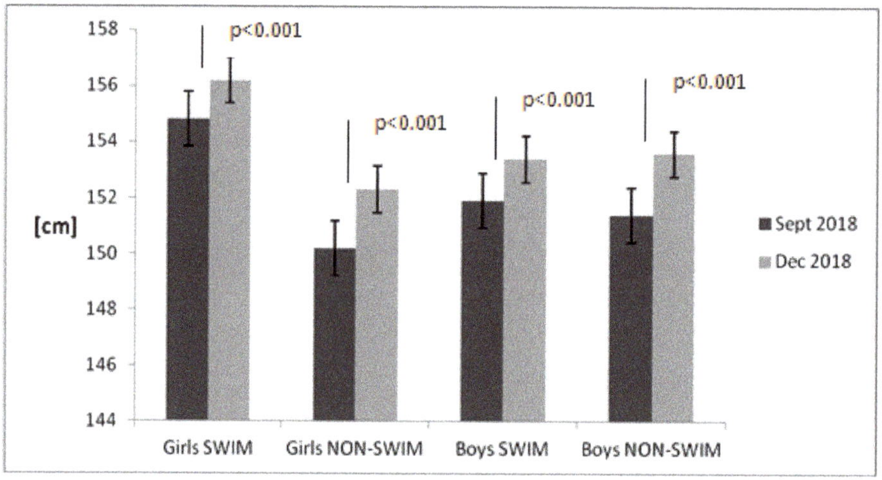

Figure 1. Body height changes in the examined groups of children. Data are presented as mean value and standard error. The non-swimmers group consisted of children participating in team sports (soccer, basketball), athletics, martial arts and dance.

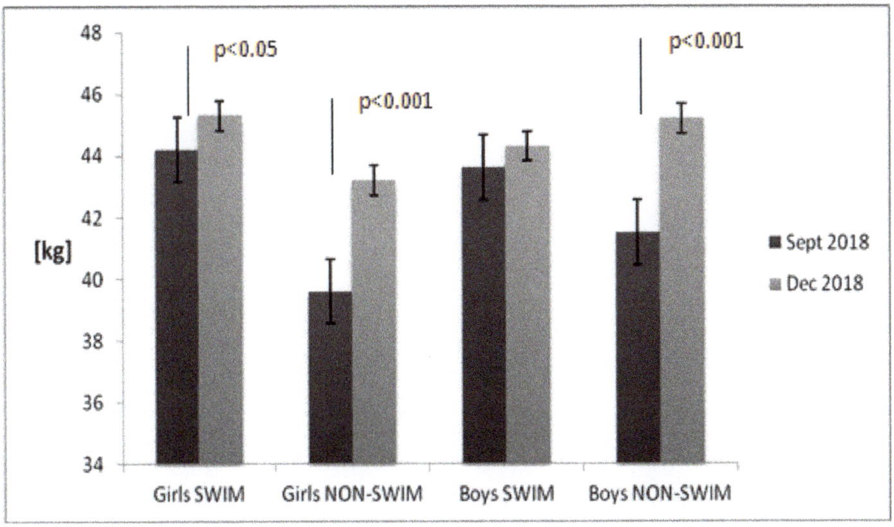

Figure 2. Body mass changes in the examined groups of children. Data are presented as mean value and standard error. The non-swimmers group consisted of children participating in team sports (soccer, basketball), athletics, martial arts and dance.

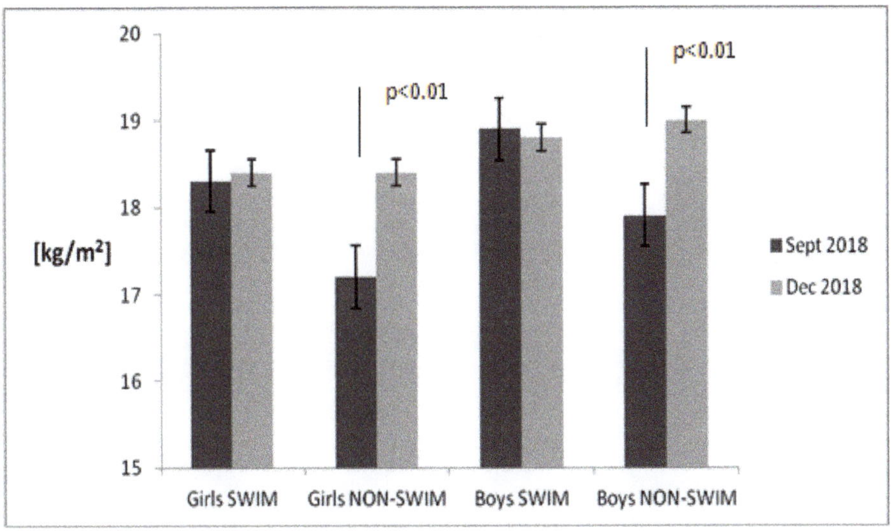

Figure 3. Body mass index (BMI) changes in the examined groups of children. Data are presented as mean value and standard error. The non-swimmers group consisted of children participating in team sports (soccer, basketball), athletics, martial arts and dance.

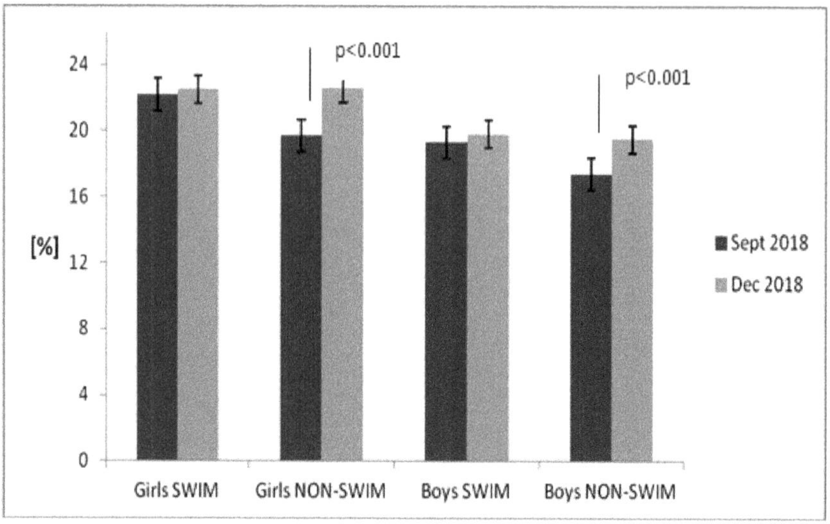

Figure 4. Body fat percentage changes in the examined groups of children. Data are presented as mean value and standard error. The non-swimmers group consisted of children participating in team sports (soccer, basketball), athletics, martial arts and dance.

4. Discussion

The aim of our study was to compare the effects of swimming and participating in other sports on the body composition of 11- to 12-year-old children. Our results indicate that, whereas the BMI and body fat content of swimmers may not change significantly over 12 weeks, they can increase significantly in children participating in other sports. It should be noted, however, that the indicators in both groups were within the national norms for this age group [21]. As in other studies [22,23], the correlation between body mass index and body fat percentage was strong in boys and girls ($r > 0.95$). Moreover, on average, swimmers exercised 12 h per week, and non-swimmers exercised for 5 h per week.

The lack of a significant change in the BMI and adipose tissue content of the swimmers might be due to the mostly aerobic nature of their activities. Laframboise and de Graauw [24] completed an analysis of the literature on the relationships between aerobic exercise and overweight schoolchildren. According to them, aerobic exercise might be recommended for children as a means of reducing body mass. The findings from our research are consistent with the remarks made by Ness et al. [25]. Those authors took anthropometric measurements and analyzed the physical activity of 5500 British 12-year-old children over a 3-day period. They came to the conclusion that unstructured physical activity influences body mass and BMI to a smaller extent than structured moderate-to-vigorous intensity exercise. Similarly, beneficial changes in body composition (i.e., fat mass and free fat mass) were larger in Spanish 11-year-olds that performed vigorous exercise for 12 weeks than in those that performed moderate exercise over this time period [26].

We believe that the weekly number of training sessions in swimmers and non-swimmers may also have affected the results of their body composition. Similar conclusions can be found in the literature. For example, in 10-year-old Brazilian children, the amount of vigorous physical activity taken during 1 week was inversely correlated with their BMI and body fat percentage [27]. Similarly, Pastuszak et al. [28] monitored sports-school students and non sports-school students for a period of over 3 years. Sports-school students participated in 15–25 h of sports activities a week, whereas non sports-school students participated in 4 h of physical education classes at school and 4 h of extra-curricular sports activities a week. Compared to the control group, the sports-school students had lower body mass and BMI values, and lower body fat percentages. In contrast, Kubusiak-Slonina et al. [29] did not find any statistically significant correlations between the BMI and the level of physical activity of 92 12-year-old Polish children. Similarly, BMI and body fat values did not differ significantly between 10-year-old South African children involved in team sports and their peers not participating in extra-curricular sports activities [30].

In the present study, the body composition indicators of the swimmers were very similar to those of swimmers from other European countries [31,32]. It is noteworthy that, in our study, the BMI values of swimmers and non-swimmers were within the national population norms [21]. In other words, the children examined in this study displayed ratios of height to body mass that are typical for their age in their regional and national context. However, studies in non-European contexts have reported noticeable differences in anthropometrics between children of different nationalities who engage in a similar number of training sessions per week with similar training loads [33,34].

The effects of maturation may also have affected the results presented here, particularly with regard to the girls. On the one hand, children involved in regular strenuous swimming workouts may exhibit delayed puberty and not put on as much body fat as children participating in sports requiring less intensive training [35]. For example, Damsgaard et al. [36] found that early pubescent female swimmers had a sum of skinfolds that was lower than that of handball players but higher than that of gymnasts. On the other hand, Opstoel et al. [37] reported that prepubertal swimmers had higher values of body fat and BMI than their peers engaged in athletics, ball sports, gymnastics and dance. This is similar to what we observed, as our swimmers also had greater BMI values and body fat percentage at the baseline than their non-swimming peers. We believe that this could be due to the early phase of the swimmers' training phase. The first measurements were taken in mid-September, just 2 weeks after starting the new school-year season. At that time, the swimmers were not training at high intensities or high volumes, and thus their training sessions were not physiologically demanding.

The results obtained here, based on the PCERT scale, indicate that the swimmers perceived greater mean exertion during training than their peers involved in other sports (swimmers, 7.1; non-swimmers, 5.8). These differences may be caused by physiological factors. A study by Yelling et al. [19] indicated that 11–12-year-old girls and boys had an average heart rate of 170 beats per minute at stage 5 and 180 beats per minute at stage 7 on the PCERT scale, respectively. It is also possible that this perception is due to psychological factors. The swimmers started training every day at 6:00 a.m. and attended a second session in the afternoon three to five times a week. The non-swimmers usually started their exercise sessions in the afternoon and did not attend a second exercise session.

Strengths and Limitations

This study makes a contribution to the literature by providing a comparison of the body composition of prepubescent children participating in different sports. Whereas the literature contains many comparisons of athletes and non-athletes in this regard, we are aware of only one other comparison of the body composition of prepubescent swimmers and that of children of the same age participating in other sports [37].

There are several limitations that should be kept in mind while assessing the results of this study. First, our study contained fewer participants than those in most sport science studies because we used only children who attended the same school. On the one hand, this helped to reduce the confounding effects of socioeconomic status and daily routine on our results. Moreover, because all of these children ate their midday meal in the school canteen, their diets would have been more similar than otherwise. On the other hand, due to the relatively small sample size, our study may have lacked the statistical power to detect smaller differences between groups. Although the power is low, we provide valuable data which can be combined with that of other studies via statistical meta-analysis to provide more precise estimates and higher power, as is commonly done in medical studies of children [38]. Second, heart-rate monitors were not used to measuring training intensity in the groups of swimmers and non-swimmers; thus, it was not possible to collect objective data on exercise intensity. Therefore, we were not able to verify whether the swimmers' workouts were indeed more intense than the training sessions performed by subjects in the other sport disciplines. Third, our analysis of body composition focused on adipose tissue content, which in our view, is the most easily comparable indicator of risk. We considered the remaining variables (e.g., body height or body mass index) to be of less importance for the issue in question. Finally, we decided not to create a control group consisting of children not involved in structured extracurricular sports activities, as many studies have presented significant differences in body composition between physically active children and their inactive peers [39–41]. We also decided not to present the findings from our survey of nutritional habits, because the substantial differences between swimmers and non-swimmers were not revealed. Moreover, the literature provides examples of similar eating habits of youth swimmers and non-swimmers [42].

5. Conclusions

Our results indicate that physical activity perceived as "hard" (7 or more on the PCERT scale) that is engaged in for at least 10 h per week may stop the growth of body fat in prepubescent children. In contrast, activity perceived as "starting to get hard" (5 or less on the PCERT scale) performed for 5 h a week may not be sufficient to stop the growth of body fat in children of this age. Importantly, however, adipose tissue content in both groups remained within the national norms for children at this age. It would be interesting to independently confirm these results with a larger sample of children. Moreover, a confirmatory study could be conducted over a longer period of time to see if the trend of an increasing content of adipose tissue in moderately active children persists.

Author Contributions: G.B.: design, planning, analysis, results interpretation and manuscript writing. A.G.: statistical analysis, results interpretation and manuscript writing. P.M.: results interpretation and manuscript writing. All authors have read and agreed to the published version of the manuscript.

Funding: This study was supported by grant 18.610.010-300 of the Polish Ministry of Science and Higher Education.

Institutional Review Board Statement: The study was conducted in accordance with the Declaration of Helsinki. The Ethics Committee of the University of Warmia and Mazury approved the study (KB-8/17).

Informed Consent Statement: Informed consent was obtained from children participating in the study and their parents.

Data Availability Statement: The data presented in this study are available on request from the corresponding author.

Acknowledgments: We thank Mark Leonard for stimulating discussions and for help with editing the English-language version of this manuscript.

Conflicts of Interest: The authors declare no conflict of interest.

References

1. Strong, W.B.; Malina, R.M.; Blimkie, C.J.; Daniels, S.R.; Dishman, R.K.; Gutin, B.; Hergenroeder, A.C.; Must, A.; Nixon, P.A.; Pivarnik, J.M.; et al. Evidence based physical activity for school-age youth. *J. Pediatr.* **2005**, *146*, 732–737. [CrossRef]
2. Dencker, M.; Thorsson, O.; Karlsson, M.K.; Linden, C.; Wollmer, P.; Andersen, L.B. Daily physical activity related to aerobic fitness and body fat in an urban sample of children. *Scand. J. Med. Sci. Sports* **2008**, *18*, 728–735. [CrossRef]
3. Ostojic, S.M.; Stojanovic, M.D.; Stojanovic, V.; Maric, J.; Njaradi, N. Correlation between fitness and fatness in 6-14-year old Serbian school children. *J. Health Popul. Nutr.* **2011**, *29*, 53–60. [CrossRef] [PubMed]
4. Carlson, J.A.; Crespo, N.C.; Sallis, J.F.; Patterson, R.E.; Elder, J.P. Dietary-related and physical activity related predictors of obesity in children: A 2-year prospective study. *Child Obes.* **2012**, *8*, 110–115. [CrossRef] [PubMed]
5. World Health Organization. *Global Recommendations on Physical Activity for Health*; WHO: Geneve, Switzerland, 2010.
6. Janssen, I. Physical activity guidelines for children and youth. *J. Appl. Physiol. Nutr. Metab.* **2007**, *32*, S109–S121. [CrossRef]
7. Kahlmeier, S.; Wijnhoven, T.M.; Alpiger, P.; Schweizer, C.; Breda, J.; Martin, B.W. National physical activity recommendations: Systematic overview and analysis of the situation in European countries. *BMC Public Health* **2015**, *15*, 133. [CrossRef] [PubMed]

8. Ridley, K.; Olds, T.S. Assigning energy costs to activities in children: A review and synthesis. *Med. Sci. Sports Exerc.* **2008**, *40*, 1439–1446. [CrossRef]
9. Ribeiro, E.H.C.; Guerra, P.H.; De Oliveira, A.C.; Da Silva, K.S.; Santos, P.; Santos, R.; Okely, A.; Florindo, A.A. Latin American interventions in Children and adolescents' sedentary behavior: A systematic review. *Rev. Saude Publ.* **2020**, *54*, 59. [CrossRef]
10. Aguilar-Farias, N.; Martino-Fuentealba, P.; Carcamo-Oyarzun, J.; Cortinez-O'Ryan, A.; Cristi-Montero, C.; von Oetinger, A.; Sadarangani, K.P. A regional vision of physical activity; sedentary behavior; and physical education in adolescents from Latin America and the Carribean: Results from 26 countries. *Int. J. Epidemiol.* **2018**, *47*, 976–986. [CrossRef] [PubMed]
11. Katzmarzyk, P.T.; Denstel, K.D.; Beals, K.; Carlson, J.; Crouter, S.E.; McKenzie, T.L.; Pate, R.R.; Sisson, S.B.; Staiano, A.E.; Stanish, H.; et al. Results from the United States Report Card on Physical Activity for Children and Youth. *J. Phys. Act. Health* **2018**, *15* (Suppl. S2), S422–S424. [CrossRef]
12. Draper, C.E.; Tomaz, S.A.; Bassett, S.H.; Burnett, C.; Christie, C.J.; Cozett, C.; De Milander, M.; Krog, S.; Monyeki, M.A.; Naidoo, N.; et al. Results from South Africa's 2018 Report Card on Physical Activity for Children and Youth. *J. Phys. Act. Health* **2018**, *15* (Suppl. S2), S406–S408. [CrossRef]
13. Ruch, N.; Melzer, K.; Mader, U. Duration, frequency and types of children's activities: Potential of a classification procedure. *J. Exerc. Sci. Fit.* **2013**, *11*, 85–94. [CrossRef]
14. Van Sluijs, E.M.; Skidmore, P.M.; Mwanza, K.; Jones, A.P.; Callaghan, A.M.; Ekelund, U.; Harrison, F.; Harvey, I.; Panter, J.; Wareham, N.J.; et al. Physical activity and dietary behaviour in a population-based sample of British 10-year old children: The SPEEDY study. *BMC Public Health* **2008**, *8*, 388. [CrossRef] [PubMed]
15. Riso, E.M.; Kull, M.; Mooses, K.; Hannus, A.; Jurimae, J. Objectively measured physical activity levels and sedentary time in 7-9-year-old Estonian schoolchildren: Independent associations with body composition parameters. *BMC Public Health* **2016**, *16*, 346. [CrossRef] [PubMed]
16. Jago, R.; MacDonald-Wallis, C.; Solomon-Moore, E.; Thompson, J.L.; Lawlor, D.A.; Sebire, S.J. Associations between participation in organized physical activity in the school or community outside school hours and neighbourhood play with child physical activity and sedentary time: A cross-sectional analysis of primary school-aged children from the UK. *BMJ Open* **2017**, *7*, e017588. [CrossRef] [PubMed]
17. Wiium, N.; Safvenbom, R. Participation in organized sport and self-organized physical activity: Associations and developmental factors. *Int. J. Environ. Res. Public Health* **2019**, *16*, 585. [CrossRef]
18. Roberts, C. Tanner's Puberty Scale: Exploring the historical entanglements of children; scientific photography and sex. *Sexualities* **2016**, *19*, 328–346. [CrossRef]
19. Yelling, M.; Lamb, K.L.; Swaine, I. Validity of a pictorial perceived exertion scale for effort estimation and effort production during stepping exercise in adolescent children. *Eur. Phys. Educ. Rev.* **2002**, *8*, 157–175. [CrossRef]
20. Calyniuk, B.; Ambrozy, J.; Bester, J.; Czuchraj, W.; Dostal, K.; Dubiel, M. Eating habits and frequency of consumption of selected products among children aged 10–13 years residing in urban and rural areas. *Ann. Acad. Med. Siles* **2013**, *67*, 231–237. (In Polish)
21. Stupnicki, R.; Tomaszewski, P.; Milde, K. Body Mass Index—Proposed norms for children and youths. *Pap. Anthropol.* **2013**, *22*, 203–213. [CrossRef]
22. Janjic, J.; Baltic, Z.M.; Glisic, M.; Ivanovic, J.; Boskovic, M.; Popovic, M.; Lovrenovic, M. Relationship between body mass index and body fat percentage among adolescents from Serbian Republic. *J. Child Obes.* **2016**, *1*, 10. [CrossRef]
23. Duarte, M.O.; Ruelas, Y.F.; Lopez-Alcaraz, F.; del Toro-Equihua, M.; Sanchez-Ramirez, C.A. Correlation between percentage of body fat measured by Slaughter equation and bio impendance analysis technique in Mexican schoolchildren. *Nutr. Hosp.* **2014**, *29*, 88–93. [CrossRef]
24. Laframboise, M.A.; de Graauw, C. The effects of aerobic physical activity on adiposity in schoolaged children and youth: A systematic review of randomized controlled trials. *J. Can. Chirop. Assoc.* **2011**, *55*, 256–268.
25. Ness, A.R.; Leary, S.D.; Mattocks, C.; Blair, S.N.; Reilly, J.J.; Wells, J.; Ingle, S.; Tilling, K.; Smith, G.D.; Riddoch, C. Objectively measured physical activity and fat mass in a large cohort of children. *PLoS Med.* **2007**, *4*, e97. [CrossRef]
26. Sanchez, G.F.L.; Suarez, A.D.; Radziminski, L.; Jastrzebski, Z. Effects of a 12-week physical education program on the body composition of 10- and 11-year-old children. *Sci. Sports* **2017**, *32*, e155–e161. [CrossRef]
27. De Moraes Ferrari, G.L.; Oliveira, L.C.; Araujo, T.L.; Matsudo, V.; Barreira, T.V.; Tudor-Locke, C.; Katzmarzyk, P. Moderate-to-vigorous physical activity and sedentary behavior: Independent associations with body composition variables in Brazilian children. *Pediatr. Exerc. Sci.* **2015**, *27*, 380–389. [CrossRef]
28. Pastuszak, A.; Lewandowska, J.; Buśko, K.; Charzewska, J. Effects of elevated physical activity on changes in body composition and subcutaneous fat distribution in boys aged 10 to 16 years: A longitudinal study. *Anthrop. Rev.* **2014**, *77*, 45–55. [CrossRef]
29. Kubusiak-Slonina, A.; Grzegorczyk, J.; Mazur, A. The assessment of fitness and physical activity in school children with excessive and normal weight. *Endokrynol. Otylosc.* **2012**, *8*, 16–23. (In Polish)
30. Hendricks, S.; Uys, M.; Broyles, S.T.; Naidoo, N.; Katzmarzyk, P.T.; Lambert, E.V. Factors associated with team sport participation in South African children. *BMJ Paediatr. Open* **2019**, *3*, e000495. [CrossRef]
31. Jürimäe, J.; Haljaste, K.; Cicchella, A.; Lätt, E.; Purge, P.; Leppik, A.; Jürimäe, T. Analysis of swimming performance from physical; physiological and biomechanical parameters in young swimmers. *Pediatr. Exerc. Sci.* **2007**, *19*, 70–81. [CrossRef]
32. Nikitakis, I.S.; Paradisis, G.P.; Bogdanis, G.C.; Toubekis, A.G. Physiological responses of continuous and intermittent swimming at critical speed and maximum lactate steady state in children and adolescent swimmers. *Sports* **2019**, *7*, 25. [CrossRef] [PubMed]

33. Zhao, K.; Hohmann, A.; Faber, I.; Chang, Y.; Binghong, G. A 2-year longitudinal follow-up of performance characteristics in Chinese male elite youth athletes from swimming and racket sports. *PLoS ONE* **2020**, *15*, e0239155. [CrossRef]
34. Greco, C.C.; Denadai, B.S.; Pellegrinotti, I.L.; Freitas, A.B.; Gomide, E. Limiar anaeróbio e velocidade crítica determinada com diferentes distâncias em nadadores de 10 a 15 anos: Relações com a performance e a resposta do lactato sanguíneo em testes de endurance. *Rev. Bras. Med. Esporte* **2003**, *9*, 2–8. [CrossRef]
35. Kapczuk, K. Elite athletes and pubertal delay. *Minerva Pediatr.* **2017**, *69*, 415–426. [CrossRef] [PubMed]
36. Damsgaard, R.; Bencke, J.; Matthiesen, G.; Petersen, J.H.; Muller, J. Body proportions, body composition and pubertal development of children in competitive sport. Scan. *J. Med. Sci. Sports* **2001**, *11*, 54–60. [CrossRef]
37. Opstoel, K.; Pion, J.; Elferink-Gemser, M.; Hartman, E.; Willemse, B.; Philippaerts, R.; Visscher, C.; Lenoir, M. Anthropometric characteristics; physical fitness and motor coordination of 9 to 11 year old children participating in a wide range of sports. *PLoS ONE* **2015**, *10*, e0126282. [CrossRef]
38. Harris, K.C.; Kuramoto, L.K.; Schulzer, M.; Retallack, J.E. Effect of school-based physical activity intervenions on body mass index in children: A meta-analysis. *CMAJ* **2009**, *180*, 719–726. [CrossRef] [PubMed]
39. Martinez-Gomez, D.; Ruiz, J.R.; Ortega, F.B.; Veiga, O.L.; Moliner-Urdiales, D.; Mauro, B.; Galfo, M.; Manios, Y.; Widhalm, K.; Béghin, L.; et al. Recommended levels of physical activity to avoid an excess of body fat in European adolescents: The HELENA study. *Am. J. Prev. Med.* **2010**, *39*, 2003–2011. [CrossRef]
40. Reichert, F.F.; Wells, J.C.W.; Ekelund, U.; Menezes, A.M.; Victora, C.G.; Hallal, P.C. Prospective associations between physical activity level and body composition in adolescence: 193 Pelotas (Brazil) cohort study. *J. Phys. Act. Health* **2015**, *12*, 834–839. [CrossRef]
41. Telford, R.M.; Telford, R.D.; Cochrane, T.; Cunningham, R.B.; Olive, L.S.; Davey, R. The influence of sport club participation on physical activity; fitness and body fat during childhood and adolescence: The LOOK longitudinal study. *J. Sci. Med. Sport* **2016**, *19*, 400–406. [CrossRef]
42. Collins, A.; Ward, K.D.; Mirza, B.; Slawson, D.; McClanahan, B.; Vukadinovich, C. Comparison of nutritional intake in US adolescent swimmers and non-athletes. *Health* **2012**, *4*, 873–880. [CrossRef] [PubMed]

children

MDPI

Communication

Effects of After-School Volleyball Program on Body Composition in Overweight Adolescent Girls

Nebojša Trajković [1,*], Anja Lazić [1], Drena Trkulja-Petković [2], Valentin Barišić [2], Vladan Milić [3], Siniša Nikolić [4,5] and Goran Sporiš [2]

[1] Faculty of Sport and Physical Education, University of Niš, 180000 Nis, Serbia; anja.lazic96@hotmail.com
[2] Faculty of Kinesiology, University of Zagreb, 10110 Zagreb, Croatia; drena.trkulja-petkovic@kif.unizg.hr (D.T.-P.); valentin.barisic@kif.unizg.hr (V.B.); sporis.79@gmail.com (G.S.)
[3] Study Program of Sport and Physical Education, Department of Biomedical Sciences, State University of Novi Pazar, 36300 Novi Pazar, Serbia; vladan.milic@yahoo.com
[4] Department of Physical Medicine and Rehabilitation "Dr Miroslav Zotović", 78000 Banja Luka, Bosnia and Herzegovina; sinisamnikolicbl@gmail.com
[5] Faculty of Sport and Physical Education, University of Novi Sad, 21000 Novi Sad, Serbia
* Correspondence: nele_trajce@yahoo.com; Tel.: +38-16-5207-0462

Abstract: The current study aimed to investigate the effects of an after-school volleyball program on body composition in overweight adolescent girls. Forty-two girls were randomly divided into a volleyball group (VG) (n = 22 age: 15.6 ± 0.5 years) and control group (CG) (n = 20; age: 15.5 ± 0.7 years). Both groups continued with their usual physical education activities, while VG was included as small-sided games, two times a week, after school, on modified volleyball courts. Body mass, body mass index (BMI), body fat in kg, body fat percentage, and muscle mass were analyzed by a bioelectrical impedance method. There was a significant interaction of group (VG vs. CG) × time (pre-vs. post) for weight [$F_{1,40} = 7.933$; $p = 0.004$] and BMI [$F_{1,40} = 5.764$; $p = 0.015$]. Additionally, a significant main effect of time was found for body fat (kg) [$F_{1,40} = 17.650$; $p < 0.001$] and body fat (%) [$F_{1,40} = 18.721$; $p < 0.001$]. The results of the current study show that a twelve-week after-school volleyball program, including two sessions a week, can improve body composition in overweight adolescent girls.

Keywords: girls; vigorous intensity; body weight status; ball game

Citation: Trajković, N.; Lazić, A.; Trkulja-Petković, D.; Barišić, V.; Milić, V.; Nikolić, S.; Sporiš, G. Effects of After-School Volleyball Program on Body Composition in Overweight Adolescent Girls. *Children* **2022**, 9, 21. https://doi.org/10.3390/children9010021

Academic Editors: Odysseas Androutsos and Antonis Zampelas

Received: 23 November 2021
Accepted: 27 December 2021
Published: 29 December 2021

1. Introduction

Nowadays, in a time of accelerated technological development, obesity in children and adolescents is taking on the character of an epidemic. The number of overweight and obese individuals worldwide has increased from 30 years ago, and, now, almost a third of the world population is defined as obese or overweight [1]. Concerningly, the majority of adolescents do not meet physical activity recommendations despite well-documented health benefits [2]. Although adolescence is a period when long-term healthy eating and exercise habits are formed, more than 340 million young people in the world were overweight or obese in 2016, and this negative trend continues to progress [3]. Accordingly, this age has been recognized as the period that may play a crucial role in the buildout and persistence of obesity and for leading to negative consequences during adulthood. Therefore, an effective approach to engage with youth and ameliorate their body composition status is of great importance. Children and adolescents spend most of their day in school. Therefore, the school setting has been recognized as a potentially effective environment for different exercise interventions [4]. Even though school interventions are effective and practical [5], previous studies [6,7] claimed that most adolescents are inactive during physical education classes and that the intensity of exercise is low. In accordance with all the above, there is a need for additional or different programs.

A previous review concluded that high-intensity interval training (HIIT) could be an effective method in accordance with the short duration of the activity for improving health-related indicators in adolescents [8]. However, there is no clear consensus on the positive effects of HIIT on body composition. Volleyball is considered to be one of the most played team sports worldwide. It is a sport that attracts all structures of the population; in particular, it involves no direct contact with the opponent but requires speed of movement and thinking. Volleyball, as an intermittent and complex activity where high-intensity elements are interrupted by a short period of low-intensity activities [9], could be included in work with obese female adolescents. Taking into account the previous statements that the type, configuration, and popularity of physical activity are required in both childhood and adolescent periods [10], volleyball appears to be an appropriate solution. Additionally, volleyball is one of the most popular physical activities in the aforementioned population [11]. Moreover, the total intensity of movements, according to the ratio of working metabolic rate (MET), can be measured at 6 METs in volleyball [12], which is considered to be vigorous [13]. Consequently, it is important to determine whether or not including volleyball as the only additional physical activity in PE classes can lead to an adjustment in body composition parameters and thus provide considerable health benefits.

Considering that volleyball has its popularity predominantly among the female population, it was of great interest to determine how it affects the body composition of female adolescents. To the best of the authors' knowledge, to date, no study has focused on investigating the effects of small-sided volleyball games on body composition in overweight female adolescents.

Therefore, this study aimed to investigate the effects of a volleyball after-school program on body composition in overweight adolescent girls. We hypothesized that the exercise intervention would cause greater changes in body composition in overweight adolescent girls compared to physical education classes only.

2. Materials and Methods

2.1. Participants

This was a randomized study comparing an after-school volleyball program with traditional physical education classes in overweight female adolescents. Forty-two adolescent female participants from various classes of schools across Southern Serbia were included in the study. The participants were randomly divided into a volleyball group (VG) (n = 20; height: 158.97 $\pm$ 5.39; age: 15.6 $\pm$ 0.5 years) and control group (CG) (n = 22; height: 159.96 $\pm$ 10.71; age: 15.5 $\pm$ 0.7 years). Both groups continued with their usual physical education activities. During the volleyball program, there were no time-loss injuries.

The participants did not have any known medical problems. To be included, the participants had to be aged between 14 and 16 years, present a body mass index (BMI) ranging from the 85th to 95th percentile according to gender and age [14], be free of any medications that could impact the results, and not be involved in any systematic physical activities at the time of study or at least six months before the study (besides regular physical education classes at school, lasting up to 90 min/week).

All participants and parents/guardians were fully informed of the possible risks associated with the experimental procedures and gave their consent to participate in the study. Moreover, the participants and parents/guardians were informed that they could leave the experimental program at any time. The study protocol was approved by the ethics committee of the Faculty of Sport and Physical Education in Niš (Ref. No. 20/2020) and carried out in accordance with the Declaration of Helsinki.

2.2. Procedures

Measurement of body composition of all participants at the initial and final measurement was performed in the same school hall. The measurements were performed early in the morning, before 10 a.m., on both occasions. The testing was conducted by the same researchers on the initial and final measurement and in identical order.

Participants were measured in minimal clothing and barefoot for height and weight. Body height was assessed to the nearest 0.1 cm using a stadiometer (SECA Instruments Ltd.,

Hamburg, Germany). Body mass, BMI, body fat %, body fat in kg, and muscle mass were assessed via a bioelectrical impedance method using a standardized body composition analyzer, the InBody230 (InBody Co., Ltd., Cerritos, CA, USA). BMI was calculated using the following formula: BMI = (Weight in kilograms)/(Height in meters) squared. The InBody230 is a reliable device for women and men with high ICC for BF% ($\geq$0.98) and has a low standard error of measurement [15]. Before the measurements, the participants were advised to excrete and refrain from drinking excessive amounts of liquids and not to deviate from their typical breakfast habits.

2.3. Exercise Intervention

It is widely accepted that children and adolescents should be engaged in strenuous or moderate physical activity for at least 60 min a day, which represents 420 min/week [16]. We included volleyball activities, played as small-sided games, two times a week after school. The experimental program lasted for 12 weeks. Volleyball was played on a court (4.5 to 6 m $\times$ 9 to 12 m), and the number of players was modified from 3 vs. 3 to 5 vs. 5 players. Each session started with a warm-up lasting 10–12 min. Moderate-intensity jogging was performed at the beginning of the warm-up session (3 min), along with dynamic stretching (4 min), and a specific volleyball warm-up with the ball (5 min). The main part of the session consisted of 40 min of volleyball. The session ended with a 5 min cool-down period. The intensity of training was monitored using the rate of perceived exertion (RPE) with a 10-point scale, collected during the training period. The CG undertook their regular PE classes involving team ball games (handball, basketball, or soccer) and individual sporting activities most common in many countries (track and field, gymnastics, or table tennis). In both groups, volleyball was excluded from regular PE classes. Both groups were not engaged in other organized physical activities besides organized volleyball activities and PE classes. The participants who had 70 > percent of attendance during the 12 weeks were taken for further analysis. There was no sample dropout during the experimental program and all participants were included in further analysis.

2.4. Statistical Analysis

Statistical analysis was performed with SPSS statistical program version 22 (SPSS Inc., Chicago, IL, USA). To test the normality, we performed the Kolmogorov–Smirnov test, which showed that the data were normally distributed ($p > 0.05$). Furthermore, Levene's tests were determined for each dependent variable. The effects of the experimental treatment were analyzed using a two-way analysis of variance (ANOVA) in which the main and interaction effects (group $\times$ time) were calculated. Moreover, we used the Cohen's d effect size (ES) with the use of the following criteria: trivial 0.0–0.2; small 0.2–0.6; moderate 0.6–1.2; large 1.2–2.0; and very large > 2.0 [17]. In addition to the statistical significance, which was set at $p < 0.05$, a partial eta squared (η^2) was used for the differences between groups using 0.01, 0.06, and 0.14, determined as a small effect, medium effect, and a large effect [18].

3. Results

3.1. Adherence to the Exercise Program

The average attendance for the experimental program was 92%. The main reasons for skipping the exercise sessions were menstrual cycle, cold, and fatigue.

3.2. Body Composition

The results for body composition are presented in Table 1. There was a significant interaction of group (VG vs. CG) $\times$ time (pre- vs. post) found for weight [$F_{1,40} = 7.933$; $p = 0.004$] and BMI [$F_{1,40} = 5.764$; $p = 0.015$]. Additionally, a significant main effect of time was found for body fat (kg) [$F_{1,40} = 17.650$; $p < 0.001$] and body fat (%) [$F_{1,40} = 18.721$; $p < 0.001$].

Table 1. Results for body composition; results are presented as mean ± standard deviation (SD).

Variable	Group	Pretest	Posttest	ES	% Change	p-Value, η^2_p
Body mass (kg)						
	VG	62.55 ± 5.18	58.20 ± 5.58	−0.54	−6.9%	Group: $p = 0.596$, η^2_p: 0.006
						Time: $p < 0.001$, η^2_p: 0.693
	CG	61.55 ± 5.29	60.51 ± 5.74	−0.18	−1.7%	Interaction: $p = 0.004$, η^2_p: 0.164
BMI (kg/m^2)						
	VG	24.52 ± 1.37	23.22 ± 1.46	−0.98	−5.3%	Group: $p = 0.458$, η^2_p: 0.013
						Time: $p < 0.001$, η^2_p: 0.697
	CG	25.19 ± 1.20	24.99 ± 1.66	−0.21	−0.8%	Interaction: $p = 0.015$, η^2_p: 0.211
Muscle mass (kg)						
	VG	20.42 ± 3.46	20.66 ± 3.60	+0.07	+1.2%	Group: $p = 0.425$, η^2_p: 0.014
						Time: $p = 0.393$, η^2_p: 0.016
	CG	19.68 ± 3.79	19.71 ± 3.69	0.0	0.1%	Interaction: $p = 0.393$, η^2_p: 0.016
Body fat (kg)						
	VG	13.94 ± 4.44	12.79 ± 4.59	−0.16	−5.5%	Group: $p = 0.885$, η^2_p: 0.000
						Time: $p < 0.001$, η^2_p: 0.280
	CG	13.35 ± 4.49	13.15 ± 4.24	−0.05	−1.5%	Interaction: $p = 0.090$, η^2_p: 0.065
Body fat (%)						
	VG	25.74 ± 5.27	24.64 ± 5.14	−0.14	−4.3%	Group: $p = 0.425$, η^2_p: 0.026
						Time: $p < 0.001$, η^2_p: 0.312
	CG	24.35 ± 5.32	23.83 ± 5.20	−0.10	−2.1%	Interaction: $p = 0.124$, η^2_p: 0.064

Abbreviations: VG, volleyball group; CG, control group; BMI, body mass index; ES, Cohen's d effect size.

4. Discussion

The number of overweight and obese children worldwide has risen alarmingly in the last twenty years [19], partly due to insufficient physical activity and unhealthy eating habits. The school environment is the setting where adolescents usually spend most of their day. Therefore, the current study aimed to investigate the effects of a 12-week school-based volleyball intervention on body composition in overweight female adolescents.

The most important results of the current study were the decrease in body mass and BMI after a relatively short volleyball program. Furthermore, the volleyball program over 12 weeks reduced body fat over time but showed no difference between the groups.

Obesity during childhood and adolescence, low cardiorespiratory fitness, and decreased physical activity have a strong association with various cardiovascular diseases [20]. Strategies to improve body composition status include diet, exercise interventions, or both. A recent study indicated that there is a general trend towards larger changes in body composition following HIIT protocols [8]. Volleyball is a sport that belongs to high-intensity activities [21], and it is predominantly popular in the female population; since it involves less high-intensity activity than other sports [22–24], it could be a suitable and effective activity for both obese and overweight adolescents. The current results, although among overweight adolescents, showed that the volleyball after-school program was more effective than those of the CG, with a 6.9% reduction in body mass after the experimental treatment (medium ES). Considering the baseline differences between CG and VG at the initial assessment, after the experimental program, there was a decrease in body mass index (BMI) in both groups. The reduction was greater in the VG (−5.3%) compared to CG (−0.8%). However, BMI is not the most adequate parameter to describe real differences in body composition, because a change in BMI does not mean that there has been an increase in muscle mass and a reduction in fat mass at the same time [25].

Although the current results showed a higher percentage decrease in body fat (4.3% in the VG and 2.1% in the CG group) and body fat in kg (5.5% in the VG and 1.5% in the CG) in the VG compared to the CG, the changes were not statistically significant ($p > 0.05$). A possible reason for the non-significant changes could be that both the CG and VG were exposed to physical activity for a period of 12 weeks. However, even the lowest level of increased physical activity leads to positive changes in the overweight and obese population. Moreover, adolescence is a period related to increased hormonal work and increased biological growth and development [26], where participants can be exposed to a longer duration of physical activity [27]. Another reason may be that, during the experimental program, the participants did not have any kind of educational content or control of nutrient intake that would affect the outcome [28].

It should be acknowledged that the study had some limitations. Firstly, physical activity was not measured during and outside school, which could be recognized as a limitation. Secondly, the small sample size of participants could also be considered as a limitation, particularly given that only overweight girls were included in the study. Third, dietary habits and total calorie intake were not controlled. However, they were instructed to maintain their usual calorie intake. Finally, after the experimental program and the final measurement, there was no further monitoring of the participants in the parameters of body composition, in order to determine whether there was a return of values to the initial level. Nevertheless, this study justifies the use of volleyball in the planning of physical education classes and after-school activities, showing that playing volleyball could significantly improve the body composition of overweight girls.

5. Conclusions

Overall, our findings show that an after-school volleyball program over twelve weeks is suitable for improving body composition in overweight adolescent girls. This study also shows that playing volleyball, added to regular PE classes, has the potential to reduce or even compensate for the limitations of the exercise program in a regular class in a relatively short time.

Author Contributions: Conceptualization, N.T. and S.N.; methodology, G.S.; software, D.T.-P.; validation, A.L. and S.N; formal analysis, G.S. and V.M.; investigation, N.T.; resources, V.B.; data curation, G.S.; writing—original draft preparation, N.T.; writing—review and editing, A.L.; visualization, D.T.-P.; supervision, V.B.; project administration, G.S.; funding acquisition, V.B. All authors have read and agreed to the published version of the manuscript.

Funding: This research received no external funding.

Institutional Review Board Statement: The study was conducted according to the guidelines of the Declaration of Helsinki and approved by the Institutional Review Board (or Ethics Committee) of the Faculty of Sport and Physical Education, University of Niš (20/2020, Date 23 May 2020).

Informed Consent Statement: Informed consent was obtained from all subjects involved in the study.

Data Availability Statement: Not applicable.

Conflicts of Interest: The authors declare no conflict of interest.

References

1. Chooi, Y.C.; Ding, C.; Magkos, F. The epidemiology of obesity. *Metabolism* **2019**, *92*, 6–10. [CrossRef] [PubMed]
2. Ekelund, U.; Tomkinson, G.; Armstrong, N. What proportion of youth are physically active? Measurement issues, levels and recent time trends. *Br. J. Sports Med.* **2011**, *45*, 859–865. [CrossRef] [PubMed]
3. W.H.O. *Report of the Commission on Ending Childhood Obesity*; World Health Organization: Geneva, Switzerland, 2016. Available online: https://apps.who.int/iris/bitstream/handle/10665/204176/9789241510066_eng.pdf (accessed on 13 September 2021).
4. Dobbins, M.; Husson, H.; DeCorby, K.; LaRocca, R.L. School-based physical activity programs for promoting physical activity and fitness in children and adolescents aged 6 to 18. *Cochrane Database Syst. Rev.* **2013**, *2013*, CD007651. [CrossRef]
5. Kriemler, S.; Meyer, U.; Martin, E.; van Sluijs, E.; Andersen, L.B.; Martin, B.W. Effect of school-based interventions on physical activity and fitness in children and adolescents: A review of reviews and systematic update. *Br. J. Sports Med.* **2011**, *45*, 923–930. [CrossRef] [PubMed]
6. Kirkham-King, M.; Brusseau, T.A.; Hannon, J.C.; Castelli, D.M.; Hilton, K.; Burns, R.D. Elementary physical education: A focus on fitness activities and smaller class sizes are associated with higher levels of physical activity. *Prev. Med. Rep.* **2017**, *8*, 135–139. [CrossRef]
7. Nettlefold, L.; McKay, H.A.; Warburton, D.E.R.; McGuire, K.A.; Bredin, S.S.D.; Naylor, P.J. The challenge of low physical activity during the school day: At recess, lunch and in physical education. *Br. J. Sports Med.* **2010**, *45*, 813–819. [CrossRef] [PubMed]

8. Eddolls, W.T.B.; McNarry, M.A.; Stratton, G.; Winn, C.O.N.; Mackintosh, K.A. High-Intensity Interval Training Interventions in Children and Adolescents: A Systematic Review. *Sports Med.* **2017**, *47*, 2363–2374. [CrossRef] [PubMed]
9. Gabbett, T.; Georgieff, B.; Anderson, S.; Cotton, B.; Savovic, D.; Nicholson, L. Changes in Skill and Physical Fitness Following Training in Talent-Identified Volleyball Players. *J. Strength Cond. Res.* **2006**, *20*, 29–35. [CrossRef] [PubMed]
10. Laroche, J.-A.; Girard, S.; Lemoyne, J. Tracing adolescent girls' motivation longitudinally: From FitClub participation to lei-sure-time physical activity. *Percept. Mot. Skills* **2019**, *126*, 986–1005. [CrossRef]
11. Glinkowska, B.; Glinkowski, W.M. Association of sports and physical activity with obesity among teenagers in Poland. *Int. J. Occup. Med. Environ. Health.* **2018**, *31*, 771–782. [CrossRef]
12. Scribbans, T.D.; Berg, K.; Narazaki, K.; Janssen, I.; Gurd, B.J. Heart rate during basketball game play and volleyball drills accu-rately predicts oxygen uptake and energy expenditure. *J. Sport. Med. Phys. Fit.* **2015**, *55*, 905–913.
13. Ainsworth, B.E.; Haskell, W.L.; Herrmann, S.D.; Meckes, N.; Bassett, D.R.; Tudor-Locke, C.; Greer, J.L.; Vezina, J.; Whitt-Glover, M.C.; Leon, A.S. 2011 Compendium of Physical Activities: A second update of codes and MET values. *Med. Sci. Sport. Exerc.* **2011**, *43*, 1575–1581. [CrossRef] [PubMed]
14. Cole, T.J.; Bellizzi, M.C.; Flegal, K.M.; Dietz, W.H. Establishing a standard definition for child overweight and obesity worldwide: International survey. *BMJ* **2000**, *320*, 1240–1255. [CrossRef]
15. McLester, C.N.; Nickerson, B.S.; Kliszczewicz, B.M.; McLester, J.R. Reliability and Agreement of Various InBody Body Composition Analyzers as Compared to Dual-Energy X-ray Absorptiometry in Healthy Men and Women. *J. Clin. Densitom.* **2020**, *23*, 443–450. [CrossRef] [PubMed]
16. Riebe, D.; Franklin, B.A.; Thompson, P.D.; Garber, C.E.; Whitfield, G.P.; Magal, M.; Pescatello, L.S. Updating ACSM's recommendations for exercise preparticipation health screening. *Med. Sci. Sports Exerc.* **2015**, *47*, 2473–2479. [CrossRef]
17. Hopkins, W.G.; Marshall, S.W.; Batterham, A.M.; Hanin, J. Progressive Statistics for Studies in Sports Medicine and Exercise Science. *Med. Sci. Sports Exerc.* **2009**, *41*, 3–13. [CrossRef] [PubMed]
18. Pallant, J. *A Step by Step Guide to Data Analysis Using SPSS for Windows*, 3rd ed.; McGraw Hill Open University Press: New York, NY, USA, 2007.
19. Marković, L.; Đorđić, V.; Trajković, N.; Božić, P.; Halaši, S.; Cvejić, D.; Ostojić, S.M. Childhood Obesity in Serbia on the Rise. *Children* **2021**, *8*, 409. [CrossRef] [PubMed]
20. Gomez, G.P.; Huffman, F.G. Risk Factors for Type 2 Diabetes and Cardiovascular Diseases in Hispanic Adolescents. *J. Adolesc. Health* **2008**, *43*, 444–450. [CrossRef]
21. Sheppard, J.M.; Gabbett, T.J.; Stanganelli, L.-C.R. An Analysis of Playing Positions in Elite Men's Volleyball: Considerations for Competition Demands and Physiologic Characteristics. *J. Strength Cond. Res.* **2009**, *23*, 1858–1866. [CrossRef]
22. Krustrup, P.; Aagaard, P.; Nybo, L.; Petersen, J.; Mohr, M.; Bangsbo, J. Recreational football as a health promoting activity: A topical review. *Scand. J. Med. Sci. Sports* **2010**, *20*, 1–13. [CrossRef]
23. Póvoas, S.C.A.; Castagna, C.; Resende, C.; Coelho, E.F.; Silva, P.; Santos, R.; Pereira, R.; Krustrup, P. Effects of a Short-Term Recreational Team Handball-Based Programme on Physical Fitness and Cardiovascular and Metabolic Health of 33–55-Year-Old Men: A Pilot Study. *BioMed Res. Int.* **2018**, *2018*, 4109796. [CrossRef]
24. Stojanović, E.; Stojiljković, N.; Stanković, R.; Scanlan, A.T.; Dalbo, V.J.; Milanović, Z. Recreational Basketball Small-Sided Games Elicit High-Intensity Exercise With Low Perceptual Demand. *J. Strength Cond. Res.* **2021**, *35*, 3151–3157. [CrossRef] [PubMed]
25. Aasheim, E.T.; Aylwin, S.J.B.; Radhakrishnan, S.T.; Sood, A.S.; Jovanovic, A.; Olbers, T.; le Roux, C.W. Assessment of obesity beyond body mass index to determine benefit of treatment. *Clin. Obes.* **2011**, *1*, 77–84. [CrossRef] [PubMed]
26. Malina, R.M. Growth and maturity status of young soccer players. In *Science and Soccer*, 2nd ed.; Routledge: London, UK, 2003; pp. 295–314.
27. Costigan, S.A.; Eather, N.; Plotnikoff, R.C.; Taaffe, D.; Lubans, D. High-intensity interval training for improving health-related fitness in adolescents: A systematic review and meta-analysis. *Br. J. Sports Med.* **2015**, *49*, 1253–1261. [CrossRef] [PubMed]
28. Nemet, D.; Barkan, S.; Epstein, Y.; Friedland, O.; Kowen, G.; Eliakim, A. Short- and Long-Term Beneficial Effects of a Combined Dietary-Behavioral-Physical Activity Intervention for the Treatment of Childhood Obesity. *Pediatrics* **2005**, *115*, e443–e449. [CrossRef] [PubMed]

MDPI
St. Alban-Anlage 66
4052 Basel
Switzerland
www.mdpi.com

Children Editorial Office
E-mail: children@mdpi.com
www.mdpi.com/journal/children

www.ingramcontent.com/pod-product-compliance
Lightning Source LLC
Chambersburg PA
CBHW061557210926
51850CB00009B/2128